A TECHNICAL MANUAL FOR PARASITIC WEED RESEARCH AND EXTENSION

A Technical Manual for Parasitic Weed Research and Extension

Edited by

Jürgen Kroschel

University of Kassel, Germany

Deutsche Gesellschaft für Technische Zusammenarbeit (GTZ) – University of Hohenheim

KLUWER ACADEMIC PUBLISHERS

DORDRECHT / BOSTON / LONDON

Library of Congress Cataloging-in-Publication Data

A technical manual for parasitic weed research and extension / edited by Jürgen Kroschel.
p. cm.
ISBN 0-7923-6880-0 (hb : alk. paper)
1. Parasitic plants. 2. Weeds. 3. Parasitic plants--Control. 4. Weeds--Control. 5. Parasitic plants--Research. 6. Weeds--Research. I. Kroschel, Jürgen.

SB611 .T43 2001
632'.52--dc21

2001029163

ISBN 0-7923-6880-0

Published by Kluwer Academic Publishers,
P.O. Box 17, 3300 AA Dordrecht, The Netherlands.

Sold and distributed in North, Central and South America
by Kluwer Academic Publishers,
101 Philip Drive, Norwell, MA 02061, U.S.A.

In all other countries, sold and distributed
by Kluwer Academic Publishers,
P.O. Box 322, 3300 AH Dordrecht, The Netherlands.

Printed on acid-free paper

Printed in the Netherlands.

Dedicated to Prof. Dr. Werner Koch

Contents

Foreword

Parasitic weeds are known to be difficult to control in agricultural production. Despite an enormous increase in knowledge of their biology and the development of different control options, no substantial reduction of infestation has been achieved in the past either with regard to *Orobanche* spp. in the Mediterranean region or with regard to *Striga* spp. in sub-Saharan Africa.

The specific biology of parasitic weeds is still a great obstacle for the development of control strategies. Most of the host plant damage is already done before the parasites' emergence aboveground. This means that control tools are needed which interfere with the underground stages in order to protect the present crop. Until now, we have only few control methods available which fulfil these requirements. In addition, research has to deal with various parasites, which attack a wide range of crops in different agro-ecosystems. This implies that solutions specific to parasites, crops and farming systems must be developed. In many countries and regions it has been shown that the knowledge of extension workers related to the biology and control of parasitic weeds has to be improved as a prerequisite to assist farmers adequately in their control efforts. In addition, suitable extension methods and strategies have to be developed with and for extension services in order to support extension workers in the transfer of possible solutions.

This **"Technical Manual for Parasitic Weed Research and Extension"** provides methodologies for various aspects of research and extension related to parasitic weed species of the genera *Striga, Alectra, Orobanche* and *Cuscuta*. It summarises procedures, which have been practically applied or developed either by researchers of the supra-regional project "Ecology and Management of Parasitic Weeds" of the Deutsche Gesellschaft für Technische Zusammenarbeit (GTZ) GmbH, which was carried out in co-operation with the University of Hohenheim from 1988 to 1998, or by researchers from International Agricultural Research Centres and Advanced Research Institutes.

It is not claimed to list all possible techniques applied so far in studies on parasitic weeds, nor can the respective descriptions of each procedure cover each and all aspects in full detail. It is merely intended to offer a palette of possibilities and to give hints on practical aspects related to the specific requirements of a parasite or a technique. When required, further in depth information on any given subject might be drawn from the specific reports in the literature of which a selection is provided at the end of each chapter.

One intention regarding the compilation of methods into this manual was to provide easy handling of the techniques. Hence, wherever possible and appropriate, throughout the manual the following structure was applied:

INTRODUCTION: background, overview, principle of the method.
OBJECTIVE(S): aim(s) of the method.
MATERIAL(S): list of items including chemicals, equipment, time requirement.
PROCEDURE(S): implementation, preparation, sample size.
EVALUATION: hints for data collection, calculations, evaluation and statistical aspects.

COMMENT(S): specific/general remarks on the method, validation, criticism.

LITERATURE: as cited in the text plus further recommended literature.

It is hoped that this manual will benefit researchers and extension workers of international and national agricultural research and extension institutes and universities, who either intend to start in this subject or want to apply further techniques they are not yet familiar with. Ultimately, it will improve the research capacity and will help to provide a better understanding of these outstandingly interesting plants. Consequently it may lead to the development of control methods, which are more suited to be used by small scale farmers and therefore help to improve their food security and daily lives.

PD Dr. J. Kroschel
University of Kassel, Witzenhausen

Acknowledgements

The idea and the need to compile a "Technical Manual for Parasitic Weed Research and Extension" has been arisen during the years the International Training Course on "Biology and Control of Parasitic Weed" (from 1993 to 1998) has been performed at the University of Hohenheim, Germany. In 1996, the first chapters have been prepared and it has been a pleasure gathering scientists from different disciplines of parasitic weed research to contribute to this publication.

For the use of original slides I thank Dr. Karl-Heinz Linke (plate 4: 2a,b; plate 8: 1, 2, 6; plate 9: 1; plate 10: 2a-c; plate 11: 1a,b, 2a,b), Olaf Klein (plate 12: 1-6; plate 13: 1, 3, 4, 5), Dr. Andrea Jost (plate 15: 1a, 2a,b), Dorette Müller-Stöver (plate 16: 1, 2), and Dr. Stefan Kachelriess (plate 17: 4). For the use of the key to the main *Striga*, *Alectra*, *Orobanche* and *Cuscuta* species I am grateful to Dr. Chris Parker and Dr. Charlie Riches.

The book would not have been accomplished without the financial and infrastructural support of the supra-regional project "Ecology and Management of Parasitic Weeds", which was implemented by the Deutsche Gesellschaft für Technische Zusammenarbeit (GTZ) GmbH on behalf of the German Federal Ministry for Economic Cooperation and Development (BMZ). I am thankful to Uta Hübler, Charlotte Spengler, Jutta Müther and Sigrid Haber for supporting the editing of the publication. I am indebted, in particular, to Dr. Chris Parker and Dr. Bettina Haussmann for proof-reading the manuscript, and to Olaf Klein preparing most of the line drawings.

List of Contributors

ABBASHER, Awad Abbasher
Faculty of Agriculture and Natural Resources-Abu Haraz
University of Gezira
P.O.Box 42, Wad Medani, SUDAN

DEIMLING, Sabine
Institute of Plant Breeding, Seed Science, and Population Genetics (350)
University of Hohenheim
D-70593 Stuttgart, GERMANY

FISCHER, Elke
Institute of Plant Breeding, Seed Science, and Population Genetics (350)
University of Hohenheim
D-70593 Stuttgart, GERMANY

FISCHER, Patrick A
Hans Klemm Str. 7, 50739 Köln, GERMANY

HAUSSMANN, Bettina IG
Institute of Plant Breeding, Seed Science, and Population Genetics (350)
University of Hohenheim
D-70593 Stuttgart, GERMANY

HESS, Dale E
International Crops Research Institute for the Semi-Arid Tropics (ICRISAT)
B.P. 320, Bamako, MALI

HUMMLER, Klaus
Im Kranich 3, D-72202 Gündringen, GERMANY

JOEL, Daniel M
Department of Weed Research, Agricultural Research Organization
Newe-Ya'ar Research Center
P.O. Box 1021, Ramat-Yishay 30095, ISRAEL

JOST, Andrea
Deutsche Gesellschaft für Technische Zusammenarbeit (GTZ) GmbH
Integrated Pest Management Project
P.O. Box 1004, Arusha, TANZANIA

KACHELRIESS, Stefan
Breiteweg 1, D-35415 Pohlheim, GERMANY

KATZIR, Nurit
Department of Weed Research, Agricultural Research Organization
Newe-Ya'ar Research Center
P.O. Box 1021, Ramat-Yishay 30095, ISRAEL

KLEIN, Olaf
Friedrich-Liststr.16, D-70596 Stuttgart, GERMANY

KRANZ, Brigitte
Remppisstr.14, D-70599 Stuttgart, GERMANY

KROSCHEL, Jürgen
Institute of Crop Science
University of Kassel
Steinstr. 19, D-37213 Witzenhausen, GERMANY

LINKE, Karl-Heinz
Bernhard-Riedmüllerstr. 35, D-88450Berkheim-Illerbachen, GERMANY

MÜLLER-STÖVER, Dorette
Institute of Plant Production and Agroecology in the Tropics and Subtropics (380)
University of Hohenheim
D-70593 Stuttgart, GERMANY
OSWALD, Andreas
Institute of Crop Science
University of Kassel
Steinstr. 19, D-37213 Witzenhausen, GERMANY
PORTNOY, Vitaly
Department of Weed Research, Agricultural Research Organization
Newe-Ya'ar Research Center
P.O. Box 1021, Ramat-Yishay 30095, ISRAEL
RANSOM, Joel K
Centro Internacional de Mejoramiento de Maiz y Trigo (CIMMYT)
P.O. Box 5186 Lazimpat, Kathmandu, NEPAL
THOMAS, Heiko
Steigerwaldstr. 18, D-41462 Neuss, GERMANY
VURRO, Maurizio
Istituto Tossine e Micotossine da Parassiti vegetali
Consiglio Nazionale delle Ricerche
Viale Einaudi 51, 70125 Bari, ITALY
WELZ, Günter
Aventis Crop Science USA
Midwest Field Research Station
P.O. Box 219, Seymour, IL 61875, USA

1 Introduction

Parasitic weeds belong to a group of plants which have lost their autotrophic way of life during their developmental process (evolution). They are nutrition specialists to the disadvantage of their host plants. With special organs, the haustoria, they penetrate into the vessels of other plants in order to supply themselves with water and nutrients. Different developmental steps in the process of evolution can be understood by the existence of different degrees of parasitism. Facultative parasites like species within the genus *Rhinanthus* or *Melamyprum* represent those at the beginning of the process while obligate parasites represent more advanced plants. Obligate parasites are either hemi-parasites with chlorophyll containing species of the genus *Striga* and *Alectra* (Scrophulariaceae), or holoparasites without chlorophyll, e.g., of the genus *Orobanche* (Orobanchaceae). Instead of being root parasites like species of the genus *Orobanche*, *Striga* and *Alectra*, species of the genus *Cuscuta* (Cuscutaceae) are obligate shoot parasites.

1.1 Parasitic weed species and their hosts

Parasitic weeds have evolved specificity to crops and plants in the natural vegetation. In addition to the normal germination requirements of angiosperm seeds, seeds of parasites of the genus *Striga*, *Alectra* and *Orobanche* require a suitable chemical stimulant, which is released by roots of hosts. But also some non-hosts release root exudates which can trigger the germination of parasite seed without being parasitized. These plants can be used as trap crops to reduce the seed bank of parasites in the soil (→ annex 11.6). *Cuscuta* spp. do not require a host germination stimulant.

Striga asiatica and *S. hermonthica* are almost entirely specific to grasses and therefore important parasites of cereals like maize, sorghum, pearl millet, rice and others. *S. gesnerioides* prefers dicot hosts, mainly cowpea and tobacco (Table 1). The most important *Orobanche* spp. are *O. crenata*, *O. cumana*, *O. cernua*, *O. ramosa*, *O. aegyptiaca*, *O. minor* and *O. foetida*. Most are restricted to dicot hosts. They represent a large threat to crops of the families Asteraceae, Fabaceae and Solanaceae. The most important *Alectra* spp. is *Alectra vogelii* parasitizing mainly cowpea, bambara groundnut and groundnut. *Cuscuta* spp. have a very low host specificity and can parasitize different host species simultaneously. Although dicots are preferred, attack on monocots has been observed.

1.2 Geographical distribution

40 *Striga* species are reported worldwide: 33 in Africa, 8 in Asia and 3 species in Australia. Eleven species of *Striga* are parasites on agricultural crops. *S. asiatica* is the most widespread *Striga* species. *S. asiatica* was

brought from India to Eastern Africa about 2000 years ago from where it started to spread to southern Africa. This species was also introduced unintentionally to North Carolina in the USA. The origin of *Striga hermonthica* is thought to be the northern part of Ethiopia (Tigray) and the Nuba mountains of Sudan, areas where *S. hermonthica* can be found on native vegetation, particularly on *Eragrostis* species. These regions are also the regions of origin of sorghum. Therefore, it can be assumed that the distribution of *S. hermonthica* was associated with the trade of sorghum seeds to other parts of Africa. *S. gesnerioides* is found in Africa, the Arabian Peninsula and the Indian subcontinent; in the 1970s it was unintentionally introduced to Florida.

Table 1. Economic important parasitic weed species and their main hosts

Parasitic weed species	Main crops
***Striga* species**	
S. asiatica &	Maize, sorghum, pearl millet, rice, finger millet,
S. hermonthica	sugarcane
S. gesnerioides	Cowpea, tobacco
S. aspera	Maize, sorghum, finger millet, fonio, rice, sugarcane
S. forbesii	Maize, sorghum, sugarcane
***Orobanche* species**	
O. crenata	Faba bean, pea, lentil, chickpea
O. ramosa / O. aegyptiaca	Tomato, tobacco, hemp, eggplant, lentil, curcubit
O. *cumana /O. cernua*	Sunflower, tobacco, tomato, potato
O. minor	Alfalfa, clover, trefoil
O. foetida	Faba bean, alfalfa, trefoil
***Alectra* species**	
A. vogelii	Cowpea, bambara groundnut, groundnut
A. picta	Groundnut, cowpea
A. sessiliflora	Noog
A. orobanchoides	Tobacco, sunflower
A. fluminensis	Sugarcane
***Cuscuta* species**	
C. australis	Soybean
C. campestris	Alfalfa, sugarbeet, faba bean, carrot
C. chinensis	Alfalfa, soybean
C. indecora	Alfalfa
C. epithymum	Clover
C. epilinum	Flax
C. gronovii	Vine, citrus, cranberry
C. planiflora	Berseem
C. monogyna	Vine, citrus
C. pedicellata	Forage legumes
C. palaestina	Forage legumes
C. reflexa	Citrus, coffee, peach, litchi

S. hermonthica is a species particularly well adapted to the semi-arid tropics which are characterised by high temperature throughout the year and a long dry season. The dissemination of *S. hermonthica* into higher altitudes of tropical zones or into sub-tropical regions is likely where temperatures of 21°C occur over a period of 90 to 120 days. With regard to rainfall, *S. hermonthica* occurs mainly in regions with precipitation of 450-1000 mm per annum. This is obviously not true for *S. asiatica*. In Madagascar, *S. asiatica* also occurs in regions with a precipitation of more than 2000 mm and a short dry period.

The species in the genus *Alectra* are found mainly in tropical Africa and subtropical southern Africa. *A. sessiliflora* is also found in subtropical Asia. *A. fluminensis* is a species of local importance in Latin America as a parasite of sugarcane.

Orobanche spp. are found in temperate regions of the northern hemisphere as well as in the sub-tropics and tropics, especially in arid and semi-arid regions. Their main centre of distribution is the Mediterranean region with an annual precipitation between 250 and 500 mm. Areas of origin can be found in Turkey, Italy, Spain and Morocco. International trade has contributed to the spread of *Orobanche* spp. into areas with similar climatic conditions. *O. minor* and *O. ramosa* have been introduced to North and South America (Chile) by infested crop seeds.

Cuscuta spp. are distributed world-wide and attack many different host plants. Especially international trade, mainly with contaminated crop seeds, has led to the wide distribution of this parasite.

1.3 The life cycle of parasitic weeds

After seed shed, seeds of *Orobanche, Striga* and *Alectra* are dormant and require a period of after-ripening in warm, dry storage, followed by conditioning in a warm, moist environment, before they will respond to a germination stimulant (Fig. 1). Seeds of parasites in the dry state can remain dormant and viable for many years. A water-conditioning period of several days readies the seeds for germination. The necessity to receive a germination stimulant which is released by the root tip of host plants ensures that a host is available and near enough to be reached by a germ tube, which elongates and grows in direction of the origin of the signal (chemotropism). Most of the seeds in the soil will not be reached by the stimulant. They will remain viable until the following cropping periods. After the attachment to the host root, the parasite starts to penetrate the host root epidermis.

Cuscuta seeds are hard-coated and may stay dormant for several years even immediately after shedding. Germination occurs when temperature and moisture conditions are favourable, even immediately after the vegetation period. Upon germination, the epicotyl (3-6 cm long) searches for a host by circumnutation. The seeds contain food reserves for 4 to 9 days and the epicotyl is even able to move for some distance in order to find a host (Fig. 2).

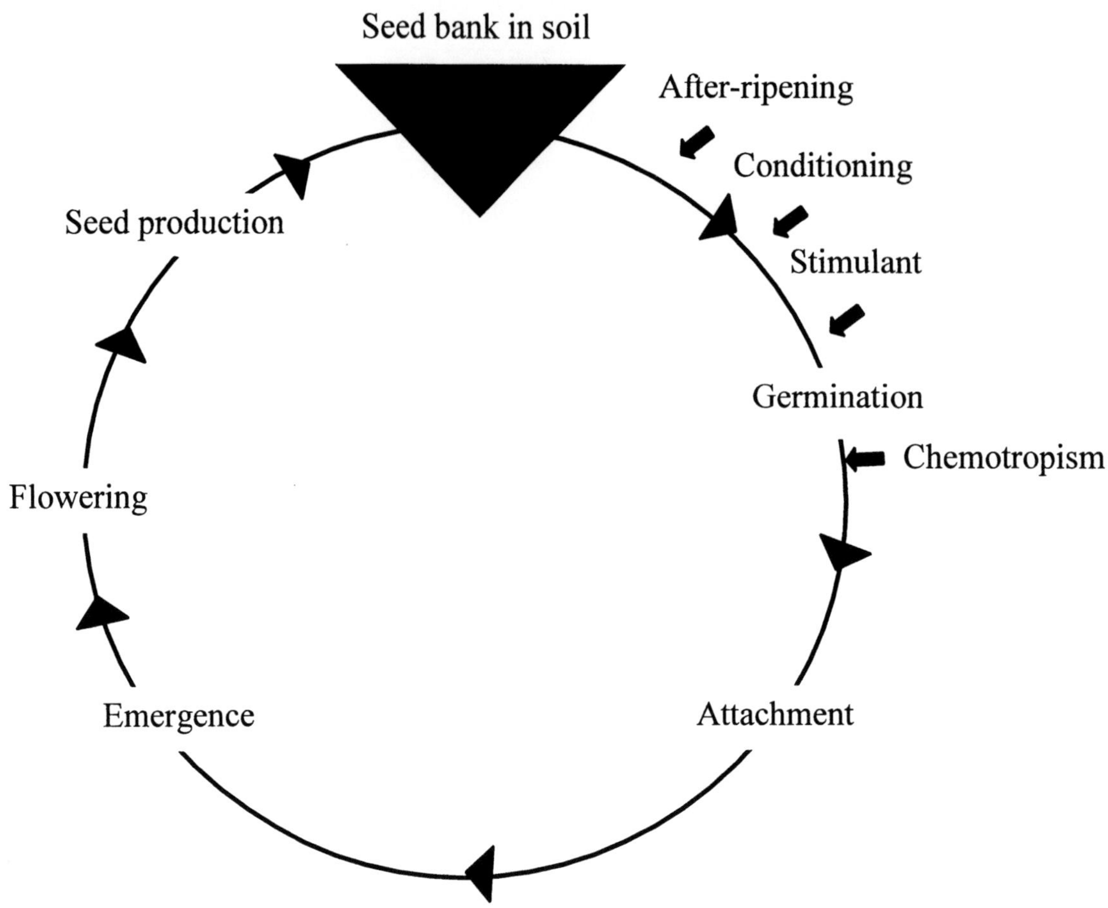

Fig. 1. The life cycle of root parasites of the genera *Striga*, *Orobanche* and *Alectra*

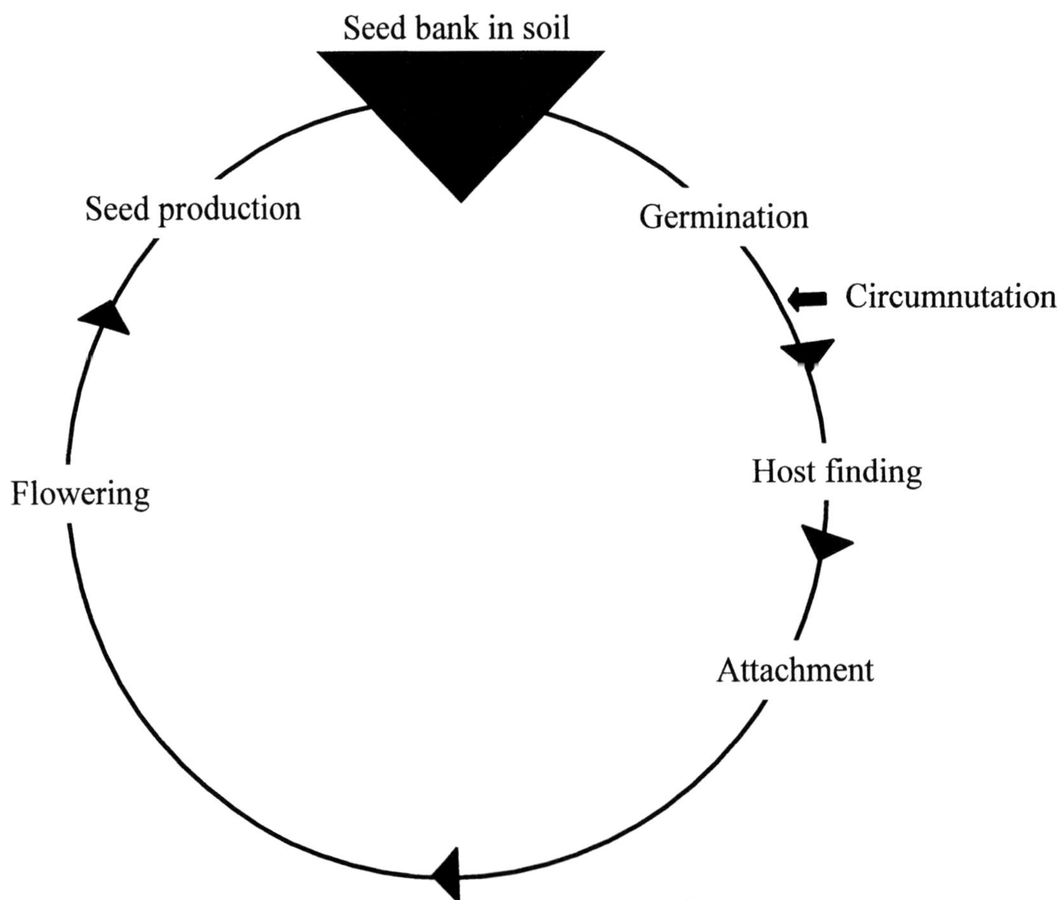

Fig. 2. The life cycle of shoot parasites of the genera *Cuscuta*

The bottleneck in parasite development is the haustorium. The haustorium represents the physical and morphological contact between parasite and host in order to supply the parasite with water, mineral nutrients and organic materials from its host. The penetration of haustorial cells into host tissue (xylem and/or phloem system) is carried out mechanically by pressure on the host endodermal cells and by hydrolytic enzymes. The primary haustorium is often supplemented by a large number of secondary haustoria. This results in an extensive and intricate system of points of attachment between parasite and host. Secondary haustoria are not restricted to the roots of one host, but different host plants of the same species (intra-specific parasitism), or even different host species may be parasitized simultaneously (inter-specific parasitism).

1.4 Major books, symposium and workshop proceedings, field guides and manuals

Major books

Kuijt J 1969. The Biology of Parasitic Flowering Plants. University of California Press, Berkeley, 346 pp.

Musselman LJ 1987. Parasitic Weeds in Agriculture. Volume I. *Striga*. CRC Press, Boca Raton, 317 pp.

Parker C & CR Riches 1993. Parasitic Weeds of the World - Biology and Control. CAB International, Wallingford, UK, 332 pp.

Press, MC & JD Graves 1995. Parasitic plants. Chapman & Hall, London, UK, 292 pp.

Sand PF, RE Eplee & RG Westbrooks 1990. Witchweed Research and Control in the United States of America. Weed Science Society of America, Champaign, 154 pp.

Sauerborn J 1991. Parasitic flowering plants - Ecology and Management. Margraf Verlag, Weikersheim, Germany, 127 pp.

Symposium and workshop proceedings

Ayensu ES, H Doggett, RD Keynes, J Marton-Lefèvre, LJ Musselman, C Parker & A Pickering 1984. *Striga* Biology and Control. IRL Press Ltd Oxford, UK.

Cubero, JI, MT Moreno, D Rubiales & J Sillero 1999. Resistance to *Orobanche*: The State of the Art. Congresos y Jornadas 51/99. Junta de Andalucia, Consejeria de Agricultura y Pesca, Spain, 199 pp.

EWRC 1973. Proceedings Symposium on Parasitic Weeds, Malta, 1973. European Weed Research Council, Wageningen, 295 pp.

Haussmann BIG, DE Hess, ML Koyama, L Grivet, HFW Rattunde & HH Geiger 2000. Breeding for *Striga* Resistance in Cereals. Proceedings, Workshop held at IITA, Ibadan, Nigeria. Margraf Verlag, Weikersheim, Germany, 376 pp.

Kroschel J, H Mercer-Quarshie & J Sauerborn 1999. Advances in Parasitic Weed Control at On-farm Level. Vol. I. Joint Action to Control *Striga* in Africa. Margraf Verlag, Weikersheim, Germany, 324 pp.

Kroschel J, M Abderabihi & H Betz 1999. Advances in Parasitic Weed Control at On-farm Level. Vol. II. Joint Action to Control *Orobanche* in the WANA Region. Margraf Verlag, Weikersheim, Germany, 347 pp.

Moreno MT, JI Cubero, D Berner, D Joel, LJ Musselman & C Parker 1996. Advances in Parasitic Plant Research. Proceedings 6th International Parasitic Weed Symposium, Cordoba, Spain, 929 pp.

Musselman LJ, AB Worsham & RE Eplee 1979. Proceedings 2nd International Symposium on Parsitic Weeds, Raleigh, 1979. North Carolina State University, Raleigh, 296 pp.

Parker C, LJ Musselman, RM Polhill & AK Wilson 1984. Proceedings 3rd International Symposium on Parasitic Weeds, ICARDA/International Parasitic Seed Plant Research Groupe, 7-9 May 1984, Aleppo, Syria. ICARDA, P.O.Box 5466, Aleppo, Syria, 265 pp.

Pieterse AH, JA Verkleij & S-J ter Borg 1994. Biology and management of *Orobanche*. Proceedings of the 3rd International Workshop on *Orobanche* and related *Striga* research, Amsterdam, The Netherlands, November 8-12, 1993. Amsterdam. Royal.Tropical Institute.

Ransom JK, LJ Musselman, AD Worsham & C Parker 1991. Proceedings 5th International Symposium on Parasitic Weeds, Nairobi, 1991. CIMMYT, Nairobi, 550 pp.

Ter Borg SJ 1986. Proceedings, Workshop on Biology and Control of *Orobanche*. LH/VPO Wageningen, The Netherlands, 206pp.

Weber HC & W Forstreuter 1987. Parasitic Flowering Plants. Proceedings 4th International Symposium on Parasitic Flowering Plants, Marburg, Germany. Philipps-Universität, Marburg, 848 pp.

Wegmann K & LJ Musselman 1991. Progress in *Orobanche* Research. Eberhards-Karls-Universität, Tübingen, Germany, 360 pp.

Wegmann K, LJ Musselman & DM Joel 1998. Current Problems of *Orobanche* Research. Proceedings of the 4th International *Orobanche* Workshop, September 23-26 1998, Albena, Bulgaria. Institute of Wheat and Sunflower Dobroudja, Bulgaria, 452 pp.

Field guides and manuals

Dembélé B, A Raynal-Roques, G Sallé & C Tuquet 1994. Plantes Parasites des Cultures et des Essences Forestières au Sahel. Institut du Sahel, Bamako, Mali, CTA, Wageningen, The Netherlands, 43 pp.

Berner DK, MD Winslow, AE Awad, KE Cardwell, DR Mohan Raj & SK Kim 1997. *Striga* research methods – a manual. International Institute of Tropical Agriculture. PMB 5320, Ibadan, Nigeria.

Linke KH, J Sauerborn & MC Saxena 1989. *Orobanche* field guide. University of Hohenheim, Germany/ICARDA, Aleppo, Syria, 42 pp. (also available in Arabic).

Ramaiah KV, C Parker, MJ Vasudeva Rao & LJ Musselman 1983. *Striga* identification and control handbook. ICRISAT Information Bulletin No. 15, ICRISAT, Patancheru, India, 52 pp (also available in french).

Author: J Kroschel

2 Ecological studies

2.1 Preparation of herbarium samples

INTRODUCTION

Herbarium samples of collected plants represent a fundamental basis for any detailed ecological investigation. For certain botanical studies it is obligatory to certify the described plants by herbarium samples. This is furthermore of great value where the survey and collection was done in poorly documented ecosystems. A collection of herbarium samples is also very useful in teaching about these plants. Due to their fleshy nature, however, some parasitic plants are rather difficult to preserve.

OBJECTIVES

- To compare plants of different origin,
- to keep samples of plants as reference or for teaching purposes,
- to further identify plants.

For parasitic plants with their specific morphological features the following three types of herbarium collections are recommended:

1. Regular herbarium (stationary herbarium)

MATERIALS

✓ Sheets of paper (any porous paper, e.g., newspaper), plant press, cardboard, glue or paper strips.

PROCEDURES

☞ The fresh plants are placed between sheets of paper.

☞ The paper sheets containing the plants are pressed between a stiff flat board (e.g., wood) or a commercially available plant press.

☞ Within the first day of pressing the plants should be controlled and plant parts rearranged where required.

☞ Depending on the thickness and water content of the plants, changing of the paper sheets after one day, once or several times, might be necessary.

☞ Fast drying of plant samples at 40 to 50°C is possible and may result in better conservation of plant colours.

☞ Control of moulds can be achieved by moistening samples with pure alcohol.

☞ For the final preparation dry plants are fixed on card-board (international size: 28 x 44 cm) with glue or strips of paper.

2. Plastic sheet herbarium (especially for use in the field)

MATERIALS

✓ Thin cardboard (e.g., 15 x 21 cm, Din A5), sticky (adhesive) transparent plastic sheets which are on each side about 1 cm larger than the cardboard (e.g., 17 x 23 cm).

PROCEDURES

☞ The fresh plants are arranged as required on the plastic sheet where they get stuck.

☞ The cardboard is placed over the plant and sticks also to the plastic, whereas the exceeding four sides of the plastic are bent on the backside of the cardboard.

☞ Pressing of the samples (preferably in a dry room) is necessary, and with fleshy plant samples it is recommended to use newspaper sheets between them and to renew them until the sample is dry.

3. Dry-preparation in the plants natural habit (for stationary use and exhibition)

MATERIALS

✓ Fine sand, needles, flat wooden or cardboard box for drying, flat wooden box with transparent lid (glass or plastic) for final preparation.

PROCEDURES

☞ The very fresh plant is mounted in a box using needles; the natural habit should be maintained.

☞ The box with the plant needs to be carefully filled with sand. All plant parts needs to be completely covered with sand.

☞ Store box in a dry room to expedite drying process (2 to 4 weeks according to plant sample and relative humidity of the room). For the drying of fleshy plants (*Orobanche*) a change of the sand may be required.

☞ After drying carefully tip out the sand and finally fix the plant in a box which is closed with a lid made of glass to protect the plant from being touched.

COMMENTS

- If possible, plant species should be identified before pressing or fixing. If not possible, good notes should be made when collecting and preparing herbarium specimens.
- Labelling should include common name, species, family, site and date of collection, flower colour, host plant, crop and further information depending on the aim of collection.
- A plant is dry when it feels crisp and not cool when touched.
- Herbarium samples should be stored in the dark to maintain the colours.
- Protection against insects and fungi to be done if required.

LITERATURE

Lanjouw J & Stafleu FA 1964. Index Herbariorum. Utrecht.

ADDRESSES

The Herbarium, Royal Botanic Gardens, Kew, Richmond, Surrey, TW9 3AB, England.

Forschungsinstitut Senckenberg, Botanische Abteilung, Senckenberganlage 25, 60325 Frankfurt, Germany.

Author: K-H Linke

2.2 Morphological studies

INTRODUCTION

Most parasitic weeds are highly variable plants. They display an extraordinary plasticity in terms of variation of morphological characters. As a result of this, taxonomy of parasitic weeds of the genera *Striga* and *Orobanche* is a difficult task. Many „species“ just turned out as morphotypes that can be crossed with each other (→ chapter 2.3.2). Accordingly, these plant families are still at a relatively early developmental stage of their evolution. For breeding purposes and for preventive measures (quarantine) it is of high relevance whether such morphotypes represent biotypes with different virulence or represent diverse species.

An important factor influencing morphology is the host. In general, vigorous host plants allow development of ample parasites. *Striga hermonthica* plants on sorghum are much taller than on less vigorous weedy hosts. *Orobanche cre-*

nata on faba bean is much taller and may develop 10 times more seed than on a lentil plant. Specific environmental conditions like herbicide application or solarization may induce or select morphological changes on *Orobanche crenata*. Besides factors like different altitudes also temperature regimes influence the morphology in terms of changes in flower colour. With lower temperature during flowering *O. crenata* produces corolla tubes which are more or less entirely pink to violet, whereas with high temperatures during the flowering period the corolla is nearly white. Intermediate types may exhibit yellow colours.

OBJECTIVES

- To identify different species within one genus,
- to identify and describe morphological types within one species,
- to develop a background for taxonomic studies,
- to differentiate genetically fixed morphological variation from environmentally induced variation.

MATERIAL

✓ Magnifying glass, dissecting microscope, light microscope, scalpel and needles, line drawings, photocopy and/or camera.

✓ Relevant taxonomic literature.

PROCEDURE

☞ Collect plants at various natural sites and fields.

☞ Label samples, indicating host, location, short site description, nutritional status and development of the host.

☞ Carry out a preliminary evaluation *in situ* and the more specific one in the lab.

1. Characters to be analysed in *Striga* (→ annex 11.1)

- Separation into section Pentapleureae (calyx with five ribs) or section Polypleureae (calyx with 10 or more ribs).
- Type of leaves (scale-like or more developed), length of bracts, size, colour and pubescence of corolla, width of the limb, size of haustorium (especially in *Striga gesnerioides*), and others.

2. Characters to be analysed in *Orobanche* (→annex 11.3)

- In a first step usually distinction is made concerning the section Trionychon (*Orobanche aegyptiaca* type) and section *Orobanche* (*O. crenata, O. cernua and O. minor*). The second step is concerned with taxonomic work on the species level.
- Important characters are e.g., calyx (regular or zygomorphic), number of lobes, branching of stem, presence of bracteoles, tip of calyx (filiform or acuminate), arrangement of flowers (solimentary or in racemes), corolla (colour, length, shape), hairiness of filaments, colour of stigma. The study of the relevant taxonomic literature is a prerequisite for further taxonomic work.

3. Characters to be analyzed in *Cuscuta* (→annex 11.4)

- Type of inflorescence, number of flowers/inflorescence, size of flowers, presence or proportion of pedicel, stem diameter, calyx shape, calyx texture, calyx length vs. corolla tube length, degree of calyx split, calyx lobe outline, position of corolla lobes (upright or spreading), degree of corolla split, corolla lobe shape, corolla texture, infrastaminal scale shape, scale position in the corolla tube, scale size, anther shape, colour and size, stigma length vs. style length, stigma shape and colour, stigma and style length vs. ovary size, dehiscence of capsules.

EVALUATION

- ∗ Diligent synoptic comparison of characters.
- ∗ Description of morphotypes on the basis of differences of relevant characters.
- ∗ Useful material (also for later evaluations) are line drawings, photocopies, photographs and herbarium samples (→ chapter 2.1).

COMMENTS

- The taxonomy is of increasing complication starting from *Alectra* (→ annex 11.2) over *Striga*, *Cuscuta* to *Orobanche*.
- To differentiate between genetically fixed morphological and environmentally induced variation, it is essential to grow the parasites under modified conditions (e.g., other host plants) and to perform studies on interfertility and genetics (→ chapter 2.3).
- Due to their fleshy nature *Orobanche* and *Cuscuta* specimens lose many of their features when dried.

LITERATURE

Chater AO & DA Webb 1972. *Orobanche*. In: TG Tutin, VH Heywood, NA Burgess, SM Walters & DA Webb (eds.), Flora Europaea, III. University Press, Cambridge, England, 286-393.

Hepper FN 1984. Species of *Striga* in East Africa. In: C Parker, LJ Musselman, RM Polhill & AK Wilson (eds.), Proceedings 3rd International Symposium on Parasitic Weeds, ICARDA/International Parasitic Seed Plant Research Groupe, 7-9 May 1984, Aleppo, Syria. ICARDA, P.O.Box 5466, Aleppo, Syria, 262-265.

Hussain HSN & EI Aigbokhan 1991. Growth stages of *Striga gesnerioides* on cowpea roots. In: JR Ransom, LJ Musselman, AD Worsham & C Parker (eds.), Proceedings of the 5th International Symposium of Parasitic Weeds, CIMMYT, Nairobi, Kenya, 279-286.

Linke KH 1989. *Orobanche* studies. In: Food legume improvement program - Annual report for 1990. ICARDA, Aleppo, Syria, 214-242.

Musselman LJ 1986. Taxonomy of *Orobanche*. In: SJ ter Borg (ed.), Proceedings of a workshop on biology and control of *Orobanche*, LH/VPO, Wageningen, 2-10.

Musselman LJ 1987. Taxonomy of witchweeds. In: LJ Musselman (ed.), Parasitic weeds in Agriculture. CRC Press, Boca Raton, Florida, 3-12.

Parker C & CR Riches 1993. Parasitic Weeds of the World - Biology and Control. CAB International, Wallingford, UK, 332 pp.

Severova E 1991. New taxonomic characters in the genus *Cuscuta* (*Cuscuta*ceae). In: JR Ransom, LJ Musselman, AD Worsham & C Parker (eds.), Proceedings of the 5th International Symposium of Parasitic Weeds; CIMMYT, Nairobi, Kenya, 514-518.

Wettstein R 1893. Scrophulariaceae. In: A Engler & K Prantl (eds.), Die Natürlichen Pflanzenfamilien 4(36), 39-107.

Author: K-H Linke

2.3 Population-genetic studies

The ability of a parasitic weeds to adapt to new ecological niches - or to the resistance of a host - depends on the amount of genetic variation of the entire population, i.e., the global populations, as well as the local populations or subpopulations. Important aspects are the level of genetic variation and the population's temporal and spatial subdivision. To assess the potential of adaptation to new environmental conditions, knowledge is also important about the micro-evolutionary forces like mutation (source of new variation); migration (transport of genes between subpopulations, enriching local variation); recombination (intermixing the genome, thereby creating new genotypes); selection (survival of the fittest), and drift (random loss of alleles if populations are small).

An understanding of the patterns of variability of parasitic weed species within and between different regions, and the extent of host selection effects, is essential if breeding programs are to target sources of resistance at different areas and to realize the nature of the resistance required. Many cross-inoculation experiments have been carried out to study host specificity but little research has gone into studying geographical variability of the parasitic weeds themselves.

Studies of the molecular variability of populations have another application in the fields of taxonomy and systematics. Identification of species is a major problem in *Orobanche* as well as *Striga* and *Cuscuta* due to high morphological variability within each species. The differentiation of species like *Orobanche aegyptiana*, *O. ramosa*, and *O. mutelii* on the basis of morphological characters alone is almost impossible. Molecular markers represent a tool that is independent from environmental and developmental factors. It has been successfully applied for the estimation of genetic relationships within and between species.

2.3.1 Molecular markers

INTRODUCTION

There are many techniques available today for the study of the genetic variability of populations of organisms from DNA sequence analysis to the study of proteins / isoenzymes. They all share the objective of assessing levels and patterns of genetic diversity on the molecular level, but differ in factors such as resolution, focus, difficulty and expense.

Enzyme electrophoresis was the first technique to permit the assessment of genetic variation in populations at the molecular level. Although this method provides relatively low resolution, it is simple, fast and the most inexpensive way to study variation in natural and artificial populations. It is by far up until now the most widely used method.

The techniques that have advanced the direct analysis of DNA have been revolutionary in their impact, from restriction fragment length polymorphism (RFLP) to DNA sequencing to polymerase chain reaction (PCR)-based technologies and beyond. The random amplified polymorphic DNA (RAPD)-technique was one of the most modern PCR-based techniques in the early 1990s, but since then a number of new protocols have found wide applications in population genetic studies, such as microsatellite and amplification fragment length polymorphism (AFLP) technologies. Other techniques that have been described involve DNA conformational and denaturing gradient methods and the study of mitochondrial or chloroplastic.

OBJECTIVES

- To assess the genetic variation within and between parasitic weed populations,
- to enhance breeding by providing information on parasitic variability, genetics of virulence, and host selection,
- to reveal genetic relationships within and between species, and to enable and /or facilitate differentiation of species.

PROCEDURES

➔ Note: because of the complex and detailed requirements in terms of materials, methods, and also data analysis, it is recommended to conduct appropriate studies with the support of advanced research institutes and working groups already experienced in the subject.

LITERATURE

Borchardt D 1998. Population genetic structure of *Setosphaeria turcica*, the causal agent of Northern corn leaf blight, investigated with molecular markers. Ph. D. Thesis, University of Hohenheim, Stuttgart, Germany.

Dietrich WF, J Miller, R Steen, MA Merchant, D Damron Boles, Z Husain, R Dredge, MJ Daly, KA Ingalls, TJ OConnor, CA Evans, MM DeAngelis, DM Levinson, L Kruglyak, N Goodman, NG Copeland, NA Jenkins, TL Hawkins, L Stein, DC Page & ES Lander 1996. A comprehensive genetic map of the mouse genome. Nature 380: 149.

Doyle FJ & JL Doyle 1990. Isolation of plant DNA from fresh tissue. Focus 12: 13-15.

Fulton TM, J Chunwongse & SD Tanksley 1995. Microprep protocol for extraction of DNA from tomato and other herbaceous plants. Plant Molecular Biology Reporter 13: 207-209.

Hamrick JL 1989. Isozymes and the analysis of genetic structure in plant populations. In: DE Soltis & PS Soltis (eds), Isozymes in Plant Biology. Chapman and Hall Ltd., 87-105.

Jeffreys AJ, V Wilson & SL Thein 1985. Hypervariable minisatellite regions in human DNA. Nature, 314: 67.

Joel DM, V Portnoy, G Tzuri, R Greenberg & N Katzir 1996. Molecular markers for the identification of *Orobanche* species. In: MT Moreno, JI Cubero, D Berner, D Joel, LJ Musselman & C Parker (eds.), Advances in Parasitic Plant Research; 6th International Parasitic Weed Symposium, Cordoba, Spain, 151-160.

Koyama ML 1998. Genetic variability of *Striga hermonthica* and stability of resistance in sorghum. Ph.D. thesis, Wye College, University of London, UK, 30 pp.

Koyama ML 2000. Molecular markers for the study of pathogen variability: implications for breeding resistance to *Striga hermonthica*. In: Haussmann BIG, DE Hess, ML Koyama, L Grivet, HFW Rattunde & HH Geiger (eds.), Breeding for *Striga* resistance in cereals. Proceedings of a workshop held at IITA, Ibadan, Nigeria. Margraf Verlag, Weikersheim, Germany, 227-246.

Lewontin RC & JL Hubby 1966. Genetics 54: 595.

Melchinger AE 1990. Use of molecular markers in breeding for oligogenic disease resistance. Plant Breeding 104: 1-19.

Millan T, S Cobos & AM Torres 1996. The use of molecular markers in *Orobanche*. In: MT Moreno, JI Cubero, D Berner, D Joel, LJ Musselman & C Parker (eds.), Advances in Parasitic Plant Research; 6th International Parasitic Weed Symposium, Cordoba, Spain, 161-166.

Nickrent DL & RJ Duff 1996. Molecular studies of parasitic plants using ribosomal RNA. In: MT Moreno, JI Cubero, D Berner, D Joel, LJ Musselman & C Parker (eds.), Advances in Parasitic Plant Research; 6th International Parasitic Weed Symposium, Cordoba, Spain, 27-52.

Orita M, Y Suzuki, T Sekita & K Hayashi 1989. Rapid and sensitive detection of point mutations and DNA polymorphisms using the polymerase chain reaction. Genomics 5: 874-879.

Paterson AH, SD Tanksley & ME Sorrells 1991. DNA markers in plant improvement. Advanced Agronomy 46: 39-90.

Sambrook J, EF Fritsch & T Maniatis 1989. Molecular Cloning - a laboratory manual (2nd edition). Cold Spring Harbour Laboratory Press, U.S.

Shawe KG & MJ Ingrouille 1993. Isozyme analysis demonstrates host selection of parasite pathotypes in the association between cowpea and *S. gesnerioides*. Brighton Crop Protection Conference-Weeds 2: 919-924.

Theodet C, P Thalouarn, C Figureau & P Ferard 1996. A new approach to *Orobanche* species identification. In: MT Moreno, JI Cubero, D Berner, D Joel, LJ Musselman & C Parker (eds.), Advances in Parasitic Plant Research; 6th International Parasitic Weed Symposium, Cordoba, Spain, 143-150.

Thormann CE, ME Ferreira, LEA Camargo, JG Tivang & TC Osborn 1994. Comparison of RFLP and RAPD markers to estimating genetic relationships within and among cruciferous species. Theoretical Applied Genetics 88: 973-980.

Torres AM, ND Weeden & A Martin 1992. Linkage between isozymes, RFLP and RAPD markers in *Vicia faba*. Theoretical and Applied Genetics 85: 937-938.

Vos P, R Hogers, M Bleeker, M Reijans, T Van der Lee, M Hornes, A Frijters, J Pot, J Peleman, M Kuiper & M Zabeau 1995. AFLP: a new technique for DNA fingerprinting. Nucleic Acids Research 23: 4407-4414.

Williams JGK, AR Kubelik, KJ Livak, JA Rafalski & SV Tingey 1990. DNA polymorphisms amplified by arbitrary primers are useful as genetic markers. Nucleic Acids Research 18: 6531-6535.

Author: BIG Haussmann & K-H Linke

2.3.2 Use of DNA markers in the study of *Orobanche* and *Striga*

INTRODUCTION

The DNA fingerprinting method is the most accurate method for parasitic plant diagnosis when flowers are not available. This is also a useful technique for the detection of polymorphism within species and plant populations.

DNA-based markers have largely overcome the disadvantages of allozyme analysis, namely that they may be affected by environmental and developmental factors, that the number of polymorphic loci is limited and that discrimination of different genotypes is not always possible. The random amplified polymorphic DNA (RAPD) technique, based on the use of short primers of arbitrary nucleotide sequence in the polymerase chain reaction (PCR), has been shown to be useful for a wide range of applications, including the DNA fingerprinting of plants (Castiglione *et al.* 1993, Francisco-Ortega *et al.* 1993, Katzir *et al.* 1996).

This method, when applied on single seeds, allows rapid genetic analysis without waiting for seed germination and plant growth which takes weeks or months in many species. We have successfully used it for analysis of *Orobanche* species that are important in agriculture, from seven different countries. It should also help in DNA fingerprinting of the seeds of other plants like the root parasites of the genera *Striga* and *Alectra*.

OBJECTIVES

- To obtain RAPD profiles that are reproducible and unique to different genotypes,
- to identify single seeds when morphological characters are ambiguous or not available.

MATERIALS

✓ Plant material: Young flower buds or young (3-5mm) tubercles can be stored frozen, preferably at –80°C, until used. Seeds can also be stored at room temperature under dry conditions.

✓ Equipment: Thermal cycler, centrifuge (microtube), power supply, Agarose gel apparatus (small and large), autoclave, balance (analytical), camera (video, Polaroid or standard), distilled water source, UV light box (long wave), UV protective goggles, incubator (37°C), micropipetters (set), pH meter, stirring hot plate, vortex, water baths, Eppendorf tubes, PCR tubes.

✓ Chemicals: Taq polymerase kit, dNTPs (x10), homogenisation buffer, chloroform:isoamyl alcohol (24:1), cold isopropanol (-20°C), 70% ethanol, TE pH 7.6, TE pH 8.0, TBE Buffer pH 8.3 (x20), sample buffer.

✓ Homogenisation buffer: mix the following two solutions (a and b), 10 ml of each, and add 4 ml of 5% N-lauroylsarcosine. (a) 0.038 g Na bisulphite in extraction buffer containing 0.1M Tris-HCl pH 7.5, 0.35M sorbitol and 0.005M Na-EDTA. (b) 0.2 g CTAB in nucleus lysis buffer containing 0.2M Tris-HCl pH 8.0, 2M NaCl and 0.05M Na-EDTA. Keep it at room temperature.

✓ TE: 10mM Tris-HCl pH 7.6 or pH 8.0 with 1mM EDTA.

✓ TBE buffer: 45mM Tris-borate, 1mM EDTA.

✓ dNTPs: 2mM of each of the following: dATP, dGTP, dCTP, dTTP.

✓ Sample buffer: 36% glycerol, 0.4% SDS, 0.08% brom phenol blue, 0.08% xylene cyanole.

PROCEDURES

Orobanche DNA preparation (Katzir *et al.* 1996):

☞ Put a single flower bud or small tubercle, or a small portion of the stem (up to 50 mg) in the bottom of a 1.5 ml Eppendorf tube.

☞ Add 300 µl of homogenisation buffer and grind tissue with power drill and plastic drill bit as a pestle.

☞ Add another 300 µl of buffer and repeat grinding as above.

☞ Incubate at 65°C for 30 min.

☞ Add 600 µl of chloroform:isoamyl alcohol (24:1). Mix well. This can be done by shaking up and down for 5-10 min.

☞ Centrifuge tubes at 15,000 rpm for 5 min.

☞ Carefully pipette off the aqueous phase (ca. 400-450 µl) into new Eppendorf tube.

☞ Add 400 µl of cold isopropanol (-20°C). Invert tube several times. Leave the tubes at 4°C for 1 h.

☞ Spin at 15,000 rpm for 5 min., discard isopropanol and wash pellet with 70% ethanol. Dry pellet simply by opening the tube for 30 min. at room temperature.

☞ Resuspend DNA in 100 µl of TE pH 8.0 at 65°C for 10 min.

☞ Spin at 15,000 rpm for 5 sec, store at 4°C.

Orobanche and *Striga* seed DNA preparation (revised from Joel *et al.* 1996):

☞ Put a single seed of *Orobanche* or *Striga* in the bottom of a 0.5 ml PCR tube.

☞ Add 30 µl of homogenisation buffer and grind tissue with power drill and plastic drill bit as a pestle.

☞ Add another 30 µl of buffer and repeat grinding as above.

☞ Incubate at 65°C for 30 to 60 min.

☞ Add 60 µl of chloroform:isoamyl alcohol (24:1). Mix well. This can be done by shaking up and down for 5-10 min.

- ☞ Centrifuge tubes at 15,000 rpm for 5 min.
- ☞ Carefully pipette off the aqueous phase (ca. 30-40 µl) into new PCR tube.
- ☞ Add 40 µl of cold isopropanol (-20°C). Invert tube several times. Leave the tubes overnight at 4°C.
- ☞ Spin at 15,000 rpm for 5 min., discard isopropanol and wash pellet with 70% ethanol. The pellet is very small and cannot be seen. Dry pellet simply by opening the tube for 20 min. at room temperature.
- ☞ Resuspend DNA in 40 µl of TE pH 8.0 at 65°C for 10 min.
- ☞ Spin at 15,000 rpm for 5 sec, store at 4°C.

RAPD analysis:

- ☞ Mixture for the RAPD reactions contains 1-5 ng of *Orobanche* DNA (0.1-0.5 ng of seed DNA), 2mM of Mg^{2+}, 10pmole of primer, 100mM of each dNTPs, 1 x Taq Buffer, 0.5 unit of Taq DNA polymerase (1 unit for seed DNA), in a total volume of 25 µl (Williams *et al.* 1990).
- ☞ The primers, that were previously found useful for the identification of *Orobanche* (Katzir *et al.* 1996) are UBC 215, UBC 250, UBC 300, (the University of British Columbia) and OPG 6 (Operon Technologies Inc.).
- ☞ The amplification program: 1 min. denaturation at 94°C, 1 min. annealing at 36°C, and 2.5 min. extension at 72°C, for 45 cycles (60 cycles for seed DNA) on a thermocycler.
- ☞ Southern hybridisation of RAPD gels can be performed as described by Sambrook *et al.* (1989).

Table 2. Polymorphic amplification products that give unique clear bands, useful for the RAPD identification of the two sections of the genus *Orobanche* and of the agricultural important species.

Primer	UBC 300	UBC 250	OPG 6	OPG 6	UBC 215	UBC215
Amplification product (base pairs)	600	750	400	660	470	1400
sect. Osproleon Wallr.						
O. cernua Loefl.	+	-	-	-	-	-
O. cumana Wallr.	+	-	-	+	-	-
O. crenata Forsk.	+	-	-	-	-	+
sect. Trionychon Wallr.						
O. aegyptiaca Pers.	-	-	+	-	+	-
O. ramosa L.	-	+	-	-	-	-

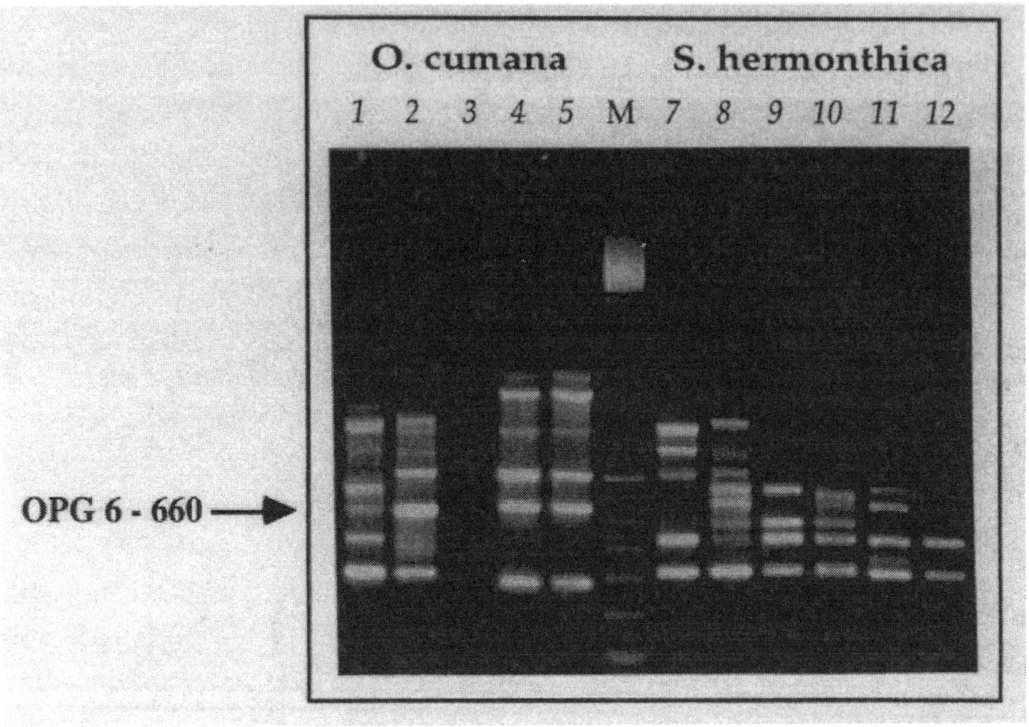

Fig. 3. RAPD pattern obtained with the primer OPG 6, of individual seeds of *Orobanche cumana* and of *Striga hermonthica.* The band OPG 6-660 that is unique to *O. cumana* occurs in all samples of this species. Note the different RAPD patterns in the *Striga* material. Lanes 1-2: *O. cumana* from Turkey. Lanes 3-4: *O. cumana* from Israel. Lane 6: DNA marker: 100 bp ladder. Lanes 7-12: *S. hermonthica* from Mali.

EVALUATION

- When the diagnostic primers (e.g., UBC 215, UBC 250, UBC300) are used, the occurrence of the respective bands serve for species identification (→Table 2, Fig. 3). Verification of the results can be done by southern hybridisation, that confirm identity of sequences.

COMMENTS

This method is very accurate and does not depend on the quality of testa cell walls, that are highly polymorphic in their morphology and rapidly deteriorate when buried in field soil. The RAPD patterns of DNA extracted from soil-borne *Orobanche* seeds is identical to that of DNA from vegetative plant material provided that the embryo and/or endosperm did not deteriorate. It is therefore recommended to examine soil borne seeds under a dissecting microscope before DNA extraction, and to select for this procedure only seeds that do not look empty.

LITERATURE

Castiglione S, G Wang, G Damiani, C Bandi, S Bisoffi & F Sala 1993. RAPD fingerprints for identification and for taxonomic studies of elite poplar (*Populus* spp.) clones. Theoretical and Applied Genetics 87: 54-59.

Francisco-Ortega J, HJ Newbury & BV Ford-Lloyd 1993. Numerical analysis of RAPD data highlight the origin of cultivated tagasate (*Chamaecytisus proliferus* ssp. *Palmensis*) in the Canary Islands. Theoretical and Applied Genetics 87: 264-270.

Joel DM, V Portnoy & N Katzir 1996. Identification of single tiny seeds of *Orobanche* using RAPD analysis. Plant Molecular Biology Reporter 14: 6-11.

Katzir N, V Portnoy, G Tzuri, M Castejon-Muñoz & DM Joel 1996. Use of random amplified polymorphic DNA (RAPD)

markers for the study of the parasitic weed *Orobanche*. Theoretical and Applied Genetics 93: 367-372.
Sambrook J, EF Fritch & T Maniatis 1989. Molecular cloning: a laboratory manual, 2nd edn. Vols. 1-3. Cold Spring Harbor Laboratory Press, Cold Spring Harbor, New York.
Williams JGK, AR Kubelic, KJ Livak, JA Rafalski & SV Tingey 1990. DNA polymorphisms amplified by arbitrary primers are useful as genetic markers. Nucleic Acids Research 18: 6531-6535.

Authors: DM Joel, V Portnoy & N Katzir

2.4 Ecological field assessments

2.4.1 Determination of ecological behaviour

INTRODUCTION

The weed flora in cultivated land is subjected to the diverse and changing aspects of soils and the interrelationships with other plants and other living organisms. Ecological studies of weed-crop interactions help to gain insight in the mechanism of distribution and adaptation of weeds to specific environmental conditions and are therefore indispensable for evolving suitable methods of weed control. In contrast to non-parasitic weeds, the occurrence of parasitic weeds is dependent on the presence of their specific host plants. The frequency of cropping host plants therefore strongly influences the reproduction of parasites and their occurrence. It is advisable to carry out ecological studies in combination with field surveys on the distribution of parasitic weeds (→ chapter 3.1.1).

OBJECTIVE

- To get a better understanding on the occurrence and distribution of parasitic weeds in agro-ecosystems in relation to environmental and anthropogenic factors.

MATERIALS

→ chapter 3.1.1, in addition

✓ indicator paper or indicator solutions (e.g., pH-meter of Hellige GmbH, Germany)
✓ HCL (10% solution), water
✓ Altimeter and compass
✓ Prepared form for a site description (→ annex 11.7)

PROCEDURE

☞ **Localisation of the site:** Selection of sites (→ chapter 3.1). The sites are localised and the area and/or fields are marked on an adequate map. This allows comparisons with, e.g., soil maps or climatic maps for data interpretation. Maps may be used to derive the required information on the locality (latitude, longitude, altitude etc.). Employment of the Global Positioning System (GPS) could largely facilitate the site localisation and its general description.

☞ **Description of the site:** For the description of the sites, all required information could be gathered in a special "multiple choice" form (→ annex 11.7), where the different aspects are provided and one has to choose the ones which apply to the field.

☞ **Relief:** Slope inclination: 0-2°-plain; 2°-5°-slight; 5°-10°-moderate; 10°-

20°-strong; 20°-30° steep; 30°-45°-very steep; >45°-extremely steep, slope direction: using compass, altitude: (m above sea level, by altimeter).

☞ **Climatic data** are usually taken from meteorological statistics.

☞ **Soil data** may be gathered from soil maps. But in most cases it is worthwhile to have some data on the actual situation in the field as well. For a thorough information laboratory studies are necessary. But data as listed in annex 11.7 can be obtained already through field observations and relevant methodologies:

Soil texture:

The soil texture can be estimated by a manual test as shown in Fig. 4.

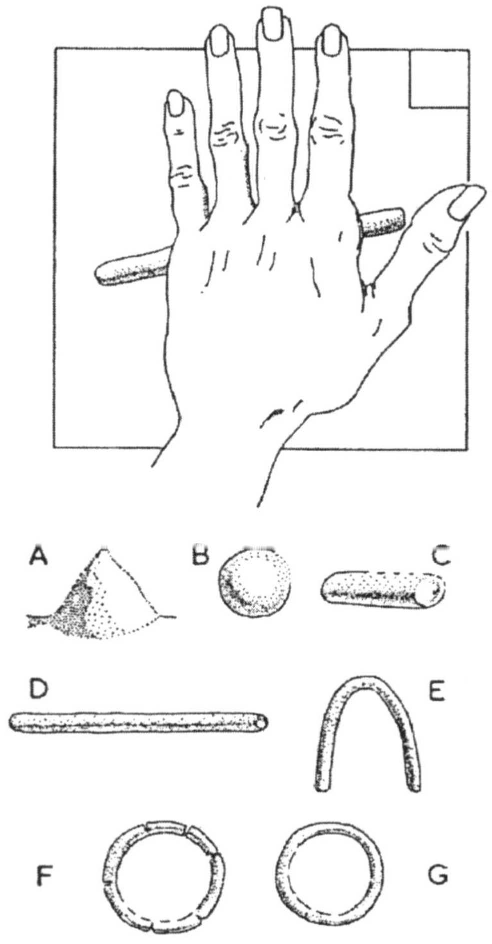

Fig. 4. Manual test to estimate the textural class of soils

A = sand, B = loamy sand, C = silt loam, D = loam, E = clay loam, F = light clay, G = heavy clay (taken from Anonymous 1981).

Content of organic matter

- Fast and simple method: Estimation of the organic matter content through the fact that the darker the soil in moist condition the higher the organic matter content.
- More exact measurement: Loss of soil weight during burning in the oven.

Soil reaction

The soil reaction is measured as the pH value. In the field pH values can be roughly assessed with the aid of indicator paper, indicator solutions or field pH-meters.

Calcium content

May be identified by judging the intensity of effervescence after HCl (10% solution) is dripped on the sample. According to the intensity of effervescence, the calcium-content is as follows: poor: <2%, moderate: 2-5%, intense: 5-10% and very intense: >10%.

For a more precise laboratory analysis, depending on the size of the farmers fields five to ten individual soil test portions (0 30 cm soil depth) have to be collected randomly using soil augers at each location. Representative soil samples are prepared thereafter by mixing these portions. Samples are air-dried and sieved (2 mm) and laboratory analysis on specific soil parameters can be carried out according to Schlichting & Blume (1966), Cottenie (1980) and USDA (1967).

EVALUATION

* Relate parasitic weed infestation intensity (→ chapter 3.1.1) with the different parameters of interest.

COMMENTS

- In general, the production system influences the development of weeds. Hence, any change in the cropping system has an impact on the occurrence of weeds. Parasitic weeds are strongly influenced by the length of fallow, the frequency of cropping host plants, the degree of resistance in the cultivated varieties, the use of trap crops (e.g., cotton) in crop rotation, and by the amount of nitrogen fertiliser applied. Therefore, during field surveys, farmers should be interviewed on the field history as precisely as possible, including cropping pattern, nutrient supply and so on. For data recording again prepared forms could be used (e.g., → annex 11.7) and kept for immediate or later evaluation with parasitic weed sampling data (→ chapter 3.1.1).

LITERATURE

Adrianaivo AP, S Kachelriess, J Kroschel & W Zehrer 1998. Biologie et gestion du *Striga* à Madagascar. Projet „Ecologie et Gestion des Plantes Parasites" / Direction de la Protection des Végétaux, Nanisana, Madagascar, 61pp.

Anonymous 1981. Agricultural compendium for rural development in the tropics and subtropics. Elsevier Scientific Publishing Company, Amsterdam, 739 pp.

Cottenie A 1980. Soil and plant testing as a basis for fertilizer recommendations. FAO Soils Bulletin 38/2 Rome.

Janetschek H 1982. Ökologische Feldmethoden. Ulmer Verlag Stuttgart, Germany.

Kroschel J 1998. *Striga* – How will it affect African agriculture in the future? – An ecological perspective. In: K Martin, J Müther & A Auffarth (eds.), Agroecology, Plant Protection and the Human Environment: Views and Concepts. PLITS 16(2): 137-158.

Pülschen L 1990. Natural site factors and agronomic aspects in the grain-plough complex of the Shewa Province, Ethiopia. PLITS 8(3), 84 pp.

Schlichting E & H-P Blume 1966. Bodenkundliches Praktikum. Parey Verlag, Berlin, Hamburg, Germany, 205 pp.

USDA 1967. Soil survey laboratory methods and procedures for collecting soil samples. Soil survey investigations rep. no. 1. Soil Cons. Serv. Washington D.C.

Vogt W, J Sauerborn & M Honisch 1991. *Striga hermonthica*, distribution and infestation in Ghana and Togo on grain crops. In: JK Ransom, LJ Musselman, AD Worsham & C Parker (eds.), Proceedings of the 5th International Symposium of Parasitic weeds, June 1991, Nairobi, Kenya, 372-377.

Walter H 1983. Weed sampling in the field and interpretation of sampling data. In: Weed management in the Philippines, Reports of seminars. PLITS 1(1): 65-80.

Authors: J Kroschel & K-H Linke

2.4.2 Toxic effects, transpiration, respiration and photosynthesis

INTRODUCTION

Host plants of parasitic weeds are damaged by the loss of carbohydrates and water. However, experiments with *Striga* demonstrated that the loss of biomass of the host is larger than it can be explained by competition for resources, indicating the occurrence of one or several toxins released from the parasite. Furthermore, *Striga* induces an increased partitioning of dry weight to the root system of the infected host at the expense of the host shoot. Hence, a better understanding of physiological aspects of the host-parasite-interaction is essential for the development of any management strategies.

The existence of toxins in parasitic plants is greatly supported by various experimental results in recent years, though the toxin(s) was (were) not identified yet. On the other hand, why should a parasite affect its host in this way, what is the biological reason? Several findings indicate serious physiological imbalance of the host after infection; would such disorder not be sufficient to already explain symptoms of toxicity?

Parasitic plants differ from normal plants due to the reduction in organs responsible for the photosynthesis. Also transpiration and respiration are different from non-parasitic plants. Studies on this topic will include measurements on the parasite as well as on its host plant, because also the photosynthesis of the infected host is severely affected.

Recommendations in „MATERIALS" and „PROCEDURES" on physiological and biochemical aspects can not be exhaustive due to the extensive field of potential analyses and investigations.

OBJECTIVES

- To quantify the effect of parasitism on host dry weight as well as partitioning between host root and shoot biomass,
- to investigate the physiological imbalance effected by parasitism,
- to investigate the effect of parasitism on physiological and biochemical characteristics and processes such as amino acids or growth regulators, especially in the context with nitrogen fertilisation, or concerning the partitioning of sugars in the host,
- to study symptoms of parasitism such as chlorotic and/or necrotic lesions, stunting or wilting even under well-watered conditions and to estimate the potential phytotoxic effect of parasites,
- to identify the potential toxin(s),
- to quantify the effect of parasitism on transpiration and photosynthesis of the host plants,
- to measure transpiration rate, respiration, stomatal conductance, water use efficiency and photosynthesis of parasitic plants,
- to study aspects of carbon fixation,
- to evaluate the efficiency of antitranspirants,
- to identify potential control targets.

MATERIALS

✓ Balance and oven for dry weight assessments.

✓ Physiological/biochemical studies will require laboratory equipment appropriate to the relevant studies (too specific to be listed here).

✓ According to the question under study experiments may be conducted in the lab, the greenhouse or in the field.

✓ Due to the complex questions under study sufficient field plots with a large level of parasite infestation within the experiment are required in field studies.

✓ Stationary or portable equipment for foliar gas exchange measurement which is available in various size and options, including microcomputer, temperature and humidity recorder, printer, printer paper, electrical power to run the printer and to load the batteries (for portable systems).

✓ The chamber should have a volume of at least 0.5 l in order to measure whole plants, as most parasites exhibit tiny leafs only.

PROCEDURE

- ☞ *Striga* and *Orobanche* plants are harvested and oven-dried shortly after most plants have finished flowering.
- ☞ Studies on partitioning require separate assessment of root and shoot biomass of infected and non-infected host plants. A rather large set of data is required to obtain statistically significant results.
- ☞ Transpiration (rate) is assessed on the basis of the release of $H_2O/m^2/s$ (stomatal conductance).
- ☞ Photosynthesis is measured by the release of $CO_2/m^2/s$.
- ☞ Measurements have to be conducted following the instructions of the respective equipment used.
- ☞ Make sure that the chamber is properly sealed.
- ☞ The transpiration rate has to be related to the temperature of tissue and surrounding air (usually measured also by the transpiration measurement).
- ☞ It is desirable for proper evaluation and discussion of the results to also measure the water capacity of the soil (pF-values). In addition it is recommended to record the prevailing weather conditions (clouds, sunshine).
- ☞ Carry out repeated measurements (at different days) under comparable weather conditions.

EVALUATION

- ✶ The number of measurement depends on the variation obtained on the single plant measurements and the type of measurement. Normally, three to five measurements per treatment and respective time are sufficient.
- ✶ Evaluation of the host root system involves length and dry weight of the main root as well as number, length and weight of lateral roots. Data from the root system can be related to data from the plant parts above ground.
- ✶ Linear regression equations with the parasite dry weight/m^2 as the independent variable and the host dry weight/m^2 as the dependent variable can be used to estimate the effect of the parasite on the host; correlation coefficients for these two variables will provide further information.
- ✶ With regard to photosynthesis, several measurements at several times (five or more) during the day are required to obtain a complete set of data.
- ✶ Collect sufficient data in order to compare photosynthesis and transpiration of the parasite with that of an infested and a non-infested host plant (each in relation to temperature and soil water conditions).
- ✶ Stomatal conductance and transpiration rate can be calculated on a leaf area basis (von Caemmerer & Farquhar 1981).

COMMENTS

- For the study on toxic effects the variation of host varieties has to be considered.
- Portable systems to measure the gas exchange are of advantage as the measurements can be conducted *in situ* under natural conditions. A limitation is sometimes the size of the chamber in which the measurement is carried out.

- Important for measurements on *Orobanche* is that the chamber can be closed tight with rubber foam sealing broad enough to exclude any air exchange between inside and outside of the chamber. This should not be taken for granted, as most systems are designed for leaves and not for a thick stem with at least 0.5 cm in diameter. Due to extreme reductions of the scaly leaves, measurements on parasites like *Orobanche*, *Cistanche*, *Cytinus*, *Lathraea* or *Cynomorium* require the inclusion of the whole parasite into the chamber for proper measurements.
- Large differences, e.g., in transpiration rate may arise from unequal water supply to a given host plant.
- Parasite dry weight/m^2 and host dry weight/m^2 usually are negatively correlated; under a high level of nutrients (basically nitrogen), however, the situation may be different.

LITERATURE

Caemmerer S von & GD Farquhar 1981. Some relations between the biochemistry of photosynthesis and the gas exchange of leaves. Planta 153: 376-387.

Gurney AL, MC Press & JK Ransom 1995. The parasitic angiosperm *Striga hermonthica* can reduce photosynthesis of its sorghum and maize host in the field. Journal of Experimental Botany 46: 1817-1823.

Knapp M 1990. [Studies on transpiration of *Orobanche crenata* Forsk. and the effect of parasitism on transpiration and photosynthesis of different host plants]. Diploma thesis (in German), University of Hohenheim, Germany.

Kroschel J 1989. [The effect of nitrogen on phanerogamic root parasites and their host plants with special refernce to *Striga hermonthica* (Del.) Benth. (Scrophulariaceae)]. PLITS 7(3).

Linke KH 1991. *Orobanche* studies. In: Food legume improvement program - Annual report for 1990. ICARDA, Aleppo, Syria, 216-242.

Parker C & CR Riches 1993. Parasitic Weeds of the World - Biology and Control. CAB International, Wallingford, UK, 332 pp.

Press MC, JJ Nour, FF Bebawi & GR Stewart 1989. Antitranspirant-induced heat stress in the parasitic plant *Striga hermonthica* - A novel method of control. Journal of Experimental Botany 40: 585-598.

Press MC, JM Tuohy & GR Stewart 1987. Gas exchange characteristics of the sorghum-*Striga* host-parasite association. Plant Physiology 84: 814-819.

Ransom JK, GD Odhiambo, RE Eplee & AO Diallo 1996. Estimates from field studies of the phytotoxic effects of *Striga spp.* on maize. In: MT Moreno, JI Cubero, D Berner, D Joel, LJ Musselman & C Parker (eds.), Advances in Parasitic Plant Research. Proceedings 6th Parasitic Weed Symposium. Cordoba, Spain, 327-333.

Stewart GR, MC Press, JD Graves, JJ Nour & A Wylde 1991. A physiological characterization of the host-parasite association between *Sorghum bicolor* and *Striga hermonthica* and its implications for *Striga* control. In: SK Kim (ed.), Combating *Striga* in Africa. Proceedings, International Workshop organized by IITA, ICRISAT and ICRC, 1988 IITA, Ibadan, Nigeria, 48-54.

Thalouarn P, MC Arnaud, C Théodet & L Rey 1991. Cytological, biochemical and genetic aspects of carbon fixation in *Striga hermonthica* and *Striga gesnerioides*. In: Ransom JK, LJ Musselman, AD Worsham, & C Parker (eds). Proceedings of the 5th International Symposium on Parasitic Weeds. CIMMYT, Nairobi, Kenya, pp. 51-57.

Tuohy JM & JS Choinski 1991. Water relations and photosynthesis in the mistletoe *Tieghemia bolusii* and its host *Dipolrhynchus condylocarpon*. In: JK Ransom, LJ Musselman, AD Worsham & C Parker (eds.), Proceedings of the 5th International Symposium of Parasitic Weeds. CIMMYT, Nairobi, Kenya, 232-236.

Tuquet C & G Sallé 1991. Biochemical features of chloropasts isolated from hemiparasitic angiosperms. In: Ransom JK, LJ Musselman, AD Worsham & C Parker (eds), Proceedings of the 5th International Symposium on Parasitic Weeds. CIMMYT, Nairobi, Kenya, pp. 262-269.

Author: K-H Linke

3 Relevance to agriculture

3.1 Quantification of infestation and yield loss

3.1.1 Field survey

INTRODUCTION

In general, field surveys are an important tool in plant protection, especially in Integrated Pest Management, to get a better understanding about the occurrence of pests (insects, diseases and weeds) and their magnitude in crop production. With regard to parasitic weeds, objectives of field surveys are as follows:

OBJECTIVES

- To provide data on the geographical distribution of parasitic weeds at the local, regional or supra-regional level,
- to obtain data on the magnitude of the infestation with regard to the parasitic weed species and the crops infected,
- to estimate a change of infestation by repeating surveys from time to time,
- to provide data on the occurrence of parasitic weeds with regard to studies on their ecological behaviour (→ chapter 2.4.1),
- to obtain basic data for the estimation of regional yield losses (→ chapter 3.1.2), and
- to identify endangered areas in order to know where to concentrate and strengthen extension work with regard to parasitic weed control.

MATERIALS

✓ Vehicle (car or motorcycle, fuel) to move from field to field

✓ Permission of relevant authorities

✓ Map with scale according to the surveyed area

✓ Identification material (field guides)

✓ Material for preparing herbarium samples (→ chapter 2.1)

✓ Data sheet (→ annex 11.7)

✓ Time requirement depends on specific conditions. It includes the preparation of the survey (several days), the survey itself (15 to 30 min per field plus transport from field to field), and the time for the evaluation (days or several weeks according to number of samples and extent of analysis).

PROCEDURE

☞ Randomized selection of fields, approx. every 5 to 10 km along roads. Use not only main roads!

☞ Select only fields were weed control, especially the weeding of parasitic weeds, has not yet been done by farmers. Give preference to fields where the crop has already reached the reproductive stage in order to include also late emerging parasites.

☞ Assess parasite and crop species.

☞ The procedure of the field inspection has to be chosen depending on the crop. In general, the fields are crossed in a diagonal line (Fig. 5). After every 20 steps, count the number of parasites/crop plant or hill (e.g., *Striga* in sorghum or maize) within a radius of 60 cm equalling $1 m^2$. If it is not possible to relate shoots of the parasite to one crop plant, it is best to count parasites/m^2 (e.g., *Striga* in rice) at different spots of the field by following always the same procedure. With regard to *Orobanche crenata* it has also been proved to estimate the infestation according to a rating scale from 0 to 6 (Table 3).

☞ Sample size: at least 20 planting holes or 20 m² selected at random.

☞ Write your assessed field data in a prepared data sheet (→ annex 11.8)

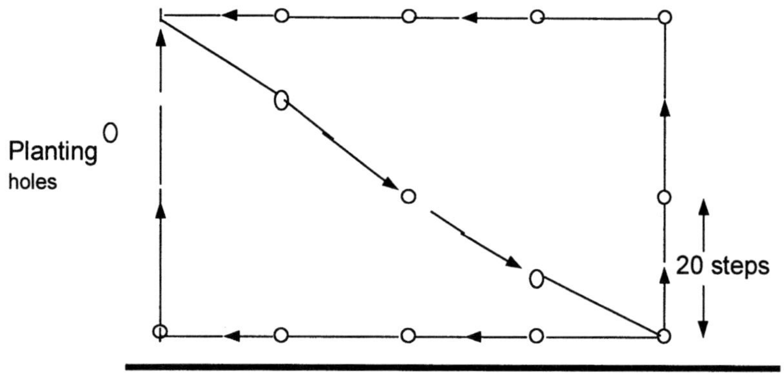

Fig. 5. Procedure of the field inspection

Table 3. Rating scale to estimate infestation of *O. crenata* in faba bean fields (Schmitt 1981)

Rating scale	Infestation	Definition
0	not infested	no *Orobanche* emerged
1	very low	*Orobanche* sporadically present
2	low	*Orobanche* present in the entire field
3	moderate	majority of host plants infected with 2 *Orobanche* shoots
4	strong	all host plants infected with more than 2 *Orobanche* shoots, but the field retains the character of a faba bean field
5	very strong	all host plants infected with more than 5 *Orobanche* shoots, field seems to be an *Orobanche*-field
6	host plants completely destroyed	field completely destroyed by *Orobanche,* no yield at all

EVALUATION

★ **Infestation rate**: Percentage of infested fields (%).

★ **Infection intensity**: Percentage of infected plants per field (sorghum, millet, maize) or number of parasites per m^2 (rice, faba bean, lentil, chickpea). The infection intensity can be grouped in five infection classes: 0%, 2.5-10%, 11-50%, 51-75% and 76-100%. A value of 2.5% is quoted, if parasitic weeds occur in a field but are not observed within the random sample.

COMMENTS

- The level of infestation can be highly variable between years. Factors like weather conditions in years of host crop cultivation, time of planting, soil fertility, intra-specific competition of the parasites etc. influence the parasites' emergence.
- A uniform sampling frame does not lead to a representative sample as population density and intensity of agricultural production are far from being uniform.

LITERATURE

Geiger U, J Kroschel & J Sauerborn 1996. *Striga asiatica*, a problem in the Middle West of Madagascar. In: MT Moreno, JI Cubero, D Berner, DM Joel, LJ Musselman & C Parker (eds.), Advances in Parasitic Plant Research. Proceedings VI Parasitic Weed Symposium. Cordoba, Spain, 479-486.

Geipert S, J Kroschel & J Sauerborn 1996. Distribution and economic importance of *Orobanche crenata* in faba bean production of Morocco. In: W Bertenbreiter & M Sadiki (1996): Rehabilitation of Faba Bean. Proceedings 1st Maghrebian Seminar on Faba Bean Research, 24.-27. Mai 1995, Rabat/Marokko, 113-120.

Reichmann S, J Kroschel & J Sauerborn 1995. Distribution and infestation of *Striga* species in Shinyanga region of Tanzania and evaluation of control methods. Brighton Crop Protection Conference Weeds, Brighton, UK, Vol. 1, 151-156.

Schmitt U 1981. Untersuchungen zur Verbreitung und Bekämpfung von *Orobanche crenata* Forsk. an Ackerbohnen in Marokko. Ph.D. Thesis, University of Bonn, Germany, 194 pp.

Wetzel T 1984. Methoden der Schaderreger und Bestandesüberwachung. In: Diagnosemethoden. Springer Verlag, Berlin, Heidelberg, New York, Tokyo, 15-17.

Author: J Kroschel

3.1.2 Yield loss assessment

INTRODUCTION

Due to the withdrawal of water, minerals and organic compounds from their hosts, the toxic effect of *Striga*, parasitic weeds are important biological constraints in agricultural production. The growth of their hosts is retarded and crop yields are lowered or reduced to zero under conditions of severe infestation. The relative total yield loss is the product of three factors:

- the infestation rate,
- the infection intensity, and
- the relative yield reduction per plant stand or m^2.

The absolute economic losses faced by the farm households and the entire regional economics further depend on (1) the potential yield of the different crops, (2) the actual acreage covered by these crops, and (3) crop prices.

To estimate regional yield losses, assessments have to be based on field surveys to determine the infestation rate and the infection intensity (→ chapter 3.1.1). In addition, yield assessments have to be carried out in single fields comparing weed-free (non-infected) plant stands with weed-infected ones (Fig. 6).

OBJECTIVES

- To quantitatively determine yield and regional production losses,
- to raise awareness of the severity of a parasitic weed problem.

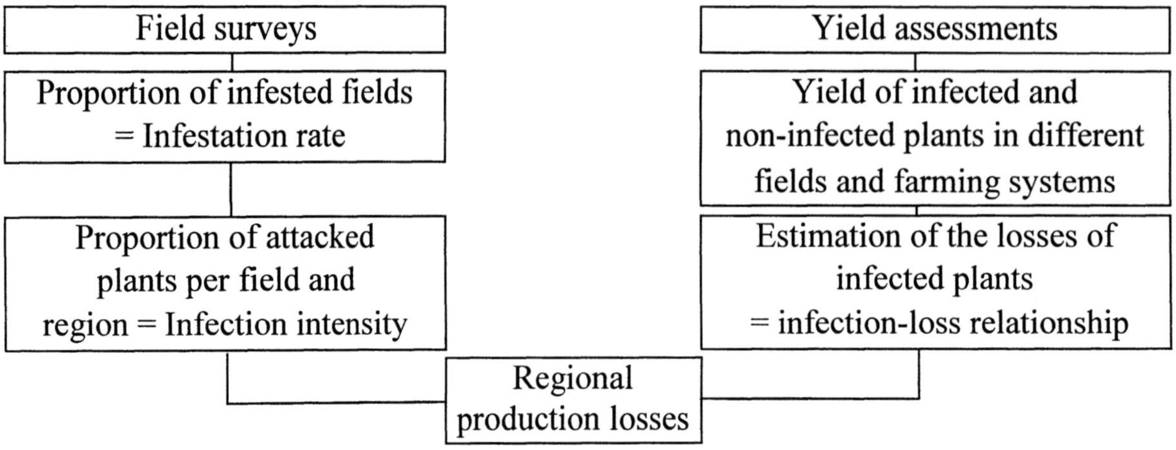

Fig. 6. Model for the estimation of regional production losses caused by parasitic weeds (Kroschel *et al.* 1996)

MATERIALS

✓ → chapter 3.1.1, and in addition

✓ a balance (for direct measurements in the field),

✓ paper-bags (for the transport of samples to the lab) and the

✓ permission of relevant farmers.

PROCEDURE

On-farm

☞ Select about 15-25 fields in regions with similar farming systems, precipitation, soils etc.

☞ Select fields with an infection intensity of 50-80% (not 100%) to allow the selection of approximately 30 plants (or 30 m²) with and without attack by parasitic weeds at random. For the unequivocal distinction between infected and non-infected plants at harvest, the randomly selected plant stands are marked already during the vegetation period (→ annex 11.8).

☞ Select fields also according to following criteria to avoid the influence of other yield limiting factors as far as possible: (1) low weed competition and no visual influence by pests and diseases, (2) no drought stress during the vegetation period, and (3) no parasitic weed control by farmers.

On-station

☞ Estimation of regional yield losses can also be derived from yield functions calculated on the basis of field trials. By using different control methods (e.g., herbicide treatments and an untreated control and/or by establishing parasite free plots → chapter 5.3.3), a relation between intensity of infestation and yield can be established (e.g., using glyphosate in faba bean to control *O. crenata*). The differences in effectiveness of the control methods applied produce variations in parasitic weed population development and infection intensity. The potential yield as well as the yield loss can be derived from yield functions (Fig. 7).

☞ Parameters for loss assessments in on-farm and on-station trials depend on the respective crop. Beside plant height this may include:

Maize: number of cobs, cob and seed weight

Sorghum: panicle and seed weight

Rice: number of panicles per m², seed weight per m²

Faba bean: number of pods/plant, seed weight, number of plants per m²

EVALUATION

* A quantified parasitic weed (or pest) density (*i*) operates on the crop yield (*y*) in a typically linear regression:

$$y = m - b(i)$$

* where (*m*) is the yield in the absence of weeds (or pests), and (*b*) is the rate of loss per unit increase in (*i*). If the yield in the presence of pests and weeds is (y_i), crop loss (*w*) may be defined as the reduction in (*m*), expressed as a percentage of it:

$$\text{Crop loss } (w)\ \% = \frac{m - y_i}{m} x100$$

* Check data for normal distribution and then conduct an analysis of variance and subsequent Tukey-test.
*

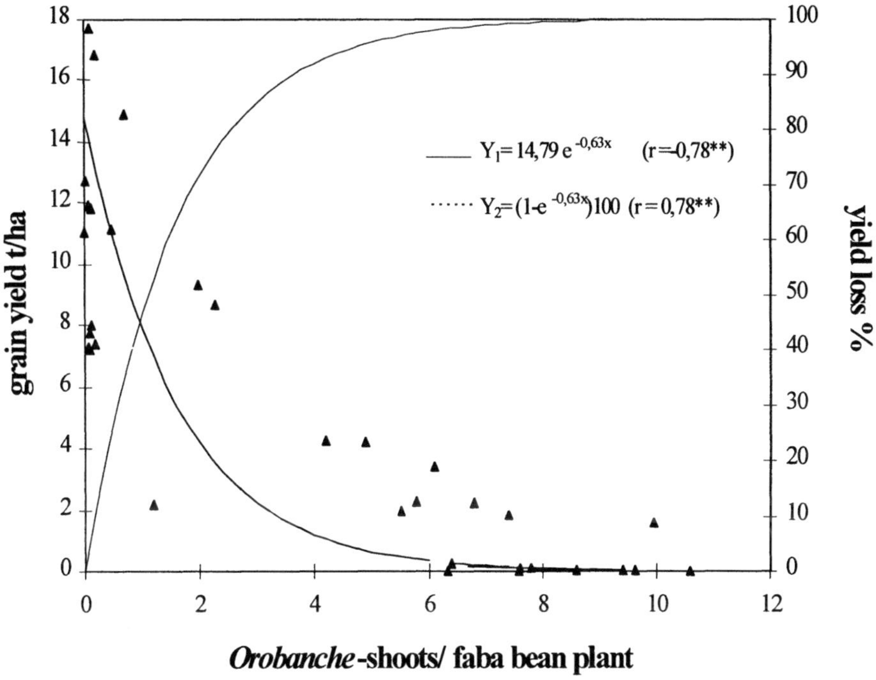

Fig. 7. Functions for absolute yield reduction (Y_1) and percentage of yield loss (Y_2) in faba bean due to *Orobanche crenata* infection (calculated on the basis of the mean faba bean plant infection in 32 plots with different treatments) (* significant at the α=0.05 level) (Geipert *et al.* 1996)

* Estimations of regional yield losses have to be derived from statistical data on crop production (cultivated area of the crop concerned, e.g., of maize or sorghum). Since these yield calculations reflect more or less the realistic cropping conditions, they include losses due to pests, diseases, and weeds including parasitic weeds. Therefore, the potential yield without parasitic weed infestation is calculated (e.g., for each district of the surveyed area) dividing the actual yield by 100-x/100 with x being the relative yield loss caused by parasitic weeds (per district).

COMMENTS

- Yield loss assessments are very time consuming and costly and help in problem identification only. Hence, a justifiable input-output relationship should be given, i.e. investigations have to be reduced to a statistically tolerable minimum of field inspections.
- Estimating regional production losses on the basis of field surveys

and yield assessments in single fields are problematic since the infestation as well as the yield is a function of several factors. The occurrence of *Striga* in the field may vary from year to year since the development of parasites is influenced by different ecological factors (e.g., soil moisture, soil temperature, rainfall distribution) which due to high interannual fluctuations. Therefore, surveys on the distribution and infestation should be carried out in several seasons to get a real impression of the situation.

- Differences in the yields of infected and uninfected crops are very variable. Factors like year of cultivation, soil fertility, weather patterns, intraspecific competition of the parasites etc. strongly influence the effect of parasitic weeds on crop yields.

LITERATURE

Agbobli CA & B Hugenin 1987. Evaluation agronomique du problème du parasitisme du *Striga* sur maïs dans le Sud Togo. In: C Weber & W Forstreuter (eds.), Parasitic Flowering Plants. Proceedings 4th International Symposium on Parasitic Flowering Plants, Marburg, Germany, 11-25.

Bebawi FF & AF Farah 1981. Effects of parasitic and non-parasitic weeds on sorghum. Experimental Agriculture 17. 415-418.

Carson AG 1988 Detailed survey of the parasitic weed *Striga hermonthica* in The Gambia. Tropical Pest Management 34(2): 162-164.

Dawson JH & JR Holstun 1971. Estimating losses from weeds in crops. In: L Chiarappa (ed.), Crop loss assessment methods: FAO Manual on the evaluation and prevention of losses by pests, diseases and weeds. FAO and C.A.B., Farnham Royal, UK.

Doggett H 1965. *Striga hermonthica* on sorghum in East Africa. Journal of Agricultural Science 65: 183-194.

Geipert S, J Kroschel & J Sauerborn 1996. Distribution and economic importance of *Orobanche crenata* in faba bean production of Morocco. In: W Bertenbreiter & M Sadiki (1996): Rehabilitation of Faba Bean. Proceedings 1st Maghrebian Seminar on Faba Bean Research, 24.-27. Mai 1995, Rabat/Marokko, 113-120.

Kroschel J 1998. *Striga* – How will it affect African agriculture in the future? – An ecological perspective. In: K Martin, J Müther & A Auffarth (eds.), Agroecology, Plant Protection and the Human Environment: Views and Concepts. PLITS 16(2): 137-158.

Kroschel J, B Mössner & J Sauerborn 1996. Estimating maize yield losses caused by *Striga asiatica* in Malawi. In: MT Moreno, JI Cubero, D Berner, D Joel, LJ Musselman & C Parker (eds.), Advances in Parasitic Plant Research. Proceedings VI Parasitic Weed Symposium. Cordoba, Spain, 335-346.

Last FT 1960. Effect of cultural treatments on the incidence of *Striga hermonthica* (Del.) Benth. and yields of sorghum in the Sudan: Field Experiments 1957/58. Annals of Applied Biology 48: 207-229.

Patterson DT 1987. Environmental factors affecting witchweed growth and development. - In: LJ Musselman (ed.), Parasitic Weeds in Agriculture, Vol. 1, *Striga*. CRC Press, Boca Raton, USA, 27-41.

Sauerborn J 1991. The economic importance of the phytoparasites *Orobanche* and *Striga*. In: JK Ransom, LJ Musselman, AD Wosham & C Parker (eds.), Proc. 5th International Symposium on Parasitic Weeds, Nairobi, Kenya, 137-143.

Vasudeva Rao MJ, VL Chidley & LR House 1989. Estimates of grain yield losses caused in sorghum (*Sorghum bicolor* (L.) Moench) by *Striga asiatica* (L.) Kuntze obtained using the regression approach. Agriculture, Ecosystems and Environment 25: 139-149.

Walker PT 1983. The assessment of crop losses in cereals. Insect Science Application, 4(1/2): 97-104.

Wetzel T 1984. Methoden der Schaderreger- und Bestandesüberwachung. In: T Wetzel (ed.), Diagnosemethoden. Springer Verlag, Berlin, Heidelberg, New York, Tokyo, 15-17.

Author: J Kroschel

3.2 Analysis of farming systems

INTRODUCTION

A number of methods to control parasitic weeds have been developed in the last decades. It has been realized that the applicability of these methods by small scale farmers differs from one country or region to another or also from farmer to farmer. An analysis of farming systems is therefore essential to develop and identify those integrated control methods, which are most appropriate for specific socio-economic conditions.

OBJECTIVES

- To assess the awareness and knowledge of farmers of parasitic weeds and their special biology,
- to assess the indigenous farming practices and their possible impact on parasitic weed population,
- to identify farmers' control strategies,
- to evaluate the possibilities of farmers to apply innovative control methods.

MATERIALS

✓ Vehicle (car or motorcycle, fuel) to move from village to village (from farmer to farmer).

✓ Permission of relevant authorities.

✓ Prepared questionnaire (→ annex 11.9).

✓ Time requirement depends on specific conditions. It includes the preparation of the questionnaire (several days), the interview itself (30 to 60 min per farmer plus the transport from farmer to farmer) and the time for the evaluation (several days or weeks according to number of interviews conducted).

PROCEDURE

☞ Compiling of a questionnaire for a standardized interview comprising open response and closed response questions on the farming system, general plant protection problems and in particular on relevant parasitic weeds (→ annex 11.9). The open response question leaves the answer entirely to the interviewee while the closed response question offers multiple choice answers.

☞ Pre-testing of the applicability of the questionnaire. If necessary rephrasing, adding new questions or dropping questions. Several tests might be needed.

☞ The length of the interview should not exceed one hour.

☞ Translation of the final questionaire into the local language.

☞ Random selection of villages and farmers. Data are collected from a relatively large number of farmers which, as a whole, are representative of the respective target group or region in order to achieve quantification, reproducibility and comparability of results. Depending on the number of farmers to be interviewed enumerators will be required with a sound training on survey techniques.

☞ Farmers' interest in the use of new methods for the control of parasitic weeds can be estimated by proposing and explaining new technologies.

EVALUATION

★ Coding answers will facilitate the survey analysis.

COMMENTS

- Instead of using a standardized interview, a non-standardized interview guided by a list of topics will be useful to get first ideas on the parasitic weed problem.
- Participatory methods, such as informal qualitative interviews and group discussions, drawing of maps and transects, ranking and scoring, etc. has also been shown to be very useful in the analysis of farming systems with regard to parasitic weeds.
- Depending on the culture and ethnic groups women in agricultural households are responsible for different agricultural activities and crops. Therefore, women farmers should also be interviewed.

LITERATURE

Bengaly MP, T Defoer & WA Stoop 1996. On-farm studies of *Striga* in South Mali: Contributions of a farmer participatory approach. In: MT Moreno, JI Cubero, D Berner, D Joel, LJ Musselman & C Parker (eds.), Advances in Parasitic Plant Research. Proceedings VI Parasitic Weed Symposium. Cordoba, Spain, 855-861.

Debrah SK 1994. Socio-economic constraints to the adoption of weed control techniques: the case of *Striga* control in the West African Semi-Arid Tropics. International Journal of Pest Management 40(2), 153-158.

Esilaba AO, F Reda, T Mulatu, JK Ransom, G. Woldewahid, A. Tesfaye, I Fitiwy & G Abate 1997. Participatory rural appraisal on *Striga* in the northern Ethiopian highlands. African Crop Science Conference Proceedings, Vol. 3.

Frost HM 1995. *Striga hermonthica* surveys in Western Kenya. Brighton Crop Protection Conference Weeds, Brighton, UK, Vol. 1, 145-150.

Gehri A, S Kachelriess, P Fischer 1998. Assessing women's needs in *Striga* control. In: J Kroschel, H Mercer-Quarshie, J Sauerborn (eds.), Advances in Parasitic Weed Control at on-farm level. Vol. I. Joint Action to Control *Striga* in Africa. Margraf Verlag, Weikersheim, Germany, 45-56.

Kroschel J & J Sauerborn 1996. Farming systems and the problems of applying *Striga* control methods - a comparison of case studies from northern Ghana, Tanzania and Malawi. In: MT Moreno, JI Cubero, D Berner, D Joel, LJ Musselman & C Parker (eds.), Advances in Parasitic Plant Research. Proceedings VI Parasitic Weed Symposium. Cordoba, Spain, 843-853.

Kroschel J, J Sauerborn, S Kachelriess, V Hoffmann & H Mercer-Quarshie 1997. Possibilities and constraints in implementing *Striga* control methods in African agriculture. In: F Heidhues & A Fadani (eds.), Food security and innovations, successes and lessons learned, International Symposium, Verlag Peter Lang, Frankfurt am Main, Germany, 121-135.

Kroschel J 1999. Analysis of the *Striga* problem, the first step towards future joint action. In: J Kroschel, H Mercer-Quarshie & J Sauerborn (eds.), Advances in Parasitic Weed Control at On-farm Level. Vol. I. Joint Action to Control *Striga* in Africa. Margraf Verlag, Weikersheim, Germany, 3-25.

Runge-Metzger A, J Kroschel, J Sauerborn 1997. The economics of *Striga* control in different farming systems in Northern Ghana. In: STO Lagoke, LE van der Straten & SS M'Boob (eds.), 3rd General Workshop of the Pan-African *Striga* Control Network (PASCON) „Integrated Management of *Striga* for the African Farmer", Oct. 18-23, 1993, Harare, Zimbabwe. FAO Accra, Ghana, 134-143.

Shaxson LJ, CR Riches & JH Seyani 1993. Incorporating farmer knowledge in the design of weed control strategies for smallholders. Brighton Crop Protection Conference Weeds, Brighton, UK, 1149-1154.

Werner J 1993. Participatory development of agricultural innovations. Procedures and methods of on-farm research. Schriftenreihe der GTZ No. 234, TZ-Verlags-Gesellschaft, Roßdorf, Germany.

Waters-Bayer A 1989. Participatory technology development in ecologicall-oriented agriculture: some approaches and tools. ODI Agricultural Administration Unit. London, U.K.

Author: J Kroschel

4 Seed features, germination and seed bank

4.1 Reproductive capacity

INTRODUCTION

Most parasitic weeds produce a large number of seeds per single plant. In context with the requirement for the stimulation of germination this high reproduction forms part of their strategy to survive. Before the last decade, little attention was paid to the reproductive capacity of these weeds, and numerous recent papers refer to non-realistic data on their seed production.

OBJECTIVES

- To determine the seed production of parasitic weeds as a basic biological feature,
- to estimate the potential threat through the reproduction of a parasite species in terms of annual seed production,
- to create information on population dynamics and to analyse a given population.

MATERIALS

✓ Bags or trays to properly transport and store the plants to be studied.

✓ Petri dishes or small trays to store capsules during drying process.

✓ Sensitive electronic balance, dissecting microscope.

✓ For the assessment of viable seeds the materials for TTC-tests are required (→ chapter 4.3).

PROCEDURE

<u>*Striga, Alectra, Orobanche*</u>

☞ The plants collected for evaluation have to be at the end of flowering but prior to seed shedding; plants in early or full flowering stage have not yet developed all capsules and the seeds are not yet viable. *Striga* plants have to be handled with special care as mature seed capsules and young flowers may occur on a single plant.

☞ Collect a representative sample of parasite shoots (e.g., *S. hermonthica* plants in a highly infested sorghum crop may reach a height of 25 cm only, whereas they are 60 cm high under other conditions; *O. crenata* shoots in a lentil crop may have an average size of 20 cm, whereas on faba beans the average size can be 45 cm).

☞ Make clear where the data collected refer to: Species of host crop, growth conditions, infestation level(s), size of field or region, date of collection.

☞ If seed capsules are collected note that seed quality varies largely from top to bottom of plant, and that immature capsules have not developed their seed potential, and ensure that capsules are collected before dehiscence.

☞ Count the number of parasite plants/m^2

☞ Count the number of seed capsules/parasite plant

☞ Count the number of seeds/capsule. For this purpose choose capsules just prior to maturity (prior to opening and seed shed); note that capsules from the top, middle and bottom of the spike may differ considerably in number and quality of seeds. Capsules have to be carefully broken or cut off from the stem and stored for further maturity under convenient conditions in a Petri dish until they open and shed seed. Count and weigh 100 seeds/capsule and estimate the number of seeds/capsule by calculation or use the method of Hammad *et al.* (1967)

☞ Use the following formula to calculate the reproductive capacity [R]:

$$\mathbf{R = S \times C \times P}$$

with S = seeds/capsule, C = capsules/plant, P = plants/m^2

☞ Count and weigh 100 seeds/capsule to obtain an average seed weight; the 1000-seed-weight is estimated by counting and weighing ten sets of 100 randomly selected seeds.

<u>*Cuscuta*</u>

☞ Count the number of capsules/m^2 (at high densities of *Cuscuta* stems/m^2 use 0.25 m^2). At later stages of maturity capsules might have fallen to the ground and should be carefully collected and included; a modified vacuum cleaner may be used for this purpose or the *Cuscuta* seed can be recovered according to the methods described in → chapter 4.5.

☞ Count the number of seeds/capsule

☞ Assess the average weight of 100 seed.

☞ For the different species of *Cuscuta* modifications of the procedure might be necessary dependent on the respective growth and flowering pattern.

☞ Use the following formula to calculate the reproductive capacity [R]:

$$\mathbf{R = S \times C}$$

with S = seeds/capsule, C = capsules/m^2

EVALUATION

* Choose the number of replications according to the variation of data, i.e. evaluate a higher number of samples if variation is high and *vice versa.*
* A correct specification of seed produced per plant includes a viability test in order to specify „viable seeds produced per plant“. This is of great importance, as the proportion of viable seeds greatly depends on the nutrition of the seed in the respective capsule. Viability of seeds in some capsules may reach only 20%. Therefore, use of the TTC-test is strongly recommended (→ chapter 4.3).

COMMENTS

- Number of seeds/capsule and capsules/plant especially in *Orobanche* is subject to large variation; hence it is necessary to notify the host plant of the collected parasite as well as the location of the capsule on the stem and the size of the parasite.

LITERATURE

Hammad S, H El-Arosi & O Al-Menoufi 1967. *Phytomyza orobanchia* Kalt (Diptera: Agromyzidae) feeding on *Orobanche crenata* Forsk. in Egypt. Bulletin de la Société Entomologique d'Egypte 51: 141-144.

Jones M 1991. Seed mass and seed production in British *Orobanche* species. In: Proceedings of the 5th International Symposium of Parasitic Weeds; JR Ransom, LJ Musselman, AD Worsham & C Parker (eds.), CIMMYT, Nairobi, Kenya, 447-453.

Linke KH 1992. Biology and control of *Orobanche* in legume crops. PLITS 10(2): 33-36.

Linke KH, H Schnell & MC Saxena 1991. Factors affecting the seed bank of *Orobanche crenata* in fields under lentil based cropping systems in Northern Syria. In: JK Ransom, LJ Musselman, AD Worsham & C Parker (eds.), Proceedings of the 5th International Symposium of Parasitic Weeds. CIMMYT, Nairobi, Kenya, 321-327.

Linke KH 1991. *Orobanche* studies. In: Food legume improvement program - Annual report for 1990. ICARDA, Aleppo, Syria, 216-242.

Lopez-Granados F & L Garcia-Torres 1991. Estimation of *Orobanche crenata* seed production. In: K Wegmann & LJ Musselman (eds.), Progress in *Orobanche* research. Eberhard-Karls-Universität, Tübingen, Germany, 92-98.

Reuter, BC 1986. The habitat, reproductive ecology and host relations of *Orobanche fasciculata* Nutt. (Orobanchaceae) in Wisconsin. Bulletin of the Torrey Botanical Club 113(2): 110-117.

Author: K-H Linke

4.2 Seed collection

INTRODUCTION

For most experiments with parasitic plants their seeds are a prerequisite. A certain amount of stored seed is therefore an essential basis for successful experimentation.

OBJECTIVE

- To obtain a standard quality and quantity of seed of parasitic plants.

MATERIAL

✓ Vehicle, paper bags, plastic bags, sieves with mesh size according to the collected parasite seed, plastic container for storage of seed, dry and cool store.

✓ The time requirements will largely depend on the density of parasite plants in the selected fields; hence, one labour can collect few hundred grams up to several kg of seed within one day. Drying of the collected plants may require a few days in *Striga* and *Cuscuta*, but up to several weeks in *Orobanche* and *Alectra*.

✓ The operation of collecting seeds can continue over several weeks until enough mature *Striga* plants are collected.

PROCEDURE

☞ The collection should be started 2-3 weeks after the parasite began to flower in the region. Prospection tours performed beforehand can help identifying the most suitable fields and the best collection time in each field.

☞ A plant is mature when there are none or only few apical flowers left and the capsules are still intact.

☞ The collection is preferably done from fields with the same host plants as will be used later in the experiment.

☞ The plants should be transported in paper bags and then sun dried at the station.

☞ When the plants are dry, either the entire spikes (*Orobanche*) or only the heads with the capsules (*Orobanche*, *Striga*) are threshed by hand in plastic bags or sacks. The bigger plant parts are separated and the remaining particles passed through several sieves. A lower mesh size of 100 µm is sufficient for *Striga*, *Orobanche* and *Alectra* seed. Smaller seeds usually possess less vitality and can be discarded (in order to know the amount of small seed in the trash, this fraction might be checked under a dissecting microscope; with substantial amount of seed in this fraction a further sieving with smaller mesh size is recommended).

☞ To obtain clean seed as required for most laboratory studies use the flotation method (→ chapter 4.5.2) with magnesium sulphate or potassium carbonate solution to further separate the parasite seed from trash and soil particles. In order to avoid any damage to the seed, any wet treatment should not exceed a few minutes and the seeds have to be dried immediately thereafter.

☞ The cleaned dry seed can be stored in plastic containers at room temperature.

☞ Correct and extensive labelling of the containers is a must, but unfortunately often neglected. It includes species, host, date and location of collection, person who collected the plants, remarks on special observations, viability within the first year of collection, weight of sample.

☞ Containers should be checked regularly for fungal infections (smell) or insect attack.

☞ The seeds remain viable for years, the cleaner the longer.

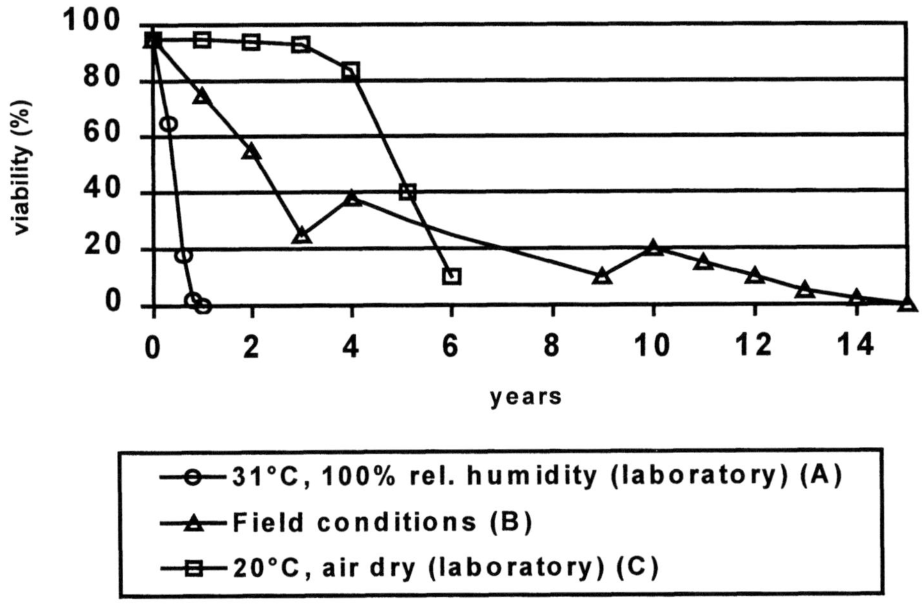

Fig. 8. The viability of seeds of *Striga asiatica*, stored under different environmental conditions. A: Kust (1963); B and C: according to Bebawi *et al.* (1984)

COMMENTS

- Seeds have to be collected at least one season prior to the set-up of the experiment due to the need for an afterripening period (e.g. *Striga*).
- The threshed seeds are more clean if the plants are collected and threshed without their root system.
- If there is evidence of races adapted to specific hosts, individual collections and explicit labelling of the seed containers is required.
- With *Alectra* plants careful evaluation of capsules is required as numerous flowers of a given shoot fail to produce seed.

LITERATURE

Bebawi FF, RE Eplee, CE Harris & RS Norris 1984. Longevity of witchweed (*Striga asiatica*) seed. Weed Science 32: 494-497.

Kust CA 1963. Dormancy and viability of witchweed seeds as affected by temperature and relative humidity during storage. Weed Science 11, 247-250.

Authors:
A Oswald, JK Ransom & K-H Linke

4.3 Testing seed viability

INTRODUCTION

Viable seeds are defined as those seeds which are capable of producing normal seedlings in a germination test under favourable conditions, after dormancy has been broken. Seed viability is assessed using 2,3,5-triphenyl tetrazolium chloride (TTC). TTC is a white salt which gives a colourless solution in water. In living cells it gets hydrated to the red formazan. Accordingly, the red staining of the seed endosperm and/or the embryo indicate the viability of the seed.

OBJECTIVES

- To test seeds of various sources before they are used in experiments to know the quality of a given seed lot and to ensure the use of seed material with a high viability,
- to compare the effect of different treatments (time of storage, temperature, chemicals...) on seed viability.

MATERIALS

✓ Glass tubes, TTC (2,3,5-triphenyl tetrazolium chloride, Clorox (NaOCl), funnel, sieve (100µm), deionized water, dissecting microscope, pipette, Petri dishes.

PROCEDURE

☞ Before seeds are used for laboratory work they should be surface sterilized to avoid the growth of microorganisms. This surface sterilisation can be done using sodium hypochlorite (NaOCl) as follows: Immerse the seeds in 1% NaOCl solution for 5 minutes (shake seeds and solution in a sealed flask to ensure full surface contact of solution on the seeds), then thoroughly wash with water through a nylon mesh (ø 100 µm). After that allow seeds to dry under room temperature or use directly for further tests.

☞ Put 1000-3000 seeds in a small glass tube.

☞ Add 5 ml of freshly prepared 1% solution of 2,3,5-triphenyl tetrazolium chloride. Close tube to prevent evaporation and place in a dark incubator adjusted at 33°C for 8 days (*S. hermonthica*) or 12 days (*Orobanche* spp.), respectively.

☞ After the incubation period filtrate the excess TTC-solution through a nylon mesh (ø 100 µm) placed in a conical flask, or remove the TTC-solution with pipette.

☞ *Orobanche* seed need a bleaching treatment in order to make the seed testa transparent. For this purpose add 5 ml of Clorox (5% NaOCl) and wait about 3 to 5 min until seed testa is discoloured. Remove Clorox solution with a pipette or wash the seeds through a sieve.

☞ Add 5 ml of water, shake well and take samples of seeds plus water with a pipette and drop on a Petri dish for further evaluation under dissecting microscope (20X).

EVALUATION

✶ Evaluate 4 x 100 seeds. As with the tiny seed of these parasites it is extremely time consuming to add exactly 100 seeds on the filter paper in a Petri dish, it is advisable to use more than 100 seeds per filter paper and to evaluate just 100 of them or to use the percentage of about 100 seeds. Evaluation of seed lots of markedly less than 50 seeds will result in undesirable high variation of the results.

Criteria for viability

★ **Viable seeds:** Endosperm red, reddish, or yellow (seed of *Orobanche ramosa* usually reveal stronger staining of the embryo than of the endosperm).

★ **Non-viable seeds:** Empty seed shells, seeds with entirely white endosperm, seeds with fungal (mostly bluish or black) or bacterial (mostly red or black) inclusions.

COMMENTS

- The standard seed testing method using TTC brings up some complications in the case of parasitic plant seed. They are relatively small and the development of formazan, especially in *Orobanche* seed, is comparably low. Higher concentrations of TTC do not produce a more pronounced staining. Reliable results require the comparison of germination and viability tests.
- Never confuse germination with viability!

LITERATURE

Aalders AJG & R Pieters 1985. *In vitro* testing with 2,3,5-triphenyl tetrazolium chloride (TTC) of *Orobanche crenata* seed metabolism. FABIS Newsletter 13: 35-37.

Bebawi FF, RE Eplee, CE Harris & RS Norris 1984. Longevity of witchweed (*Striga asiatica*) seed. Weed Science 32: 494-497.

Bischof F 1984. Untersuchung zur Bestimmung der Keimfähigkeit von Samen einiger *Orobanche*-Arten. Zeitschrift für Pflanzenkrankheiten und Pflanzenschutz, Sonderheft 10: 105-107.

Bulat H 1956. Instructions for making the tetrazolium test. Proceedings of the International Seed Testing Association 21: 52-55.

Linke KH 1987. Untersuchungen über Keimung und Jugendentwicklung von *Striga* und *Orobanche*. PLITS 5(3), 95 pp.

Mohamed AH, G Ejeta, LG Butler & TL Housley 1998. Moisture content and dormancy in *Striga asiatica* seeds. Weed Research 38: 257-265.

Moore RP 1985. Handbook on tetrazolium testing. International Seed Testing Association, Zurich, Switzerland, 99 pp.

Moore RP 1986. ISTA Handbuch für die Tetrazoliumuntersuchung. Internationale Vereinigung für Saatgutprüfung, Zürich, Schweiz, 118 pp.

Author: K-H Linke

4.4 Assessment of germination

4.4.1 Conventional germination test

INTRODUCTION

About 50% of the literature on parasitic plants is concerned more or less directly with germination studies. It is a must therefore to attribute the required attention to the respective methodological aspects.

As a prerequisite for the germination of *Striga, Alectra* and *Orobanche* seeds, the seeds have to pass a preconditioning period under moist and warm conditions. Only thereafter do they respond to a germination stimulant released e.g., from a host root. In the case of parasite seeds a seed is considered germinated when the germ tube has protruded from the seed testa (Fig. 9).

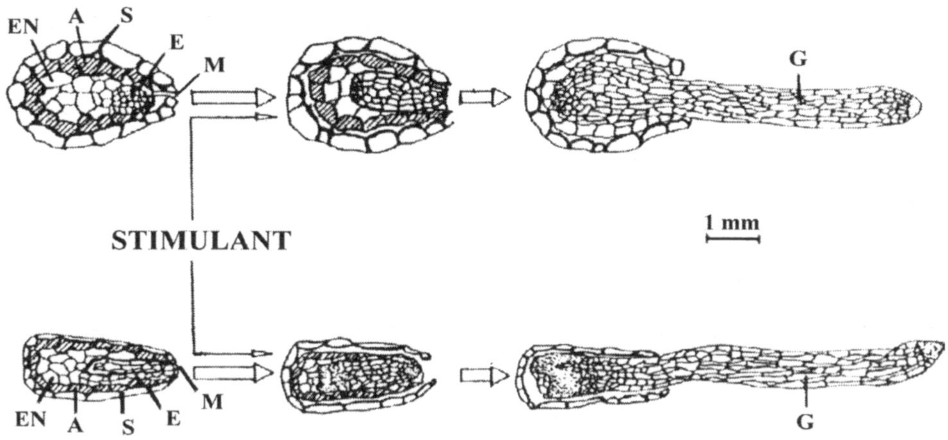

Fig. 9. Development of the germ tube of *Orobanche ramosa* (on the top) and *Striga hermonthica* (below). A = aleuron layer; E = embryo; M = microphyle; EN = endosperm; S = seed coat; G = germ tube

OBJECTIVES

- To determine the germination ability of seeds for further investigations (e.g., in root chambers, pots or field),
- to evaluate the effect of treatments on the seeds,
- to evaluate the potential of root exudates from hosts/non-hosts (e.g., for the screening of trap crops),
- to evaluate the potential of synthetic germination stimulants.

MATERIALS

✓ Filter paper, discs (1 cm Ø) of filter paper, Petri dishes, deionized water, stimulants (ethephon, GR24, GA_3, root exudates), incubator, plastic bags or comparable material to seal the Petri dishes, seeds of stimulating plants for production of exudates, washed sand, plastic pots.

PROCEDURE

1. Preconditioning of parasite seeds (prior to exposure of a stimulant)

☞ Cut small discs (1 cm Ø) from Whatman glass microfibre (Whatman GF/A) filter papers.

☞ Sprinkle the parasite seeds on each disc.

☞ Place two layers of 9 cm diameter filter papers (Whatman GF/A) in a 9.5 cm diameter Petri dish and moisten each Petri dish with 5.5 ml deionized water (to enhance preconditioning in *Orobanche* 50 to 100 ppm gibberellic acid may be added).

☞ Place the discs containing the seeds in the Petri dishes prepared above.

☞ Enclose the Petri dishes in black plastic and close tightly to prevent water evaporation.

☞ Place the Petri dishes then in a dark incubator at different temperature regimes and duration of storage depending on the parasitic weed species as follows:

Striga: 20-28°C (*S. hermonthica*) and 25-33°C (*S. asiatica*) for 10 days
Orobanche: 20-23°C for 10-14 days
Alectra: 25-33°C for 5-10 days
Cuscuta: No explicit preconditioning is required. Germination occurs under moist conditions at 10-25°C without stimulant. To overcome dormancy of *Cuscuta* seed a treatment with concentrated H_2SO_4 for 30 min. is recommended.

☞ After that time the seeds respond to a stimulant. At lower temperatures, a longer preconditioning time is required.

2. Germination stimulants
(depending on objectives of the studies):

☞ **GR24** synthetic germination stimulant.
Preparation: Dissolve 10 mg of GR24 in 10 ml of acetone. Dilute this solution 1:1000 with deionized water or with 0.3 mM buffer of 2-(N-morpholino) ethanesulfonic acid (= MES) to stabilise the solution.

☞ **Ethephon** (2-chlorethanphosphonic acid); ethylene releasing compound; ethylene is efficient for the germination of *Striga* species.
Preparation: Use commercial products of ethephon like "Ethrel" (47% a.i.). Prepare a solution of ethephon in deionized water at a concentration of 50 ppm. This solution has to be prepared freshly just prior to its usage. Storage of the solution is not advised due to its instability at a pH > 4. It is recommended to store ethephon at 8°C.

☞ **Root exudate** (water in which roots have been placed for some time)
Preparation: Grow host plants in small pots filled with pure sand. After about 10-12 days wash out the sand to clean the roots. Immerse the roots (about 20 seedlings) in 100 ml deionized water in 100 ml flask. Support the seedlings with a non-absorbent cotton, so that the roots are immersed in water, and allow them to grow under the same growth conditions for two days. The level of water in each flask is maintained by topping with distilled water to compensate for transpiration loss. Discard the seedlings thereafter and use the solution in the vessel as a fresh germination stimulant or store the germination stimulant deep-frozen (Fig. 10).

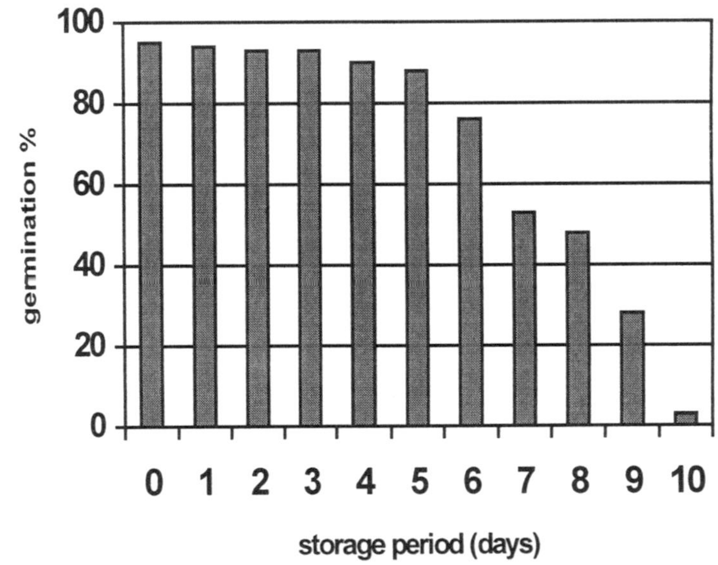

Fig. 10. Effect of the storage period of sorghum root exudate at 30°C on its activity to stimulate the germination of *Striga hermonthica* seeds

3. Stimulation of germination

☞ Add 5.5 ml of the germination stimulant to a Petri dish containing two layers of filter paper (Whatman GF/A).

☞ In these Petri dish place the discs that contained preconditioned seeds (described above) after drying on filter paper to remove any excess water.

☞ The Petri dishes are then enclosed in black plastic, firmly closed to prevent water losses and incubated in the dark at different temperature regimes and duration of storage depending on the parasitic weed species:

Striga:	30°C for 1-2 days
Orobanche:	23-25°C for 7 days
Alectra:	25-30°C for 3 days
Cuscuta:	see preconditioning of seeds.

EVALUATION

* ✶ Score a seed as germinated if the root tip (radicle) has protruded through the seed coat.
* ✶ Evaluate 4 x 100 seeds. As with the tiny seeds of parasitic weeds it is extremely time consuming to add exactly 100 seed on the filter paper in a Petri dish, it is advisable either to use more than 100 seed per filter paper and to evaluate just 100 of them or to use the percentage of about 100 seed. Evaluation of seed lots of markedly less than 50 seed may eventuate in undesirable high variation of the results.

COMMENTS

* With the root exudate method it is possible to test the capability of root exudates for the germination of parasitic weed seeds. However, mean values of germination may vary considerably from one test to another due to a very difficult to control exudate production: Root exudates, regardless of their chemical nature, are subject to rapid modification of their chemical structure and biological functions. Due to the complexity of biological reactions results may not be reliable enough for screening methods. Hence, standardisation and reproducibility may be taken precociously. Standard errors within one single test usually remain in an acceptable range. For a series of tests and if no specific testing of exudates is required, standardized stimulants like ethephon or GR24 should be preferred in germination tests.
* Sterilise Petri dishes and filter paper discs (e.g., by heat) before usage to minimise fungal and/or bacterial contamination.

LITERATURE

Botha PJ 1948. The parasitism of *Alectra vogelii*, with special reference to the germination of its seed. South African Journal of Botany 14: 63-80.

Dawoud DA 1995. Autecological studies on phytoparasites of the genera *Alectra, Orobanche* and *Striga* with special reference to their early developmental stages. PLITS 13(1), 112 pp.

Dawoud DA & J Sauerborn 1994. Impact of drought stress and temperature on *Striga hermonthica* and *Alectra vogelii* at early growth stages. Experimental Agriculture 30: 249-257.

Gaertner EE 1950. Studies of seed germination, seed identification, and host relationships in dodders, *Cuscuta* spp. Cornell University, Agricultural Experiment Station, Ithaca, New York, Memoir 294, 56 pp.

Johnson AW, G Rosebery & C Parker 1976. A novel approach to *Striga* and *Orobanche* control using synthetic stimulants. Weed Research 16: 223-227.

Kasasian L 1973. Miscellaneous observations on the biology of *Orobanche crenata* and *O. aegyptiaca*. In: Symposium on Parasitic Weeds. European Weed Research Council/ Royal University of Malta, 68-75.

Linke KH 1987. Untersuchungen über Keimung und Jugendentwicklung von *Striga* und *Orobanche*. PLITS 5(3), 95pp.

Parker C & CR Riches 1993. Parasitic Weeds of the World - Biology and Control. CAB International, Wallingford, UK, 332 pp.

Zwanenburg B, GL Mhehe, GK `t Lam, FJ Dommerholt & AA Kishimba 1986. The search for new germination stimulants of

Striga species. In: SJ ter Borg (ed.), Proceedings of a workshop on biology and control of *Orobanche*. LH/VPO, Wageningen, 35-41.

Visser JH, I Dörr & D Kollmann 1990. Compatibility of *Alectra vogelii* with different leguminous host species. Journal of Plant Physiology 135: 737-745.

Author: K-H Linke

4.4.2 Agar gel test

INTRODUCTION

Sorgoleone, the first *Striga* seed germination stimulant isolated from a natural host, is a series of four alkyl-substituted benzoquinones which are active in their reduced hydroquinone form. However, a survey of the capacity of several representative susceptible and resistant sorghum cultivars to produce sorgoleone under optimum conditions in the laboratory revealed a lack of correlation to *Striga* reaction reported from field studies. Thus the potential practical use of a screen for sorgoleone production is limited. In this situation an agar gel test was developed.

OBJECTIVES

- To screen the capacity of host root exudates to stimulate germination of *Striga* (not to assay for a particular stimulant such as sorgoleone),
- to measure germination distance rapidly and inexpensively.

MATERIALS & PROCEDURE

Surface sterilisation of *Striga* seeds

☞ Prepare Sporicidin solution: (Sporicidin cold sterilising solution available from: Sporicidin International, 12000 Old Georgetown Rd., Rockville, MD 20852, USA),

☞ mix 0.95 ml activator with 10.9 ml buffer,

☞ dilute 10-fold with distilled water (e.g., add 5 ml stock solution to 45 ml distilled water),

☞ place approximately 0.11 g of dry *Striga* seeds into a 50 ml Erlenmeyer flask,

☞ rinse 3 times with 0.1% Tween 20 to remove loose debris,

☞ add 10 ml of Sporicidin solution to the rinsed seeds,

☞ ideally an ultrasonic cleaner (e.g., Branson ultrasonic cleaner model B-12E available from Thomas Scientific, 99 High Hill Rd., PO Box 99, Swedesboro, NJ 08085-0099, USA) is used to dislodge contaminants from the seed surface. Place flask into ultrasonic cleaner and sonicate for 3 min. with occasional swirling,

☞ limit exposure to Sporicidin to less than 20 min. to retain good germination of *Striga* seeds. Rinse 3 times in sterile distilled water,

☞ prepare 0.015% aqueous benomyl solution,

☞ dissolve 0.015 g benomyl (methyl 1-(butylcarbomyl)-2-benzimidazole-carbamate) in 1 ml DMSO (dimethylsulphoxide) and add 99 ml distilled water. Swirl the suspension before use. Make a fresh solution each week,

☞ suspend seed in 0.001% benomyl solution (14 ml sterile distilled water to which has been added 1 ml of a 0.015% aqueous solution of Benomyl).

Preconditioning of *Striga* seed

→ chapter 4.4.1.

Surface sterilisation and germination of sorghum/pearl millet seed

- ☞ Soak seeds (50-100 per batch) in a 1% NaOCl solution for 15-60 min., then rinse 3 times with distilled water,
- ☞ incubate each batch of seeds in separate Petri dishes on moist (sterile distilled water) filter paper at 28°C in the dark,
- ☞ after 24 hrs vigorously germinating seeds may be used for the gel diffusion assay.

Preparation of Petri dishes

- ☞ Pipette 100-150 µl of settled *Striga* seeds into each 9 cm Petri dish,
- ☞ more seed can be used if readily available,
- ☞ pour 20 ml autoclaved 0.7% water agar (1.05 g agar in 150 ml H_2O) into each dish just before it becomes cool enough to solidify,
- ☞ with practice a uniform distribution of *Striga* seeds throughout the agar can readily be achieved,
- ☞ insert the emerging radicle of a germinating host seed (24 hr old) into the solidifying agar near one edge of the plate with the root tip pointing across the plate.
- ☞ Since the host root occasionally fails to continue to grow or grows above the medium, place two seeds on opposite sides of the plate and remove the one less favourably oriented or less vigorously growing within 12 hr.

Incubation

- ☞ Incubate plates in the dark at 28°C for 72 hr.

EVALUATION

- ✶ Observe germination of *Striga* seed through the bottom of the Petri dish with a dissecting microscope.
- ✶ Record the maximum germination distance (i.e., the distance between the host root and the most distant germinated *Striga* seed) for each plate.

COMMENTS

- Indirect selection traits are attractive to breeders given the difficulties encountered while screening for *Striga* resistance under field conditions. However, low stimulant production does not confer *Striga* resistance under all circumstances (e.g., conditions of very high *Striga* infestation) and, conversely, the high stimulant character does not necessarily result in susceptibility to *Striga* (e.g., cultivars such as N13 possessing other resistance mechanisms). The agar gel test is therefore but one tool available to those breeding for resistance to *Striga* in cereals.
- Since interaction among different resistance mechanisms could influence field reactions to *Striga* infestation, breeders will still need at some point to evaluate their materials in pots and under field conditions.

LITERATURE

Chang M, DH Netzly, LG Butler & DG Lynn 1986. Chemical regulation of distance: Characterisation of the first natural host germination stimulant for *Striga asiatica*. Journal of the American Chemistry Society 108: 7858-7860.

Doggett H 1988. Sorghum. 2nd edition, John Wiley, New York. 392-394.

Haussmann BIG, DE Hess, BVS Reddy, HG Welz & HH Geiger 1996a. Quantitative-genetic parameters for resistance to *Striga hermonthica* in sorghum. In: MT Moreno, JI Cubero, D Berner, D Joel, LJ Musselman & C Parker (eds.), Advances in Parasitic Plant Research. Proceedings of the 6th International Parasitic Weed Symposium, April 16-18, 1996, Cordoba, Spain. 681-688.

Hess DE, G Ejeta & LG Butler 1992. Selecting sorghum genotypes expressing a quantitative biosynthetic trait that confers resistance to *Striga*. Phytochemistry 31:493-497.

Netzly DH, JL Riopel, G Ejeta & LG Butler 1988. Germination stimulants of witchweed (*Striga asiatica*) from hydrophobic root exudate of sorghum (*Sorghum bicolor*). Weed Science 36: 441-446.

Okonkwo SNC & DN Okafor 1988. Stimulation of germination of seeds of root parasitic weeds by strigol analogues GR7 and GR24. In: SNC Okonkwo, IO Akobundu & AO Ayeni (eds.), Proceedings of the 3rd Biennial Conference of the West African Weed Science Society, Ibadan, 31-41.

Parker C, AM Hitchcock & KV Ramaiah 1977. The germination of *Striga* species by crop root exudates; techniques for selecting resistant crop cultivars. In: Proceedings, Asian-Pacific Weed Science Society 6th Conference, Jakarta, 67-74.

Parker C & CR Riches 1993. Parasitic Weeds of the World: Biology and Control. CAB International, Wallingford, Oxon OX10 8DE, UK, 1-74.

Ramaiah KV, VL Chidley & LR House 1990. Inheritance of *Striga* seed germination stimulant in sorghum. Euphytica 45: 33-38.

Vogler RK, G Ejeta & LG Butler 1991. Inheritance of low stimulation of *Striga* seed germination in sorghum cultivar SRN39. Agronomy Abstracts. Denver, CO, 120.

Worsham AD & GH Egley 1990. Physiology of witchweed seed dormancy and germination. In: Witchweed Research and Control in the United States, Monograph no. 5, Weed Science Society of America, Champaign, IL, 11-26.

Author: DE Hess

4.5 Extraction from soil

INTRODUCTION

The seed bank in the soil represents the prominent stage in the population dynamics of parasitic weeds. It is one of the most important factors influencing the intensity of crop infestation by parasites. It is characterised by the parasite species itself, seed density, seed distribution and age structure of the seed. The seed density is expressed as seeds/kg soil or seeds/m^2.

The principal types of dispersion pattern of seed in a given field are the homogenous (equal) distribution, and the aggregated (clumped or clustered) distribution. The statistical equivalent are the Poisson-distribution and the binomial distribution, respectively (Fig. 11).

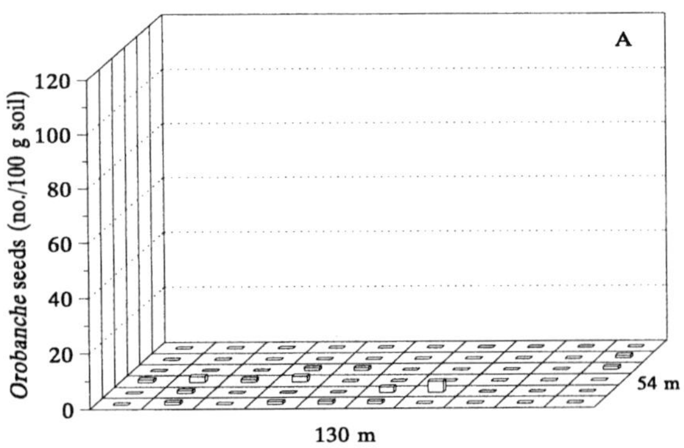

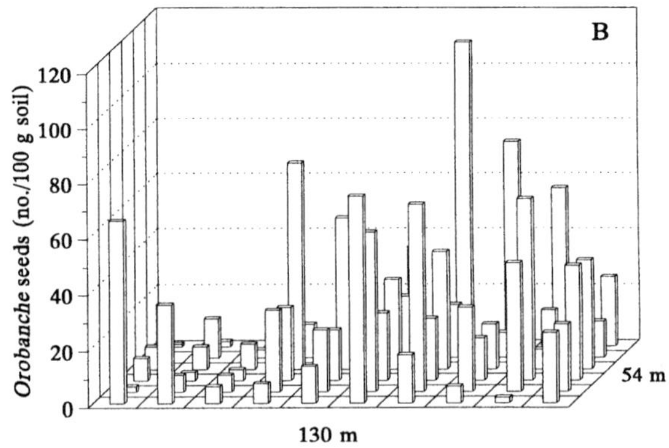

Fig. 11. Distribution of *Orobanche* seed in a field: The distribution in (A) comes close to a Poisson distribution (840 seeds/m², 20 cm depth), while in (B) it comes close to a binomial distribution (17280 seeds/m², 20 cm depth). Each field (A and B) is 54 x 130 m, represented by 60 single samples.

Depending on the aim of the respective study also the vertical dimension of seed distribution in the soil may be relevant (Fig. 12).

Principal factors influencing the seed bank in soil are previous history of infestation (e.g., crop rotation), soil tillage, soil type, soil water content, insects, nematodes and the microflora.

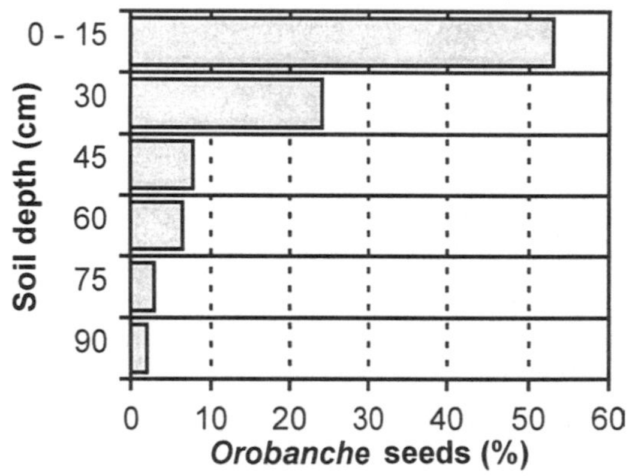

Fig. 12. Vertical distribution of *Orobanche* seed in a soil profile (Tel Hadya, Syria, chromic luvisol (60% clay) with a long *Orobanche* history).

4.5.1 Soil sampling

INTRODUCTION

The use of soil assays to investigate the presence of parasite seed in the soil provides a more reliable index of infestation than visual field surveys only. The need for accurate sampling plans to estimate seed population involves the problem of the number of cores which need to be taken to give a satisfactory estimation of the number of seeds per hectare. Principally, individual cores or mixed samples may be evaluated, aiming at a high precision or the optimal use of limited resources to obtain useful data.

OBJECTIVES
- To collect uniform and random soil samples of a given area,
- to ensure a representative collection of samples,
- to provide data for mapping the distribution of parasitic weed seed.

MATERIALS

✓ Sampling device (equipment ranges from simple to sophisticated tools): shovel, or auger of 2-3 cm diameter or hand-pulled/tractor-mounted automatic soil sampler. For soil samples from depths >50 cm an auger of 5 to 7 cm diameter is required.

✓ Plastic bags, balance, sieve of 2 to 3 mm mesh width.

✓ The time requirement largely depends on the number of samples per unit area. Higher number of samples per area provide more detailed data.

PROCEDURES

☞ To determine the seed bank of a field or plot soil samples up to 15 cm depth are taken manually with an auger. Sampling depth depends on the vertical distribution of seed in the soil, which is influenced by soil type and cultivation. For some studies it may further depend on the depth from which parasite shoots manage to emerge (usually 15-20 cm depth).

☞ Determine the desired precision of the data: maximising the precision or rough orientation on the infestation.

Hence, the aim of the study is essential for any further procedure.

☞ For highly precise data the evaluation of some single samples is required to obtain a preliminary information on the infestation level. Uniform infestation will allow less samples per unit area, whereas non-uniform infestation requires a higher number of soil samples in order to obtain the same precision. The higher the seed content of a field (mean) or the more uneven the distribution of the seeds (variance, s^2), the higher is the necessary number of cores to be taken at a given absolute confidence interval. Hence it is apparent that sampling plans must be custom made for many different situations.

☞ Make bulk samples: e.g., of a 20 m^2 plot take 10-15 sub-samples which are all mixed to form a single bulk sample.

☞ When taking soil samples on farmers' fields, each field is walked through on a diagonal course and at every seventh to tenth step a sample should be taken. The number of samples will be proportional to the size of the field.

☞ The soil samples are sieved through a 2000 µm sieve (if necessary) to remove the gravel and then they are air dried.

☞ The individual samples have to be well mixed, before a sub-sample is drawn. It has to be avoided to damage seed in the soil by crushing or grinding dry soil particles.

☞ Samples of 100 g are further processed.

EVALUATION

(→ chapter 4.5.2)

COMMENTS

- For a proper description of the experiment, the data and the results it is vital to indicate the precision of the sampling; this includes number of single samples/area, the kind of bulk preparation and the confidence interval.
- With any bulk-sampling or sub-sampling be aware that the tiny parasite seed tend to get separated from the soil within dry samples; on the bottom or top of bags higher concentration of parasite seed may occur. This requires mixing of a sample before taking sub-samples.
- Mixed sampling does not show aggregation or clumped distribution, but the precision by mixed sampling is in any case high.

LITERATURE

Ashworth LJ 1976. Quantitative detection of seed of branched broomrape in California tomato soils. Plant Disease Reporter 60: 380-383.

Barralis G, R Chadoeuf & JP Gouet 1986. Essai de détermination de la taille de l' échantillon pour l' étude du potentiel semencier d' un sol. Weed Research 26: 291-297.

Dessaint F 1991. La répartition spatiale du stock semencier: comparaison de techniques statistiques. Weed Research 31: 41-48.

Sauerborn J, G Wyrwal & KH Linke 1989. Soil sampling for broomrape seeds in naturally infested fields. In: K Wegmann & LJ Musselman (eds.), Progress in *Orobanche* research. Eberhard-Karls-Universität, Tübingen, Germany, 35-43.

Wilhelm S & LC Benson 1954. Vertical distribution of broomrape seed in tomato field soils. Plant Disease Reporter 38: 553-554.

Author: K-H Linke

4.5.2 Methods to separate seeds from soil

INTRODUCTION

The use of soil assays for the presence of parasite seeds is a more reliable index of infestation than visual field surveys only. The size of parasitic weed seeds reduces the options of their separation from soil. Hence, the available separation techniques are all based on the specific weight of the parasite's seed.

OBJECTIVES

- To survey the occurrence and the extent of infestation,
- to use the seed bank as an index to monitor effects of control measures,
- to investigate the quantitative relationship between seed bank and resulting attack by parasites,
- to know the extent and degree of infestation for optimal planning of crop rotations.

Three methods have been developed:

1. Flotation

MATERIALS

✓ Glass columns (glass separating cylinder) of 2 to 10 cm in diameter and 30 to 100 cm in length, balance, sieves of 2000 µm, 300 µm, 250 µm and 90 µm mesh width, funnel, round filter papers (150 mm diameter), dissecting microscope, aerometer for density measurement, Petri dishes, potassium carbonate or sulphate of magnesium or ammonium (fertiliser).

✓ The time requirement largely depends on the number of flotation columns and number of samples processed (with more samples the waiting times become less). Hence it may require 5 to 30 min. for the flotation itself plus the time for counting the seed. For counting under the dissecting microscope 3-7 samples/hour are feasible depending on the amount of seed and residues (soil type).

PROCEDURE

☞ This is a physical method for the separation of seeds. It bases on the differences in the specific weight of seeds and soil particles. Since viable seeds usually sink in pure water, a solution (the flotation medium) with a higher density than the seeds is added and mixed with e.g., 100 g of soil. The seeds then float on the surface and can be skimmed off. Best flotation media are mineral salts, since they are cheap and safe to handle (e.g., a magnesium sulphate or a potassium carbonate solution with a specific gravity). The specific gravity has to be adjusted to the respective seed under investigation and to the content of organic matter in the soil.

☞ Not only parasite seeds, but also other organic material floats on the surface of the mixture. Sieving (between 90 µm and 300 µm mesh size) further concentrates the seed containing residues.

☞ Final cleaning of residues is carried out with tap water; with this washing also the salt solution is eliminated and potential damage to the seed during a following storage period can be avoided.

☞ Then the residues are air dried and stored for further evaluation.

2. Centrifugation

MATERIALS

- ✓ Centrifuge (beakers should be able to contain minimum 500 g each at 2,500 rpm), sieves of 2000 µm, 300 µm, 250 µm and 90 µm mesh width, balance, funnel, round filter papers (150 mm diameter), stereo-microscope (16-30 fold magnification), aerometer for density measurement (range 1.000 to 1.200 g/ml), Petri dishes, potassium carbonate or sulphate of magnesium or sulphate of ammonium (fertiliser), kaolin (white clay).
- ✓ Dissolve sufficient salt in water to prepare a solution with a specific density of 1.16 g/ml. Verify with the aerometer.
- ✓ The time requirement largely depends on the capacity of the centrifuge.
- ✓ Time requirement for centrifugation is about 8-16 soil samples/hour depending on the size of the centrifuge beakers; each passage needs about 30 min. For the counting under a dissecting microscope 3-7 samples/hour are feasible depending on the amount of residues (soil type).

PROCEDURES

The procedures are similar to flotation but technically more advanced. All figures given refer to the use of a 1.000 ml centrifuge beaker (adjust values for centrifuge beakers of smaller size accordingly!)

- ☞ Put a sample of 100 g of air dry soil in the centrifuge beaker.
- ☞ Add 20 g of kaolin to facilitate decanting of the solution after centrifugation (kaolin produces a thin layer covering all inorganic soil particles settled on the bottom of the centrifugation beaker).
- ☞ Add 450 ml of an ammonium sulphate or magnesium sulphate solution with a specific gravity of 1.16 g/ml.
- ☞ Use a vibration-mixer or similar equipment to process the mixture for 2 min. until all particles are in suspension.
- ☞ Make sure that all centrifuge beakers have the same weight by adding salt solution.
- ☞ Centrifuge for 10 min. at 2,500 revolutions per min.
- ☞ After centrifugation the supernatant solution is decanted through a sieve combination of 2,000 µm on top, 300 µm for *Orobanche* and 250 µm for *Striga* between and 90 µm below (not only parasite seeds, but also other organic material float in the solution, whereas the inorganic soil material settles on the bottom of the jar).
- ☞ Decanted salt solution could be collected and recycled after sieving and re-adjusting to the required specific density.
- ☞ The residues on the upper sieves are discarded.
- ☞ The residues of the 90 µm mesh sieve are rinsed into a funnel with a micro-fibre filter paper.
- ☞ The folded filter papers are air dried and can be stored until counting.
- ☞ For counting, the dried residues can be transferred from the folded filter paper by gently rubbing it between two fingers and collecting all material on a Petri dish.
- ☞ This Petri dish (Ø 15 cm) has graduations on the bottom (draw a

grid or parallel lines on paper and fix it under the dish).

☞ Detection of the seed is facilitated after even distribution of the residues by tapping on the underside of the Petri dish so that particles do not lie on one another.

3. Elutriation

INTRODUCTION

An elutriator system separates seeds of parasitic plants from soil based on differences in the rate of sedimentation in rapidly moving water.

MATERIALS

✓ Table-top elutriator, glass separating cylinder (5 to 10 cm in diameter and 50 to 100 cm in length, large volumes of water (between 40 and 50 l/sample), potassium carbonate (K_2CO_3), three sieve systems with 850 µm, 250 µm and 90 µm openings, balance, filter paper, dissecting microscope, Petri dishes.

✓ Time requirement for one sample depends on experience and skill of staff, also availability of material (e.g., number of glass cylinders etc.). Under favourable conditions this method is faster for sample processing than the use of a centrifuge.

PROCEDURES

1. Calibration of the elutriator

Soil type

☞ The initial flow rate of water can be adjusted to the specific soil type used. Sandy soils have heavier soil particles than silty soils. Thus higher water pressure is needed to separate *Striga* seeds and keep the soil particles moving in the water. Calibration should be done, using different soil types with a known number of *Striga* seeds added. After processing, the recovery rate and the amounts of trash found in each sample indicate if the pressure is too high or too low for the specific soil type.

Striga species (in the following only *Striga* species are addressed)

☞ After the elutriator is filled with water, the "slow" flow rate should be adjusted so that the water rises in the upper centimetre of the cone is at the rate of approximately 90% of the rate at which *Striga* seeds will sink in standing water. This varies slightly between *Striga* species. As an example, if the seed sink 1 cm in 3 sec the rate of upward movement of the water should be 1 cm in 2.7 sec (3 sec x 0.9 = 2.7 sec). This flow rate allows the *Striga* seeds to float and eventually spill over the top of the device, while all other particles with a higher specific density will remain suspended or settle to the bottom.

2. Sample processing

☞ Large clods and pieces of plant debris should be removed from the soil sample before processing. After weighing out a sub-sample of the desired size (i.e. 500 g) it should be soaked in a few litres of water for 5-10 min. in order to break up and/or soften the larger aggregates. This slurry should be stirred occasionally to ensure uniformity.

☞ The slurry is then poured into the elutriator and processed as described above.

☞ After elutriation the bottom sieve is removed and the collected material washed from the sieve with a K_2CO_3 solution into a glass separating cylinder. The optimum size for the glass separating cylinder if relatively large samples are processed is 10 cm in diameter by 100 cm in length. The cylinder is open at the upper end and funnel shaped at the lower end. The lower end should be fitted with a cut-off valve. If possible the K_2CO_3 solution should be filled from the bottom and always before the sample is washed into the cylinder. The column should be filled halfway with K_2CO_3 solution.

☞ After washing the sample from the sieves, water is carefully added to the column to about 2 cm from top of the cylinder. Materials heavier than *Striga* settle to the bottom of the carbonate in the column while lighter particles float to the surface of the water. Because the specific density of the K_2CO_3 solution is greater than that of the *Striga* seeds, they will congregate at the interface of the K_2CO_3 solution and the water. The specific density of the carbonate solution has to be adjusted to the specific density of the *Striga* species under investigation. A specific density of 1.4 g/ml works well for *S. hermonthica*. Eplee & Norris (1990) suggest at least 1.2 g/ml for *S. asiatica*.

☞ After five min. the top of the column is stirred to ensure that any seeds trapped in the floating trash will settle. Ten to fifteen min. later the remainder of the debris at the surface can be sucked off with a syringe or a vacuum line.

☞ Twenty min. after addition of the soil in the column, the sample is drained via the bottom of the cylinder. The soil, which flows out first, is discarded. The K_2CO_3 can be caught and reused after filtering.

☞ The last 3-5 cm of the K_2CO_3 solution below the interface, the interface itself and all water on top of the interface are passed through examination screens to collect the final sample.

☞ These screens consists of a 90 µm mesh or monofilament cloth glued to 2 cm sections of PVC tubing with a 10 cm diameter. Usually three to four of these screens have to be used per sample to avoid overloading a single screen.

EVALUATION

- ★ The seeds are counted under a binocular at about 16-30 fold magnification; seed coats or damaged seeds are not counted.
- ★ The results are expressed as seeds/m² or seeds/kg soil.
- ★ The formula for calculating the number of seeds/m² from the number of seeds/100 g soil is as follows:*

Number of seeds/m² = Number of seeds/100g soil × 1.2 × 20 × 100

*for soil samples taken to a depth of 20 cm and a soil density of 1.2 g/cm³

COMMENTS

- Check the recovery rate for each soil type by mixing a known number of seed (e.g., 100) into non-infested soil prior to the evaluation. This is necessary to verify all methods, as they bear a number of possible errors. The recovery rate might vary between 50 and 100%, while lower recovery rates and a higher variation within replicates of the same soil sample (×) occur when the level of infestation is low.
- Replicate to know the variation between samples.

- Soils with a high clay content may need processing with a dispersing agent prior to centrifugation.
- Sandy soils might trap/bury seeds under the rapidly settling sand particles during centrifugation. The recovery rate will then be reduced.
- The 100 g samples for flotation and centrifugation have to be taken carefully from the bulked soil sample as an aggregation of seed already occurs during storage.
- If your centrifuge does only allow for smaller beakers (e.g., 500 ml but not smaller than 250 ml), the 100 g of soil are to be divided into suitable sub samples and processed separately. Quantities of kaolin and sulphate solution are to be adjusted accordingly. Residues after centrifugation can be collected on the same filter paper. The time requirements for the processing of one sample increases drastically if the centrifuge contains only very small beakers. This makes the processing of larger numbers of samples very time consuming if not impossible.
- The advantage of the elutriation method is that large soil samples (between 250 g and 500 g) can be quickly processed, thus allowing larger samples to be taken, and that seeds can be effectively separated from all types of soils (sand to clay) without loss in precision and recovery rate. The constraints are that large volumes of water and potassium carbonate are required; and the equipment has to be custom made and is relatively expensive. The recovery rate of *Striga* seeds from the soil is usually between 80% and 90% with this method.

LITERATURE

Ashworth LJ 1976. Quantitative detection of seed of branched broomrape in California tomato soils. Plant Disease Reporter 60: 380-383.

Benz W, W Koch & A Moosmann 1984. Ein Extraktionsverfahren zur Bestimmung des Unkrautsamenpotentials in Böden. Zeitschrift für Pflanzenkrankheiten und Pflanzenschutz, Sonderheft 10: 109-114.

Eplee GH 1976. Determination of *Striga* seeds in soil. USDA witchweed laboratory, P.O. Box 269, Whiteville, North Carolina 28472.

Eplee RE & RG Norris 1990. Soil sampling collection equipment and equipment to separate seeds from soil. In: PF Sands, RE Eplee & RG Westbrooks (eds.), Witchweed Research and Control in the United States. Weed Science Society of America, Champaign, IL, 136-140.

Kachelriess S 1987. Estimation of *Orobanche* seeds in soil. FABIS Newsletter 19: 17-18.

Kachelriess S 1988. Methode zur Bestimmung von *Orobanche*-Samen im Boden. Zeitschrift für Pflanzenkrankheiten und Pflanzenschutz, Sonderheft 11: 55-60.

Ndungu D, GD Odhiambo & JK Ransom 1994. Methodology for quantifying *Striga* seed numbers in soils of East Africa. In: E Adipala, MA Dekunda, JS Tenywa, MW Ogenga-Latigo, & JO Mugah (eds.), Africa Crop Science Conference Proceedings Vol. 1: 220-221.

Sauerborn J, G Wyrwal & KH Linke 1991. Soil sampling for broomrape seeds in naturally infested fields. In: K Wegmann & LJ Musselman (eds.), Progress in *Orobanche* Research. Eberhard-Karls-Universität, Tübingen, Germany, 35-43.

Tayler LR 1961. Aggregation, variance and the mean. Nature 189: 732-735.

Visser JH & LF Wentzel 1980. Quantitative estimation of *Alectra* and *Striga* seed in soil. Weed Research 20: 77-81.

Authors:
K-H Linke, A Oswald, JK Ransom & S Kachelriess

4.6 Detection from crop seed

INTRODUCTION

Seed of parasitic weeds may be disseminated among others by contaminated crop seed. Reasons for this are (1) the seed size of parasites, (2) the large amount of parasite seed produced per single plant, (3) the type of harvest (lentil seed harvested by combine may carry a lot of parasite seed, whereas Sorghum heads harvested by hand may be almost clean from *Striga* if the heads are not laid on the ground) and (4) especially in *Cuscuta* the similarity in seed size and shape with some host plants (*Trifolium, Medicago*). This way of parasite seed distribution and potential infestation of new areas is of special relevance for seed production companies and other institutions distributing seed.

OBJECTIVE

- To check crop seed samples before planting or dispatching to non-infested areas in order to avoid further spread of parasitic weeds.

MATERIALS

✓ Set of sieves: 90 µm to collect the parasite seed, several larger mesh sizes according to the crop seed (very convenient is the dry- and wet sieving unit of Retsch company, 42781 Haan, FRG), funnel, filter paper, dissecting microscope, tap water.

PROCEDURE

☞ The sieve with the mesh size of 90 µm is mounted on bottom to collect the parasite seed.

☞ The sieve on top (>400µm mesh size) is filled with 300 g of the crop seeds to be analysed.

☞ With tap water the crop seeds are thoroughly washed and the residues collected on the lower 90 µm sieve.

☞ Slightly stirring or moving of the crop seed facilitates the separation of potential parasite seed.

☞ The residues of the lower sieve are washed into a funnel and collected there on filter paper.

EVALUATION

∗ The residues are air-dried and examined under a dissecting microscope (16 x) for the occurrence of parasite seed.

∗ Results are calculated as number of parasite seed per 1 kg (or 100 kg) of crop seed.

∗ According to the desired precision of the results several replications per crop sample are analysed.

COMMENTS

- Taking the crop seed samples is the critical point: Most seed of the parasites are found on the bottom of the crop seed bags or boxes as a result of separation during filling and transport. Careful investigation of the „dust" in the bags or boxes is necessary. Due to the smooth surface of most crop seed there is little chance for the seed of parasitic plants to be attached to it.
- The size of the sieve on top (crop seed) has to be adjusted to the respective crop seed.
- A calibration of the technique with a known number of parasite seed mixed with crop seed prior to the test is recommended.
- For further information on cleaning of contaminated crop seed → chapter 6.5.1.

LITERATURE

Benz W, W Koch & A Moosmann 1984. Ein Extraktionsverfahren zur Bestimmung des Unkrautsamenpotentials in Böden. Zeitschrift für Pflanzenkrankheiten und Pflanzenschutz, Sonderheft 10: 109-114.

Berner DK, KF Cardwell, BO Faturoti, FO Ikie & OA Williams 1994. Relative roles of wind, crop seeds, and cattle in the dispersal of *Striga* species. Plant Disease 78: 402-406.

Jacobsohn R & R Marcus 1988: Quantitative determination of broomrape (*Orobanche* spp.) seeds in vetch seeds. Weed Research 28: 159-162.

Linke KH 1999. Status quo of *Orobanche* management: Preventive, cultural, and physical control. In: J Kroschel, M Abderabihi, H Betz (eds.), Advances in Parasitic Weed Control at On-farm Level. Vol. II. Joint Action to Control *Orobanche* in the WANA Region. Margraf Verlag, Weikersheim, Germany, 107-133.

Mack RN 1991. The commercial seed trade: an early disperser of weeds in the United States. Economic Botany 45: 257-273.

Author: K-H Linke

4.7 Seed bags in soil

INTRODUCTION

The behaviour of seed in natural soil differs from that of seed stored under laboratory conditions. Only limited information is available on that topic (e.g., concerning longevity of seed under various rotation systems, etc.). Monitoring the tiny seed of these plants requires a technique to recover the seed from the soil, i.e. to fill the seeds into special bags in order to find them again after some time.

OBJECTIVES

- To investigate effects of soil conditions, treatments, etc. on the seed population and seed characteristics (e.g., seed germination, viability, longevity or decay rate of seed) in *in situ* experiments.

MATERIALS:

✓ Parasite seed of one lot, plastic screen cloth (polyamid, polyethylene, polyester) with a mesh opening of about 90 µm, nylon strings, electric welding equipment for plastic bags, gas burner or candle, pots with soil or special field site.

PROCEDURE

☞ Two types of bags are presented below. Both are convenient for parasitic weed seed and were used in many experiments. The first type of bag is almost two-dimensional (flat bag), allowing more contact to the soil, the second (Eplee bag) is of typical bag shape.

Bag preparation

<u>1. Flat bags</u>

☞ Prepare little bags (~3 to 4 cm^2, i.e. 1.5 x 1.5 x 0.1 cm) out of the plastic cloth by welding the material together at three sides; electrical powered equipment which normally serves to weld polyethylene bags in the kitchen helps in producing bags.

☞ Leave the forth side open to insert the parasite seeds into the bag (sewing of bags is not recommended since cotton and poly-acrylic yarns dissolve quickly in the soil, parasite seed could get lost through the sewing holes or damaged during the sewing process).

☞ Place at least 150 parasite seeds for viability test and 500 seeds for the germination test in the bag using a funnel or fine paper.

☞ Seal the bag by welding (hot iron, Bunsen burner, candle etc.) the open side.

2. "Eplee bags"

☞ The bags are prepared by tying up the corners of a 8 cm^2 nylon cloth.

☞ Cut nylon squares (size 8 cm^2).

☞ Place the seeds in the centre.

☞ Carefully pick up the corners of the cloth, twist them together and tie the bag with a nylon string using the Millers knot (Millers Knot: hold one end of the string in hand, wrap the string around the finger twice, then tuck the string through the loop from the base of the fingers towards the tip, slip the loop over the twisted corners of the bundle and pull the string to a knot).

Burying in the soil and recovery of the bags from the soil:

☞ Tie long (according to the respective soil depth) strings to the bag.

☞ Place the seed bags in the test site by digging the packet in the soil at the respective depth(s).

☞ After incubation in the soil (for days, months or years) retrieve the seed packet from the soil.

☞ Wash the seed packets in tap water.

☞ Open the bags carefully and rinse the seeds on a nylon cloth filter.

☞ After surface sterilisation (→ chapter 4.3) of the seeds start with the germination (→ chapter 4.4) or viability test (→ chapter 4.3).

EVALUATION

★ Count only non-damaged seed (damaged seed or empty seed shells to be noted separately).

★ According to the layout of the total experiment (e.g., as a split-plot design).

★ Express germination or viability as percentage to compare between treatments and control.

COMMENTS

- The little bags with their almost two-dimensional shape compared to the relatively big "Eplee-bags" have the advantage of a more direct contact between seed and soil.
- Conduct viability/germination tests with the seed lot prior to the experiment and thereafter in order to clearly identify the treatment effect.
- For long term studies a few bags with exactly 100 seed should be included in the experiment to assess the number of seeds that might get decomposed during the course of the experiment. Without this information some seed may disappear and the remaining might reflect a high viability.
- For easy recovery of the bags it is helpful to mark the places where they are buried with a stick and to draw an exact map indicating where and at which depth the bags are buried.

LITERATURE

Eplee RE & RS Norris 1987. Field research techniques,. In: LJ Musselman (ed.), Parasitic Weeds in Agriculture, Vol. 1: *Striga*. CRC Press, Boca Raton, Florida, USA, 271-280.

Linke KH & MC Saxena 1991. Studies on the longevity of *Orobanche crenata* seeds in a lentil-wheat-barley rotation. Legume Program; research and training plans 1991/1992 season. ICARDA; Aleppo, Syria, 29-30.

Vasudeva Rao MJ 1985. Techniques for screening sorghums for resistance to *Striga*. ICRISAT Information Bulletin No. 20, Andhra Pradesh, India.

Van Hezewijk MJ, KH Linke, F Lopez-Granados, OA Al-Menoufi, L Garcia-Torres, JAC Verkleij & AH Pieterse 1994. Seasonal changes in germination response of buried seeds of *Orobanche crenata*. Weed Research 34(5): 369-376.

Authors: K-H Linke & B Kranz

5 Investigations on developmental stages

5.1 Laboratory experiments

5.1.1 Observations of the underground development

INTRODUCTION

Studies on the early development of root-parasitic weeds comprise the most difficult to monitor stages in the life-cycle of these plants. In order to monitor this host-parasite interaction different growing systems were developed which support the growth of both the host plant and the parasite at least for a certain time and which allow regular monitoring.

OBJECTIVES

- To observe the underground development of root parasitic weeds such as germination, appressorium and haustorium formation and further growth,
- to evaluate treatment effects during the early development of the parasites,
- to screen host cultivars for resistance,
- to allow follow-up and sampling of individual parasites,
- to permit synchronisation of *Orobanche* germination and massive infection by the parasite that can be frequently observed under the dissecting microscope.

1. Root chamber technique

MATERIALS

✓ A suitable root chamber has a dimension of 20 x 6 x 1 or 20 x 6 x 2 cm. It is made of non-transparent hard plastic with its removable front of transparent Plexi-glass (Fig. 13).

✓ Glass-fibre filter paper (that does not serve as substrate for microorganisms in contrast to cellulose filter paper), sterilised sand, black plastic, seed of test crops.

✓ One full observation period takes 3-6 weeks, consisting of 1 week for preconditioning and 2-5 weeks for host and parasite germination and growth observation.

PROCEDURE

☞ Cut a filter paper in strips of 6 x 20 cm.

☞ Moist the strip using deionised water on a clean bench or table.

☞ Distribute the parasite seeds on it (e.g., using dry seed on a scalpel or spoon).

☞ Lay the filter paper strip with the parasite seeds on the Plexi-glass in such a way that the parasite seeds are enclosed between the filter paper and the glass.

☞ Place this Plexi-glass with the filter paper on a root chamber and use clambers to fix it to the root chamber.

☞ Fill the root chamber space with sterilised sand.

☞ Then water the chamber to keep the filter paper and the parasite seeds moist (preconditioning).

☞ Cover the glass side with black plastic or aluminium foil to protect the root zone from light.

☞ Store the root chambers at temperature regimes and storage periods according to the requirements for preconditioning (→ chapter 4.4.1).

☞ After preconditioning sow the host seeds (2-3 seeds each which are thinned out after successful germination) between the glass and the filter paper (pre-germinated host seedlings may be used to accelerate the procedure).

☞ Transfer these chambers to an incubator or glasshouse adjusted at 12/12 hours light/dark periods and temperatures of 35/25°C for *Striga* and *Alectra* or 25/15°C for *Orobanche*, (you can also carry out these trials under natural environmental conditions if suited).

☞ Place the root chambers at an angle (ca. 45°) to avoid the growth of the host root through the filter paper.

☞ Watering is done as required (normally once a day).

☞ The application of a nutrient solution depends on the period of time required for observation; with most host-parasite relations the observation is feasible without nutrient solution for several weeks.

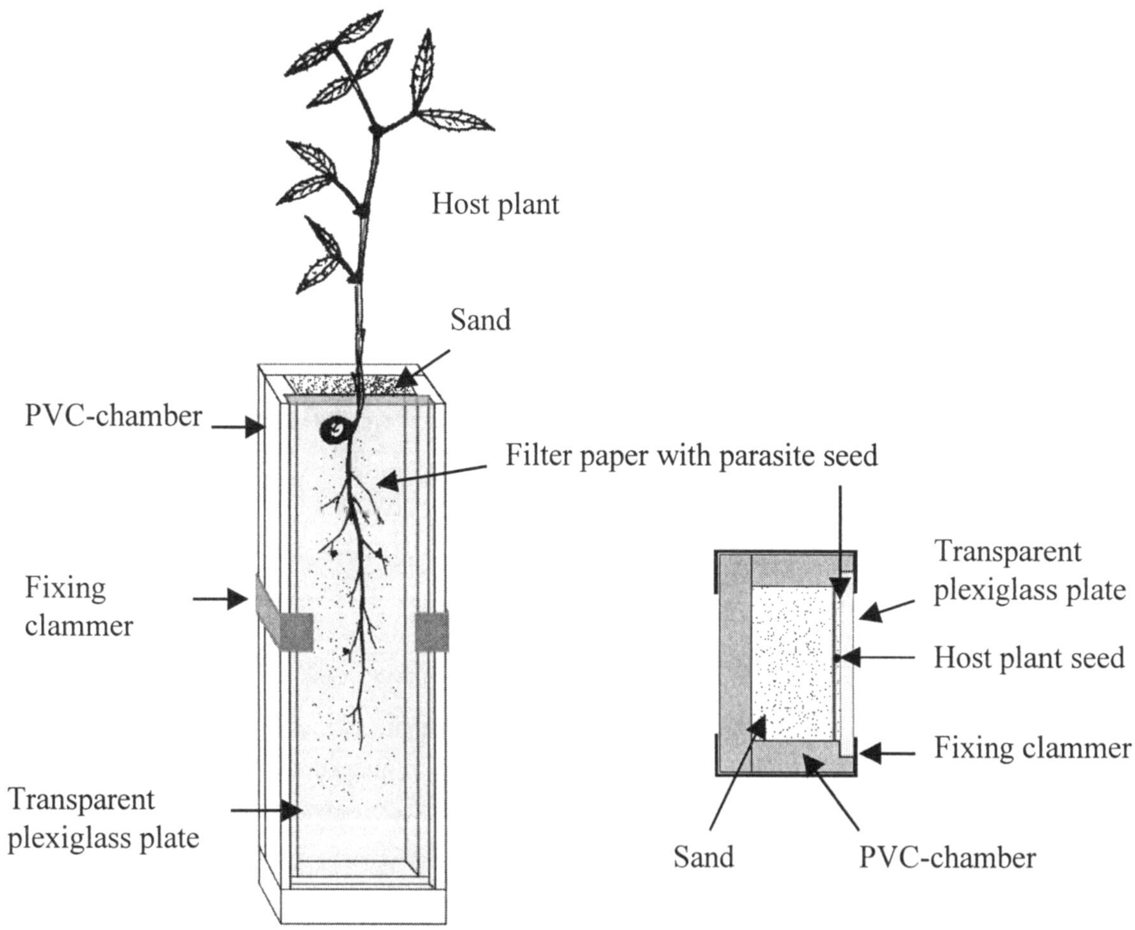

Fig. 13. Root chamber

EVALUATION

- ✶ Use a water resistant marker to mark a distance of 1.5 to 2 mm along the host root system (try to put the host root in the middle of this distance). Then divide this area in small segments of about 2 cm length. Another technique for accurate counting is the use of graduations on the plastic.
- ✶ Count the number of the parasite seeds in each segment.
- ✶ Periodically count the number of germinated parasite seeds, attachments, tubercles and shoots in each segment and record them. Markers of different colour can be used in this evaluation, in such a way that one colour is applied to each developmental stage of the parasite.

2. Petri dish method

MATERIALS

✓ Plastic Petri dishes (9 cm in diameter) with a hole drilled at one side, glass-fibre filter paper (9 cm in diameter, Whatman GF/A), incubator, sand or vermiculite as growth medium, parasite and host seed (Fig. 14).

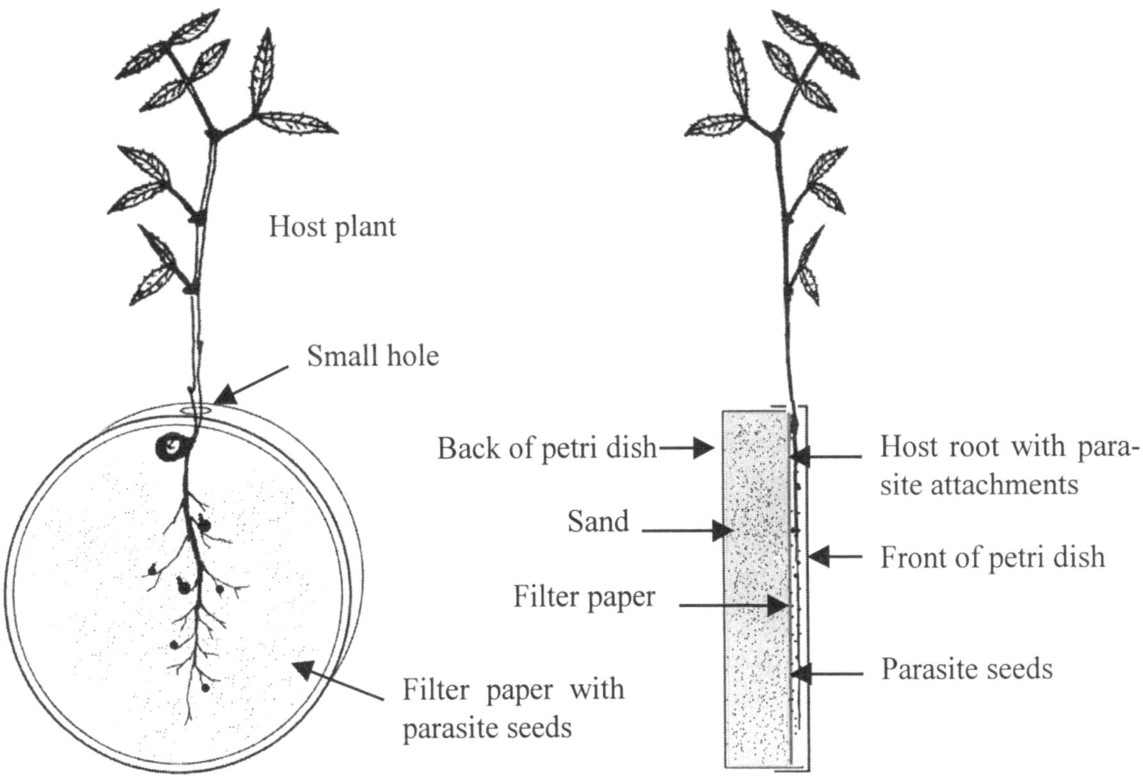

Fig. 14. The Petri dish method

PROCEDURE

- ☞ Plastic Petri dishes are filled with washed sand or vermiculite, watered with 50 ml of water and covered with fibreglass filter paper on which parasite seeds are sprinkled at a density of about 25 seeds/cm^2.
- ☞ Petri dishes are then covered with the lid and placed in an incubator in the dark for conditioning at temperature regimes and storage periods depending on the parasite species (→ chapter 4.4.1)
- ☞ Pre-germinated seedlings of the test entries are then inserted in the Petri

dish between filter paper and lid, whereby the shoot of the seedling is allowed to grow out of the Petri dish through a hole.

☞ The Petri dishes are placed in a tray and kept in an incubator with sufficient light at 25/15°C (day/night) for *Orobanche* spp. and 35/25°C for *Striga* spp. for 3 to 4 weeks, respectively. Temperature and relative humidity should be adjusted to the specific requirements of the respective host crops to be tested.

EVALUATION

★ Count the number of attachments on the root.

★ Relate the number of attachments to the host root length (root length assessment according to the method described by Tennant 1975).

COMMENTS

- Both techniques described above are also useful to screen crop varieties for their susceptibility to root parasites.
- The root chamber technique had been developed for the observation of the early development of *Striga* and *Orobanche* spp.; it can be also applied for the cultivation of *Alectra* (Lane *et al.* 1991) and other similar parasites. Other systems are required for studies on parasites such as *Cuscuta* or epiphytic parasites such as mistletoes.
- Note: Results of any laboratory screening have to be further verified in the field.
- If required, infected host plants can be carefully removed from the observation containers and planted into soil to allow further development of the parasites and/or host.

3. Polybag system

This method is based on the use of transparent polyethylene bags. The principle of this method was first published by Parker & Dixon (1983), which was modified according to the needs in *Orobanche* research. One can easily open polyethylene bags by simple cuts, and strips of adhesive tape serve as "zippers" that allow frequent opening of the bags.

MATERIALS

✓ *Orobanche* seeds, host plant seeds.

✓ NaOCl 1%, Tween 20, sterile water.

✓ Polyethylene bags, glass-fibre paper (Whatman GF/A), a metal rod longer than the width of the bag (or a piece of cane, or a plastic bar), tea-bags, a small sieve, stapler, cellotape.

✓ Rectangular boxes (e.g. 40 x 27 x 35 cm, for 30 bags).

✓ Half strength Hoagland nutrient solution.

✓ Dispenser, syringe, dissecting microscope, light-box.

PROCEDURES

☞ Seeds of host and parasite are surface-sterilised by soaking in NaOCl (1%) for 5-20 min. (the addition of a few drops of surfactant, e.g., Tween 20, is very helpful), washed twice with sterile water, and let dry. *Orobanche* seeds can be kept in tea-bags during surface-sterilisation, but make sure that the bag does not contain too many seeds otherwise the seeds inside will not come in contact with the sterilising solution.

☞ Preparation of polybag system (Fig. 15): glass-fibre paper (Whatman GF/A) is cut from large sheets to the required size (e.g., 11.5 x 28.5 cm), and inserted into transparent stationary polyethylene bags (e.g., 35 x 24 cm). 10 ml sterilised water added to wet the paper. Dry *Orobanche* seeds are evenly sprinkled over the wet glass fibre paper, using a small sieve with a reasonable seed density of 30-50 seeds per cm^2. Seeds should be present where the host root system is going to develop. The top of the bag is folded over a piece of cane, a metal rod or a plastic bar (that are longer than the width of the bag) and stapled on each side.

☞ Preparation of host plants: surface-sterilised host seeds are sown in autoclaved soil mixture or vermiculite. When host seedlings develop 2-4 leaves their root system is washed and they are transplanted into the polybags, 1-3 seedlings per bag. This is facilitated after cutting the bag open along the upper fold (slightly above the upper edge of the paper), and along one side edge of the paper. After placing the host plant in a proper position on the upper side of the system the bags are sealed with cellotape around the emergence point of the shoot, and along the side slit. Folding the ends of the tape allows easy re-opening of the bag when necessary. The bags are suspended in rectangular boxes, 20-30 bags in each, and incubated under suitable temperature and light conditions required for optimal host development.

☞ 20 ml of half strength Hoagland nutrient solution is added to each bag by an adjusted dispenser that collects sterile nutrient solution out of a bottle nearby, and applies a standard volume through a special hole made on the bag margin. To avoid removal of seeds by the liquid, the addition of nutrient solution should be done when bags stand in a vertical position.

☞ Bags should be checked daily for water level. Nutrient solution should be added only when the filter paper tends to dry down. This creates fluctuations in humidity that encourage germination and infection of *Orobanche*, and prevents the development of anaerobic environment in the bags.

☞ Synchronisation of *Orobanche* infection: When the root system of the host covers a significant part of the bag, a germination stimulant can be applied into the bag by syringe (3 ml of 5 ppm GR24, or of root exudate of a potent host, injected to the centre of the back side of the bag). This is best done when there is no nutrient solution at the bag bottom, which ensures an even and thorough spread of the stimulant to all *Orobanche* seeds. The bags are kept horizontally for a couple of hours, than replenished with nutrient solution and put back vertically in the dark box. This treatment results in a high percentage germination of *Orobanche* seeds, consequently many infections develop simultaneously.

EVALUATION

✶ <u>Examination under the dissecting microscope ("binocular"):</u>

Care should be taken that the bottom part of the polybag, that contains the nutrient solution, is always hanging in a vertical position. The upper part of the bag can however lie horizontally on a light-box under the microscope. Parasitic seed germination, attachment to host roots, and tubercle development can easily be seen under low magnifications. Germination and attachment sites can be marked, using different colours, on the bag surface, to enable follow-up of the development of specific sites.

* Sampling: The bag can be opened by simple cuts using a razor blade. Care should be taken to cut only the upper polyethylene sheet. Parasite samples and parts of the host root system can be either treated individually or taken out. Then the bag can be resealed by strips of cellotape.

* If many samples are expected to be taken, the cuts in the bags can be prepared before planting. Then adhesive tape serves as "zippers" that allow frequent opening of the bags. This is simple when the ends of the adhesive tape are folded, thus serving as small handles.

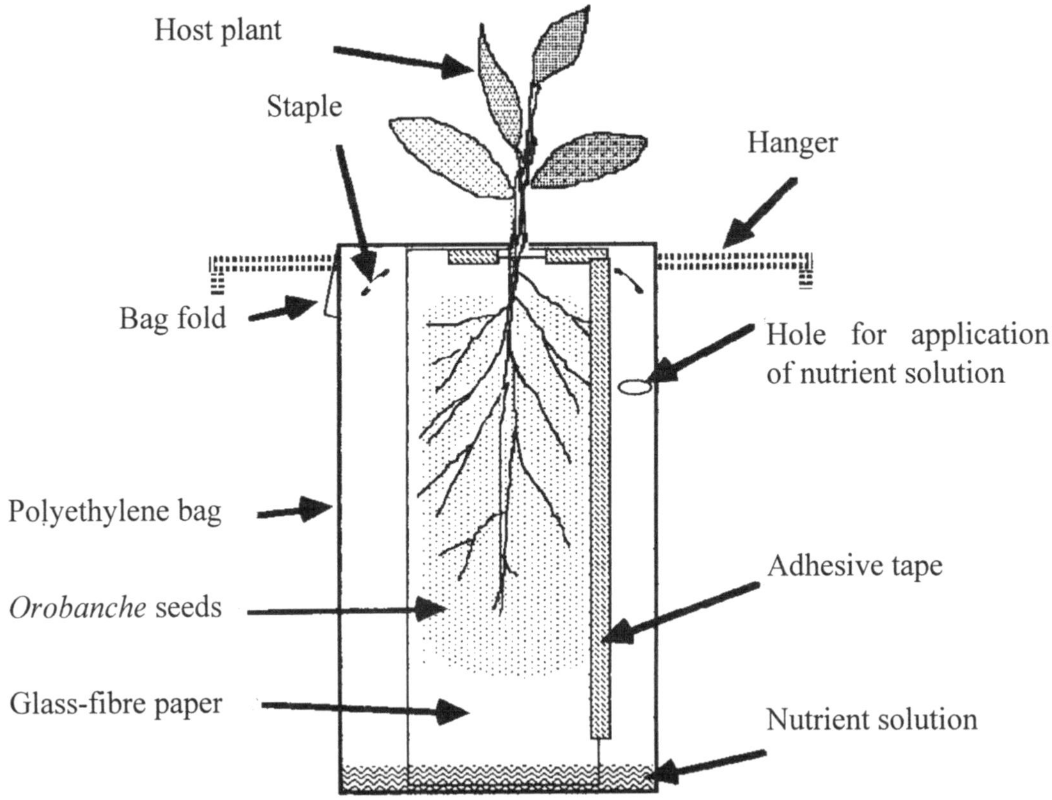

Fig. 15. A diagram of the polybag system for *in vitro* growth of *Orobanche* and similar parasitic plants.

COMMENTS

- The polybag system is a simple, convenient and space-saving method. Many different hosts and parasite species can be studied using this system.

LITERATURE

Lane JA, JA Bailey & PJ Terry 1991a. An *in vitro* growth system for studying the parasitism of cowpea (*Vigna unguiculata*) by *Striga gesnerioides*. Weed Research 32: 211-217.

Lane JA, MJ Kershaw, THM Moore, DV Child, PS Terry & JA Bailey 1991b. An *in vitro* system to investigate the parasitism of cereals by *Striga* species and resistance of cowpea to *Alectra vogelii*. In: JK Ransom, LJ Musselman, AD Worsham & C Parker (eds.), Proceedings of the 5th International Symposium of Parasitic Weeds. CIMMYT, Nairobi, Kenya, 237-240.

Linke KH & W Vogt 1987. A method and its application for observing germination and early development of *Striga* (Scrophulariaceae) and *Orobanche* (Orobanchaceae). In: HChr Weber & W Forstreuter (eds.), Proceedings, 4th Int. Symposium on Parasitic

Flowering Plants, Marburg, Germany, 501-509.
Parker C & N Dixon 1983. The use of polyethylene bags in the culture and study of *Striga* spp. and other organisms on crop roots. Annals of Applied Biology 103: 485-488.
Sauerborn J, H Masri, MC Saxena & W Erskine 1987. A rapid test to screen lentil under laboratory conditions for susceptibility to *Orobanche*. Lens 14: 15-16.
Linke KH, A Abdel-Moneim & MC Saxena 1992. Inter- and intra-specific variation in susceptibility of forage legumes to the parasitic weed *Orobanche crenata* Forsk. Field Crops Research 32: 277-285.
Tennant D 1975. A test of a modified line intersect method of estimating root length. Journal of Ecology 63: 995-1001.

Author: K-H Linke, DM Joel & J Kroschel

5.1.2 *In vitro* culture

INTRODUCTION

In vitro cultures offer some advantages for the study of biochemical and physiological features of parasitic weeds, and for the study of aspects of host-pathogen interaction. *In vitro* culture might also supply large amounts of plant material free from bacterial and fungal contaminants, which can be used for biochemical studies. In addition plant regeneration procedures have potential applications in the development of stable gene transfer systems. This technique also allows for studies about the effect of toxins on germination and seedlings of the parasite.

OBJECTIVES

- To study host parasite interactions in an environment excluding external effects,
- to study basic biochemistry of the parasite,
- to screen hosts for resistance to the parasite,
- to test toxins and herbicides.

5.1.2.1 *In vitro* culture of *Striga hermonthica*

MATERIALS

✓ Glass flasks, immersions of 1 and 10% sodium hypochlorite containing 4 drops of Tween 20 in $1l^{-1}$, ethanol (70%), deionised sterile water, pipettes, a pair of tweezers, Petri dishes (6 cm), clean parasite seeds, a medium based on N6-medium (Chu 1978) modified as follows: For one litre of medium use macro-elements: $(NH_4)_2SO_4$ (463 mg), KNO_3 (2830 mg), KH_2PO_4 (400 mg), $MgSO_4 \times 7H_2O$ (185 mg), $CaCl_2 \times 2H_2O$ (166 mg); micro-elements: $MnSO_4 \times 1H_2O$ (3.33 mg), $ZnSO_4 \times 7H_2O$ (1.50 mg), H_3BO_3 (1.60 mg), KI (0.80 mg); $FeNa_2EDTA \times 2H_2O$ (40 mg); Vitamins: thiamine HCl (1.00 mg), pyridoxine HCl (0.50 mg), nicotinic acid (0.50 mg), myo-inositol (100 mg). 20 g of sucrose is used as carbon source and 2.5 g Gelrite for gelling the media. The pH is adjusted to 5.7. This "CC-Medium" was used in former experiments for co-culture of *S. hermonthica* and sorghum. The medium is autoclaved at 121°C for 20 min. and poured into Petri dishes after cooling down.

✓ Germination of the parasite seeds is induced by stimulation with an aqueous exudate of the roots of sorghum on the medium.

PROCEDURE

Surface sterilisation of host seeds (sorghum)

☞ Soak the seeds in 70% ethanol for 10 min. and sterilise in 10% NaOCl for 20 min. After this treatment rinse the seeds three times in sterile water. Incubate seeds on CC-Medium for three days in the dark at 27°C and use uncontaminated seedlings to produce exudate.

Production of exudate

☞ The root exudate is produced by growing of uncontaminated sorghum seedlings (roots standing in water) in 10 ml water for each seedling under sterile conditions. Keep the cultures in an illuminated incubator with 10 h light at 30°C. After a seven-day culture period the water is collected and termed "exudate".

Surface sterilisation of *S. hermonthica* seeds

☞ Soak seeds in 70% ethanol for 15 min. and then in a 1% NaOCl solution for 30 min. At the beginning and the end of the sodium hypochlorite treatment the seeds are sonicated for 20 s each. Then rinse the seeds three times in sterile water. After the third time of rinsing leave a residue water with the seeds in the flask. All solutions are taken away by sterile pipettes.

Conditioning

☞ Brown & Edwards (1944) had demonstrated that *Striga* seeds did not germinate even under the influence of a host-root factor unless the seeds were pre-treated for some days with moisture. But, it has been demonstrated that germination can be induced without conditioning in water by stimulating the seeds in a moisture film of exudate on top of the media (Fischer *et al.,* 1996). To achieve germination as early as possible, preconditioning in water prior to stimulation should be omitted. Conditioning and stimulation can be carried out in one step.

Stimulation of *S. hermonthica* seeds

☞ The sterilised seeds of *S. hermonthica* are soaked up in a pipette. After sedimentation of the seeds in the tip of the pipette they can be easily delivered as concentrated drops of seeds into Petri dishes with CC-Medium. Use 3 ml of exudate given on the top of the medium as germination stimulant. Keep the cultures in the dark at 27°C. Germination starts 2 days after stimulation. Seedlings taller than 1 cm should be transferred on fresh media and cultured in an illuminated incubator (10 h light, 30°C and 2700 Lux).

Maintenance of *S. hermonthica* seedlings and plants

☞ A positive effect on the development and maintenance of seedlings and plants is achieved due to the addition of exudate instead of water to the media. A moisture film of exudate given on the top of the medium also increases plant viability.

EVALUATION

Determination of germinated seeds

✶ Evaluate 3 × 100 seeds in each Petri dish under an inverted-microscope. Each Petri dish represents one replication. Three replications per variant is the minimum for statistical analysis. This leads to 900 counted seeds per variant. The results are expressed in per cent.

Determination of surviving seedlings/plants

* Only those seedlings and plants are counted as "living", which show a green constitution and which are totally free of any necrosis. The results are expressed in per cent.

COMMENTS

- To get sterile cultures of *S. hermonthica in vitro*, it is important to use seed samples for sterilisation which are free of dirt (sand, soil) and residues of the plant (straw). Otherwise there is great risk to get contamination by bacteria or fungi. Clean samples of seeds can be produced by density gradient centrifugation prior to disinfection (→ chapter 4.2 and 4.5.2).
- Depending on the objective of the experiment above protocol may be varied as follows:

  To study host parasite interactions:
  Co-culture of host and parasite instead of using exudate as stimulant.

  To screen hosts for resistance:
  Either co-culture or the addition of exudate from each host variety as stimulant.

  To screen toxins:
  Toxin should be added to the media at different developmental stages (→ chapter 6.7).

LITERATURE

Brown R & M Edwards 1944. The germination of the seed of *Striga lutea*. I. Host influence and the progress of germination. Annals of Botany 8: 131-148.

Chu CC 1978. The N_6 Medium and its applications to anther culture of cereal crops. In: Proc. of Symp. on Plant Tissue Culture, Beijing, Science Press.

Fischer E, HH Geiger & S Deimling 1996. *In vitro*-Kultur von *Striga hermonthica*. In: International Association For Plant Tissue Culture, 5. Tagung der Deutschen Sektion, Universität Hohenheim, 10.-12. Oktober 1996.

Ejeta G & LG Butler 1996. Biotechnological approaches for understanding mechanisms of resistance to *Striga*. In: MT Moreno, JI Cubero, D Berner, D Joel, LJ Musselman & C Parker (eds.), Advances in Parasitic Plant Research; 6th International Parasitic Weed Symposium, Cordoba, Spain, 567-572.

Okonkwo SNC 1966a. Studies on *Striga senegalensis* III. *In vitro* culture of seedlings. Establishment of cultures. American Journal of Botany 53(7): 679-687.

Okonkwo SNC 1966b. Studies on *Striga senegalensis* IV. *In vitro* culture of seedlings. Effect of various sugars and Glutamine. American Journal of Botany 53(7): 687-693.

Okonkwo SNC 1970. Studies on *Striga senegalensis* V. Origin and development of buds from roots of seedlings reared *in vitro*. Phytomorphology 20: 144-150.

Okonkwo SNC 1987. Research techniques - laboratory. In: LJ Musselman (ed.), Parasitic Weeds in Agriculture Vol. I - *Striga* CRC Press. Boca Raton, Florida, 257-267.

Okonkwo SNC 1996. Aseptic culture of parasitic weeds - an overview. In: MT Moreno, JI Cubero, D Berner, D Joel, LJ Musselman & C Parker (eds.), Advances in Parasitic Plant Research; 6th International Parasitic Weed Symposium, Cordoba, Spain, 283-292.

Authors: E Fischer & S Deimling

5.1.2.2 *In vitro* culture of *Orobanche aegyptiaca*

MATERIALS

✓ Autoclave, glass flasks, sterile ELISA plates, 96 wells each, pipettes, tweezers, Petri dishes (9 cm).

✓ *Orobanche* seeds free of dirt and straw.

✓ Sodium hypochlorite 1% containing 0.1% Tween 20. Ethanol 70%. Deionized sterile water. Medium containing half-strength MS salts (Murashige & Skoog 1962), 3.6% PDA, and GA3. This PDA medium is used for seed germination, and further development up to seedlings.

✓ For further growth of *Orobanche aegyptiaca* seedlings, agar medium is used containing half strength MS, 2% sucrose, 100 mg/l mio-inositol, 0.1 mg/l thiamine, 0.5 mg/l pyridoxine and 0.5 mg/l nicotinic acid, and 8 mg/l GA3. The media is used in 0.6% (w/v) agar and adjusted to pH 5.8 with KOH. Media are autoclaved at 121°C for 60 min. After cooling down, the PDA medium is poured into Petri dishes and MS medium is used for filling ELISA plates. 130 µl medium is placed in the bottom of each well.

PROCEDURES

☞ *Orobanche aegyptiaca* seeds are placed in mesh bags as used for tea bags (8,000-10,000 seeds per bag), surface sterilized in 70% ethanol for 1 min., followed by immersion in 1% sodium hypochlorite containing Tween 20, for 15 min. The seeds are rinsed several times with sterile water and dried for 3-4 hours under sterile conditions. When seeds are dry all procedures for sterilisation are repeated for the second time.

☞ Dry sterile seeds are sown on PDA nutrient medium (about 1 000 seeds per plate). After incubation for 5-7 days in the dark, 22-25°C, massive seed germination occurs. A few days later the young radicles start to develop callii.

☞ Two distinct types of callii can be obtained: One is fragile, white, and fast growing, that soon differentiates into stunted shoots; the other is darker, more compact, and slow growing, that differentiates protrusions rather than shoots.

LITERATURE

Ben-Hod G, D Losner, DM Joel & AM Mayer 1991. *In vitro* culture of *Orobanche aegyptiaca*. Annales of Botany 68: 413-416.

Ben-Hod G, D Losner, DM Joel & AM Mayer 1991. Infection of tomato roots using callii of broomrape. Physiologia Plantarum 83: 353-356.

Harloff HJ & K Wegmann 1993. Evidence for a mannitol cycle in *Orobanche ramosa* and *O. crenata*. Journal of Plant Physiology 141: 513-520.

Kumar U & NS Rangaswamy 1977. Regulation of seed germination and polarity in seed development in *O. aegyptiaca* by growth substances. Biologia Plantarum (Praha) 19: 353-359.

Losner D, G Ben-Hod, AM Mayer & DM Joel 1996. Aseptic broomrape infection of tomato root culture. Israel Journal of Plant Sciences 44: 89-94.

Murashige T & F Skoog 1962. A revised medium for plant growth and bioassay with tobacco tissue culture. Physiologia Plantarum 15: 473-497.

Okonkwo SNC 1996. Aseptic culture of parasitic weeds - an overview. In: MT Moreno, JI Cubero, D Berner, D Joel, LJ Musselman & C Parker (eds.), Advances in Parasitic Plant Research, 6th International Parasitic Weed Symposium, Cordoba, Spain, 283-292.

Rangaswamy NS 1963. Studies on culturing seeds of *Orobanche aegyptiaca* Pers. In Maheshwari, P., Rangaswamy, NS (eds.), Plant Tissue and Organ Culture. A Symposium. 345-354.

Author: V Portnoy & DM Joel

5.2 Pot experiments

INTRODUCTION

Between the almost entire elimination of natural factors in laboratory studies and the inclusion of them in field studies the third and in between condition of research is represented by greenhouse and related studies. Most results of agricultural and related research pass the classical way from laboratory through greenhouse to the field. Accordingly, techniques in the greenhouse are numerous and have to be selected and modified according to the needs and the availability of materials. In the following, emphasis is put on pot experiments with parasitic weeds.

OBJECTIVE

- To create conditions that combine advantages of field and laboratory studies, i.e. controlled environment with nearly natural growth conditions (standard amounts of parasite seed in the soil, control of water, light and temperature, and screening almost during the whole year).

MATERIALS

✓ Soil free of parasite seed, sand, parasite seed of known origin and germinability, suitable mixing device (concrete mixer or buckets), balance, plastic pots or polyethylene bags, greenhouse with facilities for regular irrigation and shading, seed of host plant(s).

PROCEDURES

☞ Use pots with size adequate for the aim of the experiment. In many cases simple plastic pots of e.g., 13 x 13 x 13 cm will be sufficient. For studies for a longer period of time pots of bigger size are advantageous (e.g., Mitscherlich pots). The growth period will depend on the crop growth, about 12 to 20 weeks for faba bean and sorghum, respectively.

☞ Instead of pots black polyethylene bags (about 1 litre or 7 x 7 x 23 cm) filled with the same mixture of soil, sand and parasite seed can be used. Advantage: The polyethylene bag is more convenient for host roots that tend to develop a deep running root system.

☞ Estimate the quantity of growth medium required (breeding studies: kg/pot x entries x replications; other studies: kg/pot x treatments x replications).

☞ Mix the amount of (local) soil and sand as required at a ratio of 2:1 (v/v), e.g., using a concrete-mixer. Ensure to use parasite free soil only.

☞ Prepare the required amount of parasite seed. The amount may vary depending on host plant, seed germinability and purpose of the experiment: use less seed to ensure aerial development of the parasites and to avoid too much pressure on the host plant, or use more seed where final evaluation is to be conducted at an early growth stage of the host. The amount of seed (mg/kg soil) is calculated from the 1000 seed weight and the germinability of the parasite seed used: e.g., the critical amount of *O. crenata* seed to screen chickpeas ranges from 2500 to 7500 parasite seed/kg soil, i.e. with too low infestation the attack shows high variation (need for considerable increase of replications), with too high infestation the differences between test entries or treatments become marginal. Examples for parasite seed loads per kg soil in pot studies are

- *Orobanche crenata*: 0.01 g/kg (faba bean, narbon vetch, lentil); 0.03 g/kg (chickpea, safflower).
- *Orobanche aegyptiaca*, *O. ramosa*: 0.01 to 0.03 g/kg depending on host plant and objective of study.
- *Striga asiatica*: 0.02 g/kg
- *Striga hermonthica* (sorghum): 0.01 to 0.05 g/kg
- *Alectra vogelii* (cowpea): 0.06 to 0.1 g/kg
- *Cuscuta* spp.: 50 to 100 seeds spread on top of the soil, depending on seed quality and purpose of experiment. Under certain conditions higher numbers of seed might be required (e.g., when mixing the seed into the top layer of soil). Another method to infest potential host plants of *Cuscuta* is the cultivation of *Cuscuta* on a suitable host plant (e.g., *Pelargonium zonale* L.) in the greenhouse at 20 to 24°C and long-day conditions. Such culture is used as a direct source to infect test plants with already growing *Cuscuta* branches coiling to the test plant.

☞ Add the amount of parasite seed required for the respective screening into the mixer, or place parasite seed in a certain zone of the pot (→Fig. 16).

☞ Fill the pots with the prepared soil.

☞ The pots are watered to allow conditioning of parasite seed at the required temperatures (→ chapter 4.4.1), and then 3 seeds of a test entry are planted per pot and thinned afterwards to obtain 1 host plant/pot. Other objectives may require more than one host plant/pot.

☞ Fertilisation of the pots depends on the respective study. If there is no obstacle to that, granular NPK or liquid fertiliser (e.g., Wuxal at 2%) should be added for optimal growth of the host plants.

☞ Keep the pots in the greenhouse and care for regular watering and temperature conditions convenient for the crop to grow well (soil temperature 18 to 25°C for faba bean, lentil, vetch, chickpea, and 22 to 28°C for sorghum, cowpea, sunflower and tomato).

☞ According to the crop and the temperature conditions, the washing of crop roots to check the underground development (attachments) can be done 3 to 6 weeks after sowing of the host crop.

EVALUATION

The type of evaluation greatly depends on the objectives of the respective study and hence may vary considerably. In general, the following steps may be followed.

* Use 4 to 6 replications per entry or treatment; include checks without parasite seed in the soil.
* Assess the time from sowing until the first emergence of parasites.
* Record the number of emerged parasites weekly.
* Measure the height of the host plant once per week.
* Determine the host and parasite dry weight after drying for 24 h at 110°C at the end of the experiment.
* Especially for trials with *Orobanche* spp. the assessment of underground development stages is often useful. Therefore, clean the roots carefully from soil under tap water or in buckets filled with water to assess the

number of attachments, classified on a scale starting from needle stage (may be neglected due to it's size), tubercle stage, bud stage, crown roots, stem length 2 to 4 cm, stem length 4 to 6 cm and so on. Assess dry weight of the attachments.

* The evaluation of *Cuscuta* infestation has to be scored at several intervals starting from planting of the crop until harvest (or total damage). It can be expressed as % infestation of host plant or by the assessment of dry matter assessment of *Cuscuta* and host (the latter method is destructive and requires a high number of replications; it is therefore more suited for the final evaluation). The final evaluation depends on host, parasite and prevailing temperature regime; a growth period of 40 to 60 days may be expected.

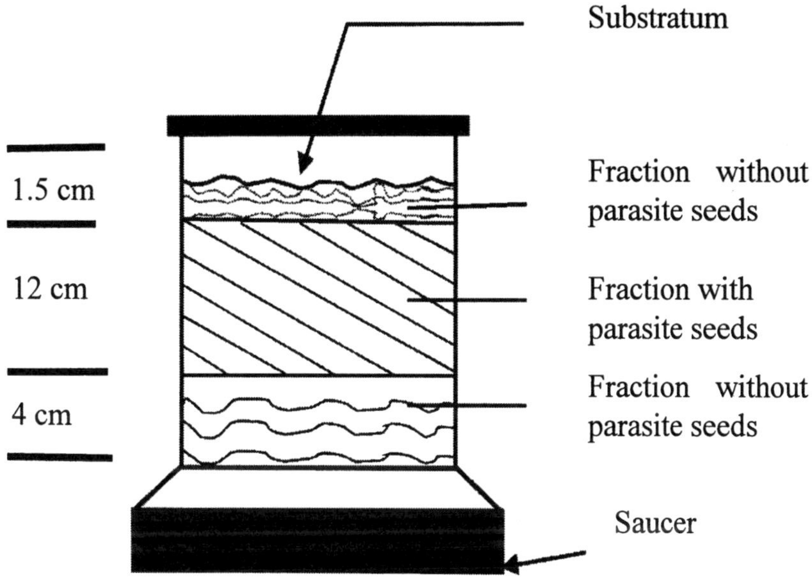

Fig. 16. Filling of a pot with substratum and parasite seed

COMMENTS

- Mixing the parasite seed with the soil requires special attention; add fine sand to the parasite seed to avoid any clumps (water content of soil should be low) or aggregation of seed.
- The evaluation of non-emerged parasites (1) may reduce the actual crop growth period to 3 to 6 weeks compared to 8 to 12 weeks necessary to see the emerged stages, (2) is more time consuming for the evaluation process itself and (3) provides more reliable results. It is, however, not feasible for *Striga* spp.

LITERATURE

Kroschel J 1989. Der Einfluß von Stickstoff auf phanerogame Wurzelparasiten und ihre Wirtspflanzen unter besonderer Berücksichtigung von *Striga hermonthica* (Del.) Benth. (Scrophulareaceae). PLITS 7(3).

Kukula S & H Masri 1985. A greenhouse technique for screening faba beans (*Vicia faba* L.) for resistance to *Orobanche* spp. FABIS Newsletter 12: 20-23.

Linke KH 1989. *Orobanche* studies. In: Food legume improvement program - Annual report for 1988. ICARDA, Aleppo, Syria. 236-239.

Linke KH 1991. *Orobanche* studies. In: Food legume improvement program - Annual report for 1990. ICARDA, Aleppo, Syria, 225.

Riches CR 1989. The biology and control of *Alectra vogelii* Benth. (Scrophulariaceae) in Botswana. Ph.D. Thesis, University of Reading, UK, 208 pp.

Saghir AR, M Kurban & B Budayr 1980. Studies on the control of *Orobanche* in Lebanon. Tropical Pest Management 26(1): 51-55.

Vasudeva Rao MJ 1985. Techniques for screening sorghums for resistance to *Striga*. ICRISAT Information Bulletin No. 20, Andhra Pradesh, India.

Vogt W 1993. Entwicklung der Wirt/Parasit-Beziehung *Sorghum bicolor/Striga hermonthica* unter dem Einfluß verschiedener Stickstofformen und Standortfaktoren. PLITS 11(3).

Wilson AK & C Parker 1984. Techniques for setting up pot experiments involving the parasitic weed *Striga*. In: C Parker, LJ Musselman, RM Polhill & AK Wilson (eds.) Proceedings of the 3rd International Symposium on Parasitic Weeds, ICARDA/ISPRG, Aleppo, Syria, 133-142.

Authors: K-H Linke & J Kroschel

5.3 Field experiments

5.3.1 Site selection and preparation

INTRODUCTION

The variation of infestation in the field and the complex interactions between the parasite, the host and edaphic and environmental factors determine the extent of seed germination and infection of the host plant. Since these processes are still not completely understood and the rate and uniformity of infection are rather unpredictable, obtaining comparable and conclusive research results can be a major constraint to efficient research on parasitic weeds and experimentation in the field. Vasudeva Rao (1985) found in experiments with a checkerboard layout conducted at five artificially infested sites and over three years a high variability in *Striga* counts between replicates of the same host variety in the same experiment each year and had different mean infestation levels each season. Therefore site selection and preparation is very critical for *Striga* experiments and several techniques have been developed to overcome these impediments.

The inclusion of farmers' fields at an early stage is often advisable to broaden the spectrum of soil and climatic conditions under which control methods or varieties are tested, as the development of parasitic weeds and their effect on the host plant depends to a large extent on these parameters, and experimental results may vary accordingly. The main difference between research stations and farmers' fields for the researcher is, that he cannot generally manipulate the level and uniformity of infestation in a farmers' field for obvious reasons. At the research station, however, this is possible and often a prerequisite to obtaining scientifically meaningful results by reducing the high variability and standard error of experiments on parasitic weeds.

OBJECTIVES

- To generate, test and evaluate new control techniques,
- to adapt already developed control techniques to the local agro-ecological conditions and farming systems,
- to assess host plant varieties for their resistance or tolerance to parasitic weeds,
- to conduct the experiments either at research stations or as researcher managed trials on farmers' fields.

MATERIALS

✓ Field experiments on parasitic weeds principally require the same technical equipment and material as other field experiments. Hence these are not listed here. Depending on the experiment, however, specific requirements with parasitic weeds may include an electronic balance for small amounts of parasite seed; also fine sand might be required to be mixed with the seed for uniform artificial infestation.

PREPARATION, COMMENTS

Site selection for on-station experiments

- ☞ First of all the experimenter should consider whether the planned experiment is aimed at short or long term control strategies. The effect of agronomic practices on parasitic weed control becomes often only evident after several seasons. Hence the experiment might have to remain for several seasons at the same location with the same design and randomisation to obtain significant treatment and interaction effects. Exceptions can be physiological studies or experiments testing the effectiveness of fertilisers or herbicides.
- ☞ The same basic rules that apply to other agricultural field experiments, apply to site selection for experiments on parasitic weeds, such as homogeneity of soil, uniform fertility status, etc. (for more detailed information see Gomez & Gomez 1984, Rohrmoser 1985). This is even more important to long term studies. Furthermore, biotic factors such as trees or termite hills might grow in time and can severely interfere with crop development and treatment evaluation.
- ☞ Special requirements of experiments on parasitic weeds are that the parasite should not spread to other adjacent sites and that the level of infestation should be uniform for the entire experimental area (unless special parasite-free areas within the experiment are required).
- ☞ To avoid further spread of parasite seed by wind (especially *Striga* and *Alectra*) the site should be big enough to plant guard rows around the experiment, which should reduce the impact of wind on soil and *Striga* seed movements. Also a few metres distance to other trials or fields greatly reduce the spread of the parasites. Run-off water in ditches or small field channels running to other fields could be another means of parasite distribution, and care should be taken to avoid this.
- ☞ For highly controlled on-station experiments, a soil with a low clay content and good drainage is desirable. Water logging may hasten the demise of parasite seeds. A high clay content makes a soil heavy and compact and may impede the development of parasitic weeds.
- ☞ When a site has been selected and there is a season to develop it, the land can be planted with a susceptible host crop to assess the level of uniformity of infestation. Areas with little or no infestation can be marked for parasite free controls or hot spots left out from further artificial infestation. In any case the assessment will help the experimenter to analyse research results once problem data occurs.
- ☞ If the experiment is to be planted straight away, methods of artificial infestation can be used.

Site selection for on-farm experiments

☞ The locations chosen for on-farm trials should be easily accessible to avoid time consuming and costly travel, and to allow for frequent visits during the growing season.

☞ The sites should be selected the previous season at a time when the respective parasitic plants are flowering or about to flower, to assess the level and uniformity of infestation.

☞ Mark areas of high or low infestation or draw a field plan.

☞ The basic requirements of a good experiment station site also apply to farmers' fields and should be checked.

☞ Quite often an on-farm site is not big or uniform enough to accommodate a whole experiment, especially if an even infestation has to be considered. In this case the number of treatments, the plot size or the number of replications can be reduced (use more sites nearby as additional replications for the analysis).

☞ Get your agreements or contract with the farmer or owner of the site.

Site preparation

☞ Soil cultivation and land preparation should be conducted as usual for the crop.

☞ Deep ploughing buries the parasite seed and will reduce infestation. If the field is plowed by tractor the working depth should not exceed 20 cm. With *Striga* land preparation with the hand-hoe is preferred.

☞ However, deep and repeated ploughing is one option to bury the seed and create a surface layer with a rather low but uniform incidence; this field could be subsequently infested artificially to achieve a uniform distribution.

☞ After ploughing, the field usually gets harrowed and eventually infested artificially.

REFERERENCES

Gomez KA & AA Gomez 1984. Statistical procedures for agricultural research. Second Edition. John Wiley & Sons, New York, USA, 680 pp.

Rohrmoser K 1985. Handbook for field trials in technical cooperation. TZ-Verlagsgesellschaft, Roßdorf, Germany.

Vasudeva Rao MJ 1985. Techniques for screening sorghums for resistance to *Striga*. ICRISAT Information Bulletin No. 20, Andhra Pradesh, India.

Authors: A Oswald & JK Ransom

5.3.2 Artificial infestation

INTRODUCTION

To rely on natural infestation alone is often not advisable, unless the characteristics of the site are well known. Methods of artificial infestation can be used to increase uniformity and decrease variability which may obstruct proper data interpretation. Different methods are available. The choice will depend on the objectives of the experiment, the amount of parasite seed, time and labour available for this operation. Generally there are two distinct procedures:

(1) In variety trials or physiological studies the response of the single plant to infection is investigated. Thus parasite seed can be applied

directly in the planting hole or in the immediate surrounding to assure a high probability of infection. This method also uses less seed than the next one described.

(2) In agronomic experiments the researcher usually wants to test farming practices, such as inter-cropping, field preparation, catch-cropping, etc., for their effect on parasite infection and damage. Therefore the soil at normal ploughing depth needs to be infested uniformly (it needs to be more like an actual farmers' field), because strategies for low infection rates of host plants or for depletion of the parasite seed bank in the soil, are under investigation. In these cases the parasite seed have to be broadcast on the entire experimental area.

Broadcasting is achieved either by direct spread of parasite seed by hand (problem of uniformity) or by mixing the seed with fine sand before spreading in order to achieve a better uniformity of parasite seed in the soil.

OBJECTIVES

- To increase uniformity of infestation with parasite seed,
- to decrease variability of infestation and of the respective data collected in the experiment.

MATERIALS

✓ Seed of parasitic plants, balance, containers of different size according to the amount of seed to be spread, fine sand as marker and as mixing agent.

PROCEDURES

1. Calculation of infestation rates

☞ The calculation of the amount of parasite seed needed to infest a certain area with a specific rate is about the same process as calculating seed rates for a crop. However, the amount of trash in the sample has to be taken into consideration.

☞ Determination of the germination rate of the seed (GR)as described in chapter → 4.4.

☞ Determination of the seed weight.

☞ Estimation of the proportion of trash (T) mixed with the seed. To do this a sample of the seed is placed in a Petri dish under a dissecting microscope and the proportion of trash is separated from the seed using, e.g., a needle. The proportion of trash as a percent of the total is expressed as follows:

T (%) = (weight of trash/weight of sample) x 100.

☞ Definition of the amount of viable seed (VS) to be used per infestation unit (IU). An IU is the smallest division of the entire area to be infested. This area will be infested with a precisely measured amount of seed. This could be a single planting hole or a subdivision (plot) of the experimental field.

☞ Calculation of the amount of seed required per IU:

Seed rate per IU = VS x weight of one parasite seed/GR x (1- T)

☞ Calculation of the amount of seed required for the whole experiment

Total amount of seed per experiment = (seed rate per IU) x (number of IUs per experiment)

Example with *Striga* seed in a sorghum field; each planting hole (= IU) should be infested with 1500 viable seeds:

GR: 70% or 0.7;

T: 20% or 0.2;

Weight of one seed: 5×10^{-6} g.

Seed rate per IU = $1500 \times (5 \times 10^{-6}) / 0.7 \times (1-0.2) = 0.013$ g

The area to be infested is 750 m^2, the spacing is 0,1875 m^2/hill:

Total amount of seed per experiment = $0.013 \text{ g} \times (750 \text{ m}^2 / 0.1875 \text{ m}^2) = 52$ g.

2. Direct host plant infection (mainly with *Striga* and *Alectra*)

☞ With *Striga*, good results are usually achieved with an application of 1000-3000 *Striga* seeds per planting hill, regardless of the number of crop plants per hill. *Striga* needs a period of 7-14 days to precondition, therefore it would be ideal to infest the hills 7 to 14 days before planting of the host crop. However, this can be a difficult and laborious operation, because all planting holes have to be marked after the infestation, to find them again for the planting of the crop. Hence there are two planting operations for one experiment. Furthermore it is often not possible to have the field prepared 7-14 days before planting, especially in regions or experimental stations which have no irrigation facilities. Since the host plant takes several days to germinate and several weeks to develop a reasonable root area, there is little need to infest before planting. Therefore direct infestation at the planting time of the experiment is used more frequently and has proved to be a reliable method.

Infestation of single plant stands

☞ Once the application rates are calculated, the *Striga* seed should be mixed with fine sand. The amount of sand required for the experiment depends on the quantity applied to each hill. One to two grams of sand/hill is sufficient to get an even distribution of the *Striga* seed in the planting hole, i.e. for the previous example with 750/0.1875 = 4000 hills, 4 to 8 kg of sand are needed.

☞ The mixing of the *Striga* seed with the sand must be very thorough to obtain a homogeneous distribution of the seeds in the mixture. Therefore the sand is divided into smaller quantities, for example according to the amount needed to infest one plot. Then the respective amount of *Striga* seed is mixed with a small portion of this sand and finally more and more sand is added while stirring continues.

☞ Using the previous example, assuming the experiment has 40 plots with 100 plant stations, and each hill receives 2 g of sand, then

100 x 2 g = 200 g of sand are used to infest one plot.

☞ The sand is mixed with

100 x 0.013 = 1.3 g of *Striga* seed.

☞ This operation is repeated 40 times to have material to infest all plots. To facilitate the application process small containers can be made, which just hold the proper amount of the *Striga*-sand mixture. Once the planting holes are dug in the field to a depth of 8-10 cm a scoop of the mixture is given and covered with soil. Then the host seed is planted and the hole covered. The best level of infection occurs when the sand-seed mixture is placed several cm below the actual planting depth of the host seed.

Infestation of plant rows

☞ For this method the IU is a whole row of a plot at a time. Prior to planting the soil is removed from strips approx. 20 cm wide and 15-20 cm deep. Then the *Striga*-sand mixture is applied and covered with soil again. It is preferable to have the *Striga* seed just above the plow layer, as the roots of the host will spread there, and infection is more likely.

☞ This operation can be done before planting, so that the *Striga* seeds can precondition, or when the host crop is sown. It requires more time and labour than the method described above but the *Striga* seed is spread over a larger area which better resembles natural infestation conditions. However, larger quantities (20 000 to 40 000 seeds per running meter) of *Striga* seed have to be used to assure good infection rates.

Infestation of single plant stand with *Striga*-water mixture

- Instead of sand, water is used as a carrier for *Striga* seeds. The mixture is applied with syringes, which hold enough water for several planting holes. The application rates should be around 5-10 ml of water/hole. The amount of water and *Striga* seed is then calculated according to the procedure described above. The *Striga*-water mixture is prepared in buckets and has to be stirred continuously. The syringes are filled and the suspension applied directly to the planting hole.

3. Infestation of entire fields (*Orobanche, Striga, Alectra*)

☞ For the uniform infestation of an entire field, parasite seed is broadcast in a seed-sand mixture. The field has to be harrowed and should be without any vegetation. Then it is subdivided in strips of 15-25 m^2. The strips are preferably 2 m wide, that a worker can cover the whole width while moving forwards or backwards. The amount of sand needed for 20 m^2 is approximately 2 kg. This quantity is mixed with the parasite seeds. The amount of seed needed is calculated on an area basis. Quantities of 30000-50000 viable seeds/m^2 should result in medium to high infestation levels, but have to be adjusted to the respective conditions (species, crop, water and soil conditions).

☞ The seed rate is calculated using the above example, an IU of 20 m^2 and an amount of viable seed (AVS) of 40000 seed/m^2 = 800000 seed/IU:

Amount of seed per IU = (80 000 x 5 x 10^{-6}) / 0.7 x (1 - 0.2) = 7.14 g

Total amount of seed = 750 m^2 / 20 m^2) x 7.14 g = 267.75 g.

☞ Considerably larger quantities of parasite seed have to be used for this method. Although the calculation was done for the exact experimental area, it is sometimes advisable to infest the whole field area including the side strips in order to have homogeneous conditions for future, perhaps larger, experiments.

☞ After broadcasting the seed-sand mixture the field should have a shallow but thorough hand harrowing to work the seed a few cm into the soil. Then the field can either be left for some time (for *Striga* to precondition) or it can be planted immediately afterwards.

LITERATURE

Ransom JK, RE Eplee, MA Langston & RS Norris 1990. Methodology for establishing witchweed (*Striga asiatica*) in research plots. Weed Technology, Volume 4, 581-584.

Authors: Ransom JK & A Oswald

5.3.3 Developing parasite free areas

INTRODUCTION

For certain experiments it is necessary to grow host plants on parasite free areas. This is achieved by eliminating the parasite seeds from the soil and eventually re-infesting the area with a known amount of seed. This is often a rather expensive operation and should only be done if absolutely necessary. The methods described below may not always clean a field completely but should often be sufficient to allow comparisons and assess the actual damage or physiological changes on the host crop inflicted by the parasite.

OBJECTIVES
- To compare treatments of infested with non-infested plots under otherwise comparable conditions,
- to assess the damage from a parasite infestation or the effect of the interaction processes.

PROCEDURES

1. Use of catch crops

☞ → chapter 6.2.1

2. Ethylene application

☞ → chapter 6.4.3.1

3. Soil solarisation

☞ → chapter 6.5.5

4. Chemical soil sterilisation

☞ → chapter 6.4.1

EVALUATION

* ★ The layout and design for these experiments should divide the field into parasite free and infested areas to avoid infestation of free plots during the growth period from seeds of maturing parasite plants or from neighbouring plots.
* ★ A split-plot design may be used rather than randomised complete block designs.

LITERATURE

Egley GH, RE Eplee & RS Norris (1990). Discovery and development of ethylene as a witchweed seed germination stimulant. In: PF Sand, RE Eplee & RG Westbrooks (eds.), Witchweed Research and control in the United States. Weed Science Society of America, Champaign, IL, USA, 56-66

Authors: A Oswald & JK Ransom

5.3.4 Experimental design for field experiments

INTRODUCTION

A main objective of any experimental design is to keep the experimental error and variability as small as possible. This is even more crucial for experiments with parasitic weeds due to the commonly found high variation in infestation levels in a field. Variability in the data obtained can mask treatment effects.

OBJECTIVES
- To keep the experimental error and variability as small as possible,
- to obtain reliable and conclusive data,
- to minimise the costs (labour, area, materials) of an experiment.

PROCEDURE, COMMENTS

- If treatments have quantitative factors it is advisable to choose rather distinct levels, because too narrow intervals might not yield statistically useful information.

- Whether to use a design with one or more error terms depends mainly on the nature of the factors under investigation. For experiments with different levels of infestation, it is better to use split- or strip-plot designs for the convenience of maintaining those levels over several seasons.
- In uniform experimental sites 3 to 4 replications should be sufficient, as the decrease in the standard error by added replications is rather small. The plot size is determined by the crop planted, the treatments applied and the variation in parasite seed distribution in a given area. Preferable are smaller plots to reduce the block size and to decrease the variability within a block.
- For on-farm experiments, where the parasite infestation in the field is probably very uneven, the number of replications should be higher than on-station to reduce the experimental error. However, other considerations, such as field size, number of visits during the growth season, etc. might create the necessity to decrease the number of replications per site to two but have more sites nearby. If this design is used the data of all sites can be pooled for statistical analysis.
- All these general considerations might or might not hold for the conditions prevailing at a specific site. Therefore, some experience is indispensable to choose the right number of replications and the adequate plot-size; but even with artificial infestation variability will be higher than for most other types of experiments.
- Specific designs and problems in breeding experiments are described and discussed in Vasudeva Rao (1985) and Berner *et al.* (1993). For more detailed information on statistics for field experiments see Gomez & Gomez (1984), Mead & Curnow (1990) and Neeley *et al.* (1991) and → chapter 5.3. More complex designs which take into account variability within a field more effective than block designs are available. However, since they require a high level of statistical understanding, they are best carried out in conjunction with a collaborating statistician.

LITERATURE

Berner DK, MD Winslow, AE Awad, KF Cardwell & DR Mohan Raj 1993. *Striga* Research Methods - A manual (Draft Edition). IITA, Ibadan, Nigeria. 71 pp.

Gomez KA & AA Gomez 1984. Statistical procedures for agricultural research. Second Edition. John Wiley & Sons, New York, USA. 680 pp.

Mead R & RN Curnow 1990. Statistical Methods in Agriculture and Experimental Biology. Chapman and Hall, London, UK. 335 pp.

Neeley D, J Matata & F Kisyombe 1991. Statistical techniques for the analysis of on-farm crop trials - a manual. CIMMYT, Harare, Zimbabwe and Nairobi, Kenya. 197 pp.

Rohrmoser K 1985. Handbook for field trials in technical cooperation. TZ-Verlagsgesellschaft, Roßdorf, Germany.

Vasudeva Rao MJ 1985. Techniques for screening sorghums for resistance to *Striga*. ICRISAT Information Bulletin No. 20, Andhra Pradesh, India.

Authors: A Oswald & JK Ransom

5.3.5 Data collection in field experiments

INTRODUCTION

With parasitic weeds not only the development and yield of the specific crop has to be followed up, but also the growth, abundance and reproductive ability of the parasite. Furthermore, secondary data like soil parameters, precipitation, weather conditions etc. are collected. Various assessments at different times during the growth period may be required.

As most of these operations are time and money consuming and require experienced labour to carry them out, they should be thoroughly planned and restricted to essentials.

OBJECTIVES

- To assess and compare the effects of treatments in an experiment,
- to understand the influence of secondary factors in an experiment,
- to better understand the relation between the occurrence of the parasite and the damage inflicted on the crop under natural or treatment-induced conditions.

MATERIALS

✓ Layout (including plot numbers, treatment numbers, description of the experiment); shovel for below-ground assessments, paper or polyethylene bags, balance, manual counter.

✓ Further equipment according to the respective objectives of the study.

PROCEDURES

1. Below-ground evaluations on the seed bank

☞ Depending on the objective the samples can be taken per plot, block or from the whole field (→ chapter 4.5.1). Usually they are drawn from the upper 20 cm of the soil. For a plot, 15-20 sub-samples which are mixed to one composite sample should be sufficient. The samples are then processed as described in chapter 4.5.2, followed by a subsequent germination and/or viability test (→ chapters 4.3 and 4.4).

☞ Another objective of below ground observations is the natural or treatment-induced demise of parasite seeds in the soil. For this purpose seed bags (→ chapter 4.7) are placed in the plots at different depths and distances from the plants exposing them to different soil conditions. They can be retrieved at any time for counts and germination and/or viability tests (→ chapters 4.3 and 4.4).

2. Below-ground evaluations on subterranean stages of the parasites

☞ According to the respective developmental size, the subterranean stages can be distinguished using phenological characters like appressorium, tubercle stage, bud stage and shoot stage, crown roots etc.

☞ For exact studies a certain number of crop plants (e.g., 4 plants/plot) are dug with the entire root system.

☞ The roots are washed carefully to remove the soil.

☞ The attached parasite plants are recorded (number, fresh and dry weight) according to their respective stages.

☞ This assessment can be carried out several times during the subterranean growth of the parasite to follow up on the development of the below-ground attack.

☞ The evaluation has to be carried out according to the field layout (e.g., as a split-plot layout).

3. Above-ground evaluation

☞ The size of the area counted in each plot depends mainly on the uniformity of the distribution of parasitic plants within the plots. With an even distribution areas used for counting can be smaller and taken at random. With a more non-uniform spread of the parasite the entire net-plot should be counted, if possible. In the literature various methods are cited, such as *Striga* counts from 1 m^2 chosen at random in each plot, counts from 5 squares of 1 m^2 at random in each plot, counts on a 0.50 m x 0.75 m quadrat around each hill of the 4 centre rows of the net plot or counts of the entire net-plot. Similar areas hold for *Alectra*. Counting *Orobanche* shoots of course depends on the uniformity of infestation; usually however, 1 m^2 per plot is sufficient in smaller crops like lentil or vetches, while 2 m^2 of a centre row in a plot sometimes is more convenient in faba beans or tomatoes.

☞ Also the number of counts per season can vary according to the objectives of the experiment, the time and labour available and the accessibility of the experimental site, e.g., in on-farm experiments. If only one count per season is possible, then it should be done when most parasites have emerged, which is around flowering time of the parasite.

☞ The intervals between counts can be reduced *ad libidum*, to two weeks after parasite emergence, to one week or three days, with a consequent gain in precision of the generated data. *Striga* plants die and degenerate quickly and after a few days the dead plants may not be distinguishable anymore. Hence the very short interval of three days might be justified. In *Orobanche* the stalks remain visible in the field until crop harvest.

☞ Furthermore the number of parasite plants in the plot can be further subdivided into emerged, flowering and mature. Especially, if the objective of the experiment is to investigate methods which interfere or inhibit the reproduction of the parasite, discerning the growth stages generates useful information.

☞ A crucial decision is whether the parasite shoots are to be uprooted before seed shed or left in the field for natural reproduction. For the first option the parasite should be removed at flowering. Only if yield parameters of the parasite such as number of capsules, seeds per capsule, etc., are to be determined, should the plants be harvested at maturity (the stage just before seed shed). This implies however, that in order to avoid seed shed the field has to be checked almost every other day. The uprooted plants can be collected in paper bags allocated to the different treatments during the growth season. At harvest of the host crop all remaining parasite plants are harvested as well and added to the bags. After drying the dry matter production per plot and season can be obtained.

COMMENTS

- The assessment of the amount of parasite seeds in the soil, i.e. of the infectious potential of a specific area, can be a powerful tool in adjusting experimental data, if used, for example, as a co-variable for crop yields. This applies particularly for experiments with a non-uniform infestation. It represents also valuable data on its own, e. g., monitoring the effectiveness of some treatments on depleting the seed bank in the soil or relating the amount of above ground parasites to the number of parasite seed discovered in the soil.

- Below-ground evaluations on subterranean stages of the parasites are of importance especially with *Orobanche*, as these stages are bigger than those of *Striga* or *Alectra*, all stages except the appressorium being distinguishable with the naked eye. The data are of importance for example in the chemical control or in ecological studies.

LITERATURE

Abayo GO, JK Ransom, J Gressel & GD Odhiambo 1996. *Striga hermonthica* control with acetolactate synthase inhibiting herbicides applied to maize seed with target-site resistance. In: MT Moreno, JI Cubero, D Joel, LJ Musselman & C Parker (eds.), Advances in Parasitic Plant Research. 6th International Parasitic Weed Symposium, Cordoba, Spain, 761-768.

Agbobli CA 1991. Effect of nitrogen rates on *Striga asiatica* emergence in maize culture in Togo. In: JK Ransom, LJ Musselman, AD Worsham & C Parker (eds.), Proceedings of the 5th International Symposium of Parasitic Weeds. CIMMYT, Nairobi, Kenya, 28-30.

Carsky RJ, L Singh & Ndikawa 1994. Suppression of *Striga hermonthica* on sorghum using cowpea intercrop. Experimental Agriculture 30: 349-358.

Hess DE & G Ejeta 1987. Effects of cultural treatments on infestation of *Striga hermonthica* (Del.) Benth. (Scrophulariaceae) on sorghum in Niger. In: HChr Weber & W Forstreuter (eds.), Parasitic Flowering Plants. Proceedings of the 4th ISPFP, Marburg, Germany, 367-375.

Yaduraju NT & MM Hosmani 1979. *Striga asiatica* control in sorghum. PANS 25(2): 163-167.

Authors: A Oswald & JK Ransom

6 Application of control methods

Note: The chapter includes a large number of techniques for the control of parasitic weeds. This may give the impression of numerous methods being available for their control. However, the opposite is true. The available methods are far from being perfect. At present, there is no technique available being both efficient and economic. However, the chance given is the careful selection of an appropriate technique (or combination of techniques) for a certain situation of agricultural system/parasite/crop.

6.1 Preventive

INTRODUCTION

Parasitic weeds have not yet infested all potential areas from the ecological point of view. Therefore, further spread has to be avoided. The seed of *Striga*, *Orobanche*, *Cuscuta* and *Alectra* are disseminated by wind, water, cattle, machinery and contaminated crop seed. One can distinguish between short distance movement and long distance movement. Without intention of doing so, dissemination is greatly supported by man. This holds true especially in the case of *Cuscuta*. Due to the high reproductive capacity of most parasitic weeds and the difficulties of their control the threshold level is set to <1 plant/field.

OBJECTIVES

- To assess the invasion of seeds of parasitic plants to non-infested fields, regions or countries,
- to assess contamination of crop seed, equipment, soil, wind, water, etc.,
- to educate and advise policy makers on effective regulations to avoid regional/international movement of parasite seeds.

PROCEDURES

☞ Survey on the potential introduction of contaminated crop seed. This includes assessment of effects of (1) buying certified seed only, buying seed only from regions known to be non-infested, (2) conduction of visual survey(s) of endangered areas from time to time to ensure the effect of preventive methods (efficient monitoring, → chapter 3.1), (3) testing of crop seeds for contamination with parasitic seed (→ chapter 4.6), and (4) on a national or international basis the placing of a certain parasite species or region under quarantine.

☞ Cleaning of crop seed (→ chapter 6.5.1) is especially essential for crop seed coming from *Cuscuta* infested areas, especially where the crop seed displays about the same size as *Cuscuta* seed. Special sieving equipment has been developed for this purpose by the seed industry. Regular control of the sieving residues and recording of respective findings will be a useful tool. But also for other parasites seed multiplication operations must ensure that only clean crop seed is offered to the farmer, e.g., in the case of *Alectra* or *Striga*.

☞ Further surveys/assessments include

- regular field inspections.
- the potential introduction of contaminated crops, straw, hay or farmyard manure from regions known to be infested.

- the examination of cattle grazing on infested land and transporting the seed via manure or skin to non-infested fields.
- the tillage of infested dry soil during windy weather to know how the seeds are transported by wind.
- the transport of seed by irrigation water from infested to non-infested parts of the field.
- the eradication (e.g., by hand) of single parasitic plants in new infested fields to prevent the build up of a large seed bank.
- the efficacy of "host avoidance" which means to kill weed hosts of the parasite as practised in the USA for the preventive control of *Striga asiatica*.
- the information transfer to farmers of yet non-infested but endangered areas (e.g., to inform them from time to time about the morphological features of a potential parasite in order to enable them to recognise an early infestation).

EVALUATION

- ★ Diligent recording of the results of survey records on new infestations or on discovery of seed in crop seeds etc. will provide useful information for further strategies.
- ★ The infested batches should be traced back to their origin.
- ★ Record findings on a map.

COMMENTS

- Preventive methods represent one of the most important and economic tools to avoid damage from parasitic weeds to crop plants. On the other hand it is probably the most neglected method due to the difficulties in assessment of effects, and because the justification of curative treatments is easier.

LITERATURE

Berner DK, KF Cardwell, BO Faturoti, FO Ikie & OA Williams 1994. Relative roles of wind, crop seeds, and cattle in the dispersal of *Striga* species. Plant Disease 78: 402-406.

Eplee R & R Norris 1996. Prevention: a management strategy for parasitic weeds. In: MT Moreno, JI Cubero, D Berner, D Joel, LJ Musselman & C Parker (eds.), Advances in Parasitic Plant Research; 6th International Parasitic Weed Symposium, Cordoba, Spain, 757-759.

Hess DE & BIG Haussmann 1999. Status quo of *Striga* control: Prevention, mechanical and biological control methods and host plant resistance. In: J Kroschel, H Mercer-Quarshie, J Sauerborn (eds.), Advances in Parasitic Weed Control at On-farm Level. Vol. I. Joint Action to Control *Striga* in Africa. Margraf Verlag, Weikersheim, Germany, 75-87.

Linke KH 1999. Status quo of *Orobanche* management: Preventive, cultural, and physical control. In: J Kroschel, M Abderabihi, H Betz (eds.), Advances in Parasitic Weed Control at On-farm Level. Vol. II. Joint Action to Control *Orobanche* in the WANA Region. Margraf Verlag, Weikersheim, Germany, 107-133.

Sand PF 1987. The American witchweed quarantine and eradication program. In: LJ Musselman (ed.), Parasitic weeds in Agriculture. CRC Press, Boca Raton, Florida, 207-223.

Author: K-H Link

6.1.1 Prevention after harvest

INTRODUCTION

The ability of parasitic plants to emerge on the stubble after crop harvest decreases from *Striga* to *Alectra*/*Orobanche* to *Cuscuta*; but *Cuscuta* grows abundantly if other suitable host plants (weeds) are available. Timely destruction of crop (post-harvest tillage) might be of special relevance in the control of *Alectra vogelii* and *Striga gesnerioides* after the harvest of cowpea (Parker & Riches 1993), and especially *S. hermonthica* can reproduce on ratoon sorghum.

OBJECTIVES
- To investigate measures conducted after crop harvest aiming at a reduction of the parasite population (seed bank and/or developmental stages of the parasite) in the field,
- to investigate measures conducted after crop harvest aiming at a reduction of the parasite's reproduction.

1. Hand weeding after crop harvest

→ chapter 6.5.2

COMMENTS
- When the parasites have not developed mature seeds until crop harvest, hand weeding of the parasite may be postponed after crop harvest. There it will be much faster and easier than in the crop stand. For example, the time required for hand pulling of *O. crenata* under low infestation was considerably less after the harvest of the crop (field pea) than before harvest.

2. Burning after crop harvest

→ chapter 6.5.3

3. Use of herbicides or germination stimulants after crop harvest

→ chapter 6.4.2 and 6.4.3

COMMENTS
- The application of herbicides on the stubble is only reasonable when sufficient parasite shoots are before or at beginning of flowering (otherwise seed production might occur); after crop harvest such stages of parasites occur more or less exclusively with *Striga*.

4. Grazing animals after crop harvest

INTRODUCTION

Stubble grazing with goats, sheep and cows is a common method to feed cattle. Seed of parasitic plants are known to survive the passage through the digestive system of sheep and goats. Thus, dissemination through such animals may occur. On the other hand, selective grazing of parasitic plants with still immature seed will reduce the number of parasitic plants/m^2 and hence attribute to the decrease of the parasite seed bank.

OBJECTIVES
- To assess the effect of grazing on number of parasites/area,
- to assess the effect of grazing on the seed bank of the parasite.

MATERIALS

✓ Field with parasitic plants emerged after harvest of the crop.

✓ Livestock for grazing (suitable are sheep, goats, ...),

PROCEDURE

☞ Select and delimit a convenient field area for grazing.

☞ Let the area be grazed by the animals for a certain time.

EVALUATION

* Type, number and age of cattle.
* Diet of cattle before grazing.
* Number and dry weight of parasites/m^2 before and after grazing.
* If mature parasite plants occur in the field, also assess number and viability of parasite seed in the excrements of the animals (→ chapter 4.3).
* Assess the seed bank before and after grazing (including the excrements).

COMMENTS

- The effect of grazing is largely influenced by factors like amount and palatability of the crop residues available in the field, by the diet exposed before grazing the respective field, by the duration of grazing, etc.
- So far, only limited information on the grazing on *Striga* plants is available.

LITERATURE

Bebawi FF & GA Elhag 1983. Nutritive value of the parasitic weed *Striga hermonthica*. Tropical Agriculture 60: 44.

Farquhar HH 1937. Witchweed - new light on the means by which it is spread. Rhodesian Journal of Agriculture 34: 563.

Jacobsohn R, D Ben-Ghedalia & K Marton 1987. Effect of the animal's digestive system on the infectivity of *Orobanche* seeds. Weed Research 27: 87-90.

Mitchell D 1985. Goats in land and pasture. In: J.W. Copland (ed.), Goat production and research in the Tropics. Canberra, Australia, ACIAR 115-116.

Parker C & CR Riches 1993. Parasitic Weeds of the World - Biology and Control. CAB International, Wallingford, UK, 332 pp.

5. Wind/water erosion

INTRODUCTION

The tiny seed of parasitic weeds are easily transported by wind and water (irrigation and erosion water). This way of seed dispersal is enhanced during periods without a crop on the field (especially wind erosion).

OBJECTIVES

- To measure the amount of seed transported by wind/water,
- to measure the distance of transportation.

MATERIALS

✓ Dissecting microscope, traps (adhesive plates, yellow traps, modified vacuum cleaner or other equipment to collect dust particles), tractor and tillage equipment; for the detection of seed in water: various sieves in dependency of water conditions.

PROCEDURE

☞ Assessment of the seed bank of the areas to be investigated (→ chapter 4.5).

☞ Assessment of soil type, composition, structure, water content.

☞ Assessment of seed transported by wind using (1) yellow traps (like those for monitoring of insects) or similar traps with adhesive or sticky surface to trap seed from the air, (2) vacuum cleaner (measure also the amount of air in m^3 passed through the machine). These traps have to be positioned at various distances from a potential source.

☞ Such source can be e.g., a tractor during tilling operations on infested dry soil under various weather condtions, a field area exposed to wind/water erosion, etc.; a potential source can be the base of a slope where eroded soil material is deposed.

☞ Assess speed of wind/water.

EVALUATION

- ✶ Calculate the number of seed trapped at various distances from a source according to the pattern detected, the duration of sampling, the distance from the source and the speed of the transport medium (wind/water).
- ✶ The development of a model might be helpful to explain and demonstrate varying conditions (wind, direction of transport, soil conditions, etc.).

COMMENTS

- The techniques have to be adjusted to the respective parasite and transport medium to be studied.

LITERATURE

Berner DK, KF Cardwell, BO Faturoti, FO Ikie & OA Williams 1994. Relative roles of wind, crop seeds, and cattle in the dispersal of *Striga* species. Plant Disease 78: 402-406.

Pavlista AD 1981. Why hasn't witchweed spread in the United States? - Weeds Today 12(1): 16-19.

Author: K-H- Linke

6.2 Cultural

6.2.1 Crop rotation with non-hosts, trap and catch crops

INTRODUCTION

Most of the parasitic weeds are obligate parasites, i.e. without a suitable host plant (= non-host) they are not able to develop. In due time their seed bank gets depleted, but it gets increased again with the cultivation of susceptible crop plants (→Fig. 17).

Trap crops (= false hosts) are plants that stimulate germination of the parasite seed but do not become infected; thus they reduce the seed bank in the soil. There are no trap crops for *Cuscuta* as its seeds do not require germination stimulants from a potential host. Catch crops (= hosts) are planted to stimulate germination and therefore they allow the development of the parasite; thus they must be destroyed (together with the parasites) before the parasites are able to produce seeds.

OBJECTIVES

- To assess the efficacy of stopping reproduction of the parasite by planting non-host crops (passive),
- to assess the efficacy of reducing the seed bank of the parasite by planting trap or catch crops (active),
- to evaluate plants for their potential value as trap or catch crops.

MATERIALS

- ✓ Infested field.
- ✓ Crop seed that is a non-host, trap or catch crop and suitable to the respective agricultural conditions (a list of trap crops can be found in → annex 11.6).
- ✓ Equipment for field experiments as usual.
- ✓ Time requirements includes (1) the usual field work for preparation, planting, management and harvest, thus depends on the size of the area,

and (2) at least one full season (or part of it in the case of catch crops), better would be the consideration of several seasons.

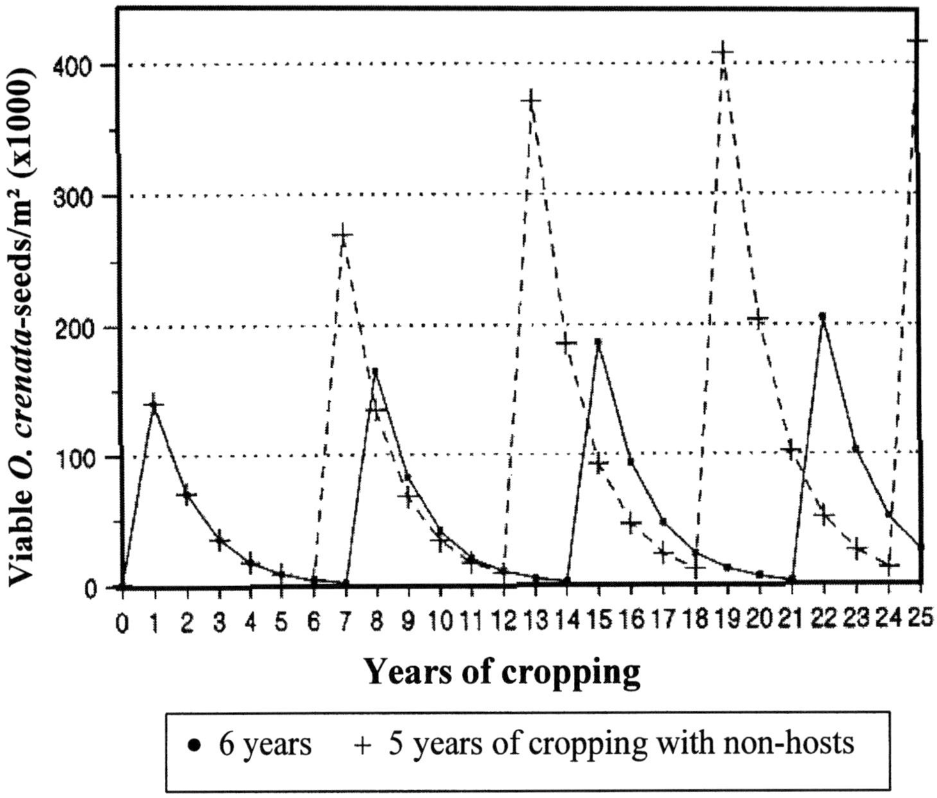

Fig. 17. Modelling of the *Orobanche crenata* seed bank in rotations of 5 and 6 years with non-hosts after 2 years of cropping of lentils (Schnell 1993).

PROCEDURE

☞ Field preparation and planting as required for the respective crop.

☞ With catch crops a high sowing density will ensure a larger root system of the crop, thus a higher effect on the parasite's seed bank.

☞ Catch crops need diligent monitoring after emergence of the crop in order to destroy it (ploughing, rotary hoe) before the parasite sets seed (usually, after 6 to 8 weeks the roots have thoroughly penetrated the upper 30 to 40 cm of soil and many parasite seeds have germinated and are attached). The planting of a catch crop may be repeated even several time within one season to deplete the seed bank of the parasite.

EVALUATION

* Soil samples before and after the season/treatment (→ chapter 4.5)
* Assessment of the emerged parasites in a subsequent host crop. Check plots are required for comparison.
* Assessment of yield parameters in subsequent host crops. Check plots are required for comparison.
* Assessment of precipitation, weather condition and general growth and adaptation of the crop.

COMMENTS

- Trap cropping is a very economic method to reduce the seed bank of parasitic weeds, but it is crucial for a trap crop to be suitable for the respective cropping system (growth requirements, management, marketing, demand, e.g., cotton as a effective trap crop for *Striga* spp.). Otherwise it will represent just a theoretical option.

- Unless repeated for several seasons the efficiency remains low (usually 15 to 40% of the parasite's seed bank might get stimulated in a single season.
- For the screening and evaluation of potential trap crops either germination tests with root exudates or the root chamber technique/Petri dish method can be used (→ chapters 4.4 and 5.1.1)
- The most efficient trap and catch crops are those crops grown for fodder production (e.g., *Trifolium alexandrinum*), as these crops are planted densely and extensively penetrate the soil with their roots.

LITERATURE

Kroschel J & Sauerborn J 1996. Farming systems and the problems of applying *Striga* control methods - a comparison of case studies from northern Ghana, Tanzania and Malawi. In: MT Moreno, JI Cubero, D Berner, D Joel, LJ Musselman & C Parker (eds.), Advances in Parasitic Plant Research. Proceedings 6th Parasitic Weed Symposium. Cordoba, Spain, 843-853.

Langston MA & TJ English 1990. Depletion of witchweed (*Striga asiatica*) in the soil. In: PF Sand, RE Eplee & RG Westbrooks (eds.), Witchweed Research and Control in the United States of America. Weed Science Society of America, Champaign, 114-125.

Last FT 1961. Direct and residual effects of *Striga* control treatments on sorghum yields. Tropical Agriculture, Trinidad 38: 49-55.

Parkinson V, Y Efron, L Bello & K Dashiell 1987. Trap crops as a cultural measure in *Striga* control in Africa. FAO Plant Protection Bulletin 35: 51-54.

Robinson EL & CC Dowler 1966. Investigations of catch and trap crops to eradicate witchweed (*Striga asiatica*). Weeds 14: 275-276.

Sauerborn J, H Sprich & H Mercer-Quarshie 1999. Parasitic weed infestation of maize and sorghum as influenced by crop rotation. In: J Kroschel, H Mercer-Quarshie & J Sauerborn (eds.), Advances in Parasitic Weed Control at On-farm Level. Vol. I. Joint Action to Control *Striga* in Africa. Margraf Verlag, Weikersheim, Germany, 161-169.

Schnell H 1993. Aspekte zur Kontrolle von *Orobanche crenata* Forsk. in Nordsyrien durch Fruchtfolgemaßnahmen unter Berücksichtigung der Dynamik der *Orobanche*-Samenpopulation im Boden. PLITS 11(2).

Sprich H 1994. Bedeutung der Fruchtfolge zur Ertragssicherung in getreidebetonten Produktionssystemen der Guinea-Savanne unter besonderer Berücksichtigung des parasitischen Unkrauts *Striga hermonthica* (Del.) Benth. PLITS 12(3).

Author: K-H Linke

6.2.2 Sowing date

INTRODUCTION

Delaying or advancing crop sowing date can reduce crop yield. However, in the presence of a certain level of infestation of parasitic plants, a shift of sowing date may be more adverse to the parasite that to the host crop, resulting in a crop yield advantage. This is the chance to study the sowing date as a means of control. The reason for less attack by parasitic weeds usually is a partial escape from an attack.

OBJECTIVES

- To provide sub-optimal growth conditions to the parasite (especially concerning germination and attack),

- to evaluate the potential of changing the sowing date for various crops.
- to assess the effect of a modified sowing date on parasite and host crop.

MATERIALS

✓ Normal crop cultivar, or

✓ Cultivar adapted to a change of the sowing date; e.g., the lentil cultivar ILL8 represents an early maturing genotype (short growth period), thus late planting does not affect yield.

PROCEDURE

☞ Planting and crop management to be conducted as usual.

☞ The period between several sowing dates has to be adjusted to the regional conditions and rainfall pattern, e.g., with four sowing dates the period may be 7 to 10 days.

☞ *Striga*: (1) early planting to obtain fast development of the crop before temperatures become high enough for *Striga* development (e.g., in Southern Africa); (2) delayed planting where the rains become more continuous to avoid intermediate drought periods which favour *Striga* (Sudan, Burkina Faso, Ethiopia). Delayed planting hampers *Striga gesnerioides* on cowpea (Niger).

☞ *Orobanche*: delayed planting of winter crops stops/reduces germination of *Orobanche crenata* due to unfavourable low temperature under Mediterranean climate (*O. crenata* germination below 8°C tends to zero).

☞ *Alectra*: Delay of cowpea planting reduced the infestation in Kenya, whereas late planted groundnut or soyabean got more severe infected in northern Nigeria (Parker & Riches 1993).

☞ *Cuscuta*: Delay of the alfalfa planting e.g., in the Pacific Northwest of the USA (late summer instead of spring/early summer) induces abundant *Cuscuta* seedlings to the crop; however, the cool temperatures in autumn inhibit the growth of the parasite more than the crop. In the following spring alfalfa is large enough to tolerate the usual herbicide treatments.

EVALUATION

★ Comparison of regular and earlier/later planted plots or fields in terms of yield parameters and number of emerged parasites per area (→ chapter 5.3.5).

★ For sufficient comparison of effects at least 3, better 5 sowing dates are required for.

★ Recording of temperature (soil, air) and precipitation for interpretation of the results.

COMMENTS

- An important limitation to this method are situations where the soil is dry and crop germinates only with the first rains. In this case irrigation is required to induce earlier germination and to ensure that the treatments can be realized.
- The advantage and required period of time for early or late planting depends on the respective host-parasite-environment relation. It has still to be proven for many relations (e.g., tobacco/*Orobanche cernua* in Eastern Europe, tomato/*Orobanche crenata* under plastic-house production, cowpea/*Alectra* at different latitudes of Southern Africa, etc.).

- Good knowledge of regional weather conditions is of advantage.

LITERATURE

Andrews FW 1945. The parasitism of *Striga hermonthica* Benth. on *Sorghum* spp. under irrigation. Applied Biology 32: 193-200.

Bebawi FF 1987. Cultural practices in witch-weed management. In: LJ Musselman (ed.), Parasitic Weeds in Agriculture, Vol. I *Striga*, CRC Press, Florida, USA, 159-171.

Dawson JH 1987. *Cuscuta* and it's control. In: HChr Weber & W Forstreuter (eds.), Proceedings,. 4th Int. Symposium on Parasitic Flowering Plants, Marburg, Germany, 137-149.

Hezewijk van MJ, AH Pieterse, MC Saxena & SJ ter Borg 1987. Relationship between sowing date and *Orobanche* development on faba been (*Vicia faba* L.) and lentil (*Lens culinaris* Medik.) in Syria. In: HChr Weber & W Forstreuter (eds.), Proceedings of the 4th International Symposium on Parasitic Flowering Plants, Marburg, Germany, 377-390.

Linke KH & MC Saxena 1991. Towards an integrated Control of *Orobanche* spp. in some legume crops. In: K Wegmann & LJ Musselman (eds.), Progress in *Orobanche* Research, Eberhard-Karls Universität, Tübingen, Germany, 248-256.

Linke KH 1991. *Orobanche* studies. In: Food legume improvement program - Annual report for 1990. ICARDA, Aleppo, Syria, 216-242.

Parker C & CR Riches 1993. Parasitic Weeds of the World - Biology and Control. CAB International, Wallingford, UK, 332 pp.

Sauerborn J 1989. The influence of temperature on germination and attachment of the parasitic weed *Orobanche* spp. on lentil and sunflower. Angewandte Botanik 63: 543-550.

Toure M, A Olivier, BR Ntare & CA St-Pierre 1996. The influence of sowing date and irrigation of cowpea on *Striga gesnerioides* emergence. In: MT Moreno, JI Cubero, D Berner, D Joel, LJ Musselman & C Parker (eds.), Advances in Parasitic Plant Research; 6th International Parasitic Weed Symposium, Cordoba, Spain, 787-794.

Author: K-H Linke

6.2.3 Mineral fertiliser and organic material

INTRODUCTION

In several *in vitro* and some field studies germination and development of parasitic plant such as *Striga* and *Orobanche* could be reduced after the application of nitrogen.

However, application of nitrogen for this purpose in the field bears some limitations: (1) Quantities required are rather high, exceeding 40 to 80 kg/ha, (2) legume host crops (as for *Orobanche*, *Alectra*) do not require high nitrogen supply, (3) in most cropping systems with *Striga* problems the input of fertiliser is limited due to financial reasons.

Application of organic material or manure has proven effective in reducing *Striga* and *Alectra*, and to some extent *Orobanche*. However, data on field experiments are scarce and in most cropping systems organic manure is preferably given to other crops (e.g., vegetables) than those attacked by parasites. In general the objective of applying fertiliser is to reduce the infestation with parasitic weeds and to support the crop plant and increase crop yield.

OBJECTIVES

- To study the effects of different nitrogen sources on the development of parasitic plants,
- to study the effects of organic compounds on the development of parasitic plants,
- to study effects of the interaction of host and parasite.

MATERIALS

✓ Various sources of nitrogen and/or organic manure.

✓ For laboratory studies Petri dishes, root chambers, filter paper, pipettes, deionised water.

✓ Various chemicals depending on the topic(s) under study; e.g., a nitrification inhibitor to prevent the first step in nitrification; germination stimulant (e.g., GR24) for germination tests (→ chapter 4.4).

✓ On a macro-molecular basis also the use of complex lab equipment may be required.

✓ For field studies equipment as usual; balance for the preparation of the fertilisation treatments; containers for the liquid (knapsack sprayer) or granular distribution of fertiliser. Tractor and vehicle for the transport and distribution of organic matter.

PROCEDURE

1. Laboratory

☞ Preparation of nitrogen solutions or mixtures/concentrations of organic material to be used either as medium during pre-conditioning and/or after stimulation of germination.

☞ Germination tests (→ chapter 4.4); agar-gel test (→ chapter 4.4.2); root-chamber technique (→ chapter 5.1.1).

2. Field and pots

☞ Collect data on the soil (literature or soil analysis), including composition, organic matter, nitrogen status, cation exchange capacity, pH.

☞ Ensure homogenous infestation with parasite seed (for artificial infestation → chapter 5.3.2).

☞ Application of fertiliser/manure according to the subject under investigation, i.e. before or after sowing the crop.

☞ Recording of weather conditions to support data interpretation.

EVALUATION

★ Laboratory: Germination tests (→ chapter 4.4).

★ Field: Analysis of soil samples at regular intervals in order to measure pH and exchangeable ammonium and nitrate contents, CO_2-production, organic components etc.; counting of emerged parasites as required; assessment of yield parameters.

★ For conclusive results it is recommended to use a series of about 5 concentrations of the medium to be tested.

COMMENTS

- Due to the complexity of the subject such studies may become extremely detailed and difficult in interpretation of the results. Hence, diligent and clear planning and conduct of the experiments is a prerequisite for useful results.
- Under certain circumstances, e.g., with *Striga*, higher nitrogen levels may increase infestation as a result of a better host development especially on poor soils; such behaviour has to be studied in detail by a series of increasing rates of nitrogen added to the medium.

LITERATURE

Farina MPW, PEL Thomas & P Channon 1985. Nitrogen, phosphorus and potassium effects on the incidence of *Striga asiatica* (L.) Kuntze in maize. Weed Research 25: 443-447.

Geipert S 1997. Potentiale und Grenzen der Bekämpfung von *Orobanche crenata* Forsk. im Ackerbohnenanbau (*Vicia faba* L.) Marokkos. PLITS 15(5).

Hunter J & JH Visser 1986. The nitrate reductase activity (NRA) of cultivated Scrophulariacean root parasites. South African Journal of Botany 52(3): 246-248.

Jain R & CL Foy 1987. Influence of various nutrients and growth regulators on germination and parasitism of *Orobanche aegyptiaca*. In: HChr Weber & W Forstreuter (eds.), Parasitic Flowering Plants. Marburg, Germany, 427-436.

Kranz B 1997. Die Entwicklung der parasitischen Blütenpflanze *Striga hermonthica* (Del.) Benth. unter dem Einfluß von organischer Düngung. Supra-regional GTZ-project "Ecology and Management of Parasitic Weeds", Internal Report, University of Hohenheim, Germany, 110pp.

Kranz B, D Fugger, J Kroschel & J Sauerborn 1998. The influence of organic manure on *Striga hermonthica* (Del.) Benth. infestation in northern Ghana. Advances in GeoEcology 31: 615-619.

Kroschel J 1995. The influence of different nitrogen sources on the development of *Striga hermonthica* and studies into the ecological behaviour of parasitic flowering plants. International Workshop on Parasitic Plants – Biology and Resistance. 30 May – 2 June 1995. Long Ashton Research Station, Bristol, UK, Book of Abstracts, 11.

Kroschel J 1989. Der Einfluß von Stickstoff auf phanerogame Wurzelparasiten und ihre Wirtspflanzen unter besonderer Berücksichtigung von *Striga hermonthica* (Del.) Benth. (Scrophulariaceae). PLITS 7(3).

Kroschel J 1998. *Striga* – How will it affect African agriculture in the future? An ecological perspective. In: K Martin, J Müther & A Auffarth (eds.), Agroecology, Plant Protection and the Human Environment: Views and Concepts. PLITS 16(2): 137-158.

Last FT 1960. Incidence of *Striga hermonthica* (Del.) Benth. on two varieties of irrigated sorghum differently manured, spaced and thinned. Tropical Agriculture, Trinidad 37: 309-319.

Linke KH 1987. Untersuchungen über Keimung und Jugendentwicklung von *Striga* und *Orobanche*. PLITS 5(3), 95 pp.

Parker C 1984. The influence of *Striga* species on sorghum under varying nitrogen fertilization. In: Parker *et al.* (eds.), Proceedings of the 3rd International Symposium on Parasitic Weeds, ICARDA/ISPRG, Aleppo, Syria, 90-98.

Pieterse AH 1996. The effect of nitrogen on *Orobanche* and *Striga* - State of the art. In: MT Moreno, JI Cubero, D Berner, D Joel, LJ Musselman & C Parker (eds.), Advances in Parasitic Plant Research; 6th International Parasitic Weed Symposium, Cordoba, Spain, 273-282.

Pieterse AH 1991. The effect of nitrogenous fertilisers on the germination of seeds of *Striga hermonthica* and *Orobanche crenata*. In: K Wegmann & LJ Musselman (eds.), Progress in *Orobanche* Research, Proceedings of the International Workshop on *Orobanche* Research, Eberhard-Karls Universität, Tübingen, Germany, 115-124.

Van Hezewijk MJ, KH Linke, JAC Verkleij & AH Pieterse 1991. The effect of ammonium and urea fertilizer in combination with nitrification inhibitors on *Orobanche crenata* infestation in faba bean and lentil - field experiments. In: JK Ransom, LJ Musselman, AD Worsham & C Parker (eds.), Proceedings of the 5th International Symposium of Parasitic weeds, June 1991, Nairobi, Kenya, 470-483.

Vogt W 1993. Entwicklung der Wirt/Parasit-Beziehung *Sorghum bicolor/Striga hermonthica* unter dem Einfluß verschiedener Stickstofformen und Standortfaktoren. PLITS 11(3).

Wilson-Jones K 1953. Relation of witchweed (*Striga*) to fertility in tropical soils. Nature 172: 128.

Author: K-H Linke & B Kranz

6.2.4 Duration of fallow period and managed fallow

INTRODUCTION

As with other seed the seed bank of parasitic plants in the soil is subject to decay. The longer a fallow period is extended, the lower will be the following infestation of a susceptible host crop. In agricultural systems with *Orobanche*, *Cuscuta* and *Alectra* fallow periods are rare and hardly exceed one season. With *Striga* the traditional fallow periods ranged from several years up to 15 and 20 years. An important aspect to be considered in this context is the half-life (longevity) of the respective seed under field conditions. Especially if the fallow period is short it becomes important to make use of the fallow period to reduce efficiently the seed bank of parasites. This can be done either by the integration of multipurpose agroforestry trees or by the promotion of plant species occurring in the fallow with an ability to induce suicidal germination. In particular legumes should be considered for improving soil fertility.

OBJECTIVES

- To evaluate the effect of various fallow periods on the occurrence of parasitic weeds,
- to investigate multipurpose agroforestry trees and fallow plant species for their efficiency to induce suicidal germination of parasite seeds in order to reduce the seed bank.

MATERIAL

✓ Equipment for germination tests (→ chapter 4.4); agar-gel test (→ chapter 4.4.2); root-chamber technique (→ chapter 5.1.1).

✓ Normal equipment for field studies.

✓ Land available for an extended period of years; durable stakes for marking the area.

PROCEDURE

1. Laboratory

☞ Test agroforstry trees and fallow plants for their capability to induce suicidal germination of parasite seeds using germination tests (→ chapter 4.4), agar-gel test (→ chapter 4.4.2) and root-chamber technique (→ chapter 5.1.1).

2. Field

☞ Subject fields/plots to fallow for a various number of seasons.

☞ The study may include (1) periods of fallow and (2) various treatments (potential plant species) aiming at a reduction of the seed bank.

☞ Diligent recording (layout, map) and marking of the area under investigation (iron or plastic stakes) is a prerequisite for a later identification of the field/plots.

☞ Fencing the area might be required in certain cases.

☞ Prior to the experiment plans have to be made for the management of the respective areas in terms of weeding (shrubs), grazing, walking, etc.; the alignment on neighbouring areas might be helpful.

EVALUATION

★ Assess the number of emerged parasites/m^2 in a susceptible crop prior to the study (→ chapters 3.1.1 and 5.3.5).

★ Assess the seed bank in the soil (→ chapter 4.5).

★ Bury seed packets in the soil for evaluation of viability after the fallow period (→ chapters 4.3 and 4.7).

COMMENTS

- Respective investigations are limited due to their requirements for long term studies. Nowadays, extremely few studies get projected over a period of more than five years.

LITERATURE

Bebawi FF 1987. Cultural practices in witchweed management. In: LJ Musselmann (ed.), Parasitic Weeds in Agriculture, Volume I *Striga*. CRC Press, Boca Raton, Florida, 159-171.

Oswald A, H Frost, JK Ransom, J Kroschel, K Shepherd & J Sauerborn 1996. In: MT Moreno, JI Cubero, D Berner, D Joel, LJ Musselman & C Parker (eds.), Advances in Parasitic Plant Research; 6th International Parasitic Weed Symposium, Cordoba, Spain, 795-800.

Portères R 1952. Linear cultural sequences in primitive systems of agriculture in Africa and their significance. Sols Africains 2: 15.

Sprich H 1994. Bedeutung der Fruchtfolge zur Ertragssicherung in getreidebetonten Produktionssystemen der Guinea-Savanne unter besonderer Berücksichtigung des parasitischen Unkrauts *Striga hermonthica* (Del.) Benth. PLITS 12(3).

Author: K-H Linke & J Kroschel

6.2.5 Flooding

INTRODUCTION

Increased soil moisture, water logging or saturated conditions of the soil infested with parasitic weed reduces the infestation of a host plant. Thereby the seeds e.g., may die from oxygen deficiency, are subject to a "wet dormancy", the germination stimulants are diluted to a non-effective concentration, soil temperature is reduced, etc. However, the extent of flooding/moisture level is difficult to compare for various situations. Furthermore, excess water may damage the crop and not be economic.

OBJECTIVES

- To investigate the effect(s) of various moisture levels on seeds and on development of a parasitic weed,
- to study the parasite's response to various osmotic potentials.

MATERIALS

1. Laboratory studies

✓ Equipment for germination tests (→ chapter 4.4).

✓ Polyethylene glycol (PEG) for the simulation of drought stress by changing the osmotic potential, balance; further equipment (gas analyser, ...) according to the specific topics of the study.

2. Field studies

✓ Normal equipment for field trials (→ chapter 5.3).

✓ Facilities for irrigation or flooding, system for assessment of soil water content, thermometer for extended recordings of soil temperature, auger for soil samples; further equipment according to the specific topics of the study.

PROCEDURE

1. Laboratory studies

☞ "Wet dormancy" is achieved by prolonged conditioning (2 to 20 weeks more than usually required) of the seed prior to the addition of a germination stimulant.

☞ Preparation of the PEG solution by dissolving different amounts of PEG in deionised water; it is recommended to prepare a series (e.g., 6) of osmotic potentials from 0 to -1.4 MPa (calculation of the osmotic potential according to Michel & Kaufmann 1973); in order to maintain the respective osmotic potential in the Petri dishes three layers of filter paper are used per Petri dish; they are sealed tightly or wrapped in plastic bags to avoid any loss of water and thereby increase the osmotic potential. Use 4 to 6 replications per treatment. The effect of the osmotic potential can be studied during conditioning and/or germination.

2. Field studies

☞ Bury seed bags at various depths in the field prior to the experiment (→ chapter 4.7). Take care that the bags can be discovered again after the treatment and that they don't get flushed away.

☞ Take soil samples for analysis of soil properties (e.g., content of parasite seed, nitrogen content).

EVALUATION

- ✶ Testing seed viability may be necessary in the laboratory (e.g., after prolonged exposure to moisture); it is obligatory in field studies.
- ✶ In field studies quantify the flooding: (1) Assess the period of excess moisture (course of pF value); (2) measure the amount of water involved (m^3 per day and area); height of water over the ground (cm).
- ✶ Assess number and dry weight of parasites before and after the treatment using a suitable check crop.
- ✶ Assess the occurrence of any erosion from the area or any accumulation of eroded material into the area.
- ✶ Seed bags may also be used to investigate whether some of the seeds get stimulated, germinated and disappear.

COMMENTS

- Field observations on the effect of flooding/excess moisture are sometimes rather contradictory (e.g., irrigated but infested tomatoes, sometimes heavy infestation of sorghum by *S. hermonthica* in the Rahad scheme, Sudan). It is vital for any serious conclusions to seriously evaluate the actual conditions (soil conditions, parasite, host, period of flooding and amount of water, pF-values, aeration of soil, etc.).

LITERATURE

Dawoud DA & J Sauerborn 1994. Impact of drought stress and temperature on *Striga hermonthica* and *Alectra vogelii* at early growth stages. Experimental Agriculture 30: 249-257.

Linke KH & W Vogt 1987. A method and its application for observing germination and early development of *Striga* (Scrophulariaceae) and *Orobanche* (Orobanchaceae). In: HChr Weber & W Forstreuter (eds.), Proceedings, 4th Int. Symposium on Parasitic Flowering Plants, Marburg, Germany, 501-509.

Michel BE & MR Kaufmann 1973. The potential use of polyethylene glycol 6000. Plant Physiology 51: 914-916.

Pieterse AH 1979. The broomrapes (Orobanchaceae) - a review. Abstracts on Tropical Agriculture 5(3): 9-35.

Vallance K 1950. Studies on the germination of seeds of *Striga hermonthica*. I. The influence of moisture treatment, stimulant dilution and after-ripening on germination. Annals of Botany 14: 347.

Author: K-H Linke

6.3 Host plant resistance and tolerance

INTRODUCTION

Resistant host plants offer a potentially durable tool for resource-poor farming conditions to control parasitic weeds. Resistant cultivars could be a major component of integrated control packages, the distinct advantage being that their cultivation does not require expensive inputs from the farmer.

With obligate parasites, interactions between them and their host plant play a crucial role in the survival of the parasites. Crop genotypes which interfere with one or several stages of the parasite's life cycle could be resistant. The following (potential) mechanisms of resistance have been suggested:

Mechanisms of resistance

- low production of germination stimulant;
- low production of haustoria inducing factor;
- mechanical barriers (e.g., lignification of cell walls or radicular cortex structure);
- inhibition of germ tube exoenzymes by root exudates;
- phytoalexin synthesis;
- post-attachment hypersensitive reactions;
- prevention of vascular connection between vessels of host and parasite;
- disturbed internal flow of nutrients to the young parasite;
- antibiosis;
- unfavourable phytohormone supply by the host;
- insensitivity to *Striga* "toxin" (e.g., maintenance of photosynthetic efficiency);
- avoidance through root growth habit (e.g., less roots in the upper 15 to 20 cm).

Definition of resistance to parasitic weeds

A crop genotype that, when grown under infestation, supports significantly fewer parasite plants and has a higher yield than a susceptible cultivar is designated resistant. The resistance is a protective action by the host which may be either passive or active. A wide range of resistance from low to high may occur within a single crop species. The converse of resistance is susceptibility.

In contrast, **tolerant** genotypes stimulate germination of parasite seeds and support as many parasites as susceptible genotypes, but without showing concomitant reduction in grain yield or overall productivity. Cultivation of tolerant cultivars can increase the parasite seed bank in the field over time. The converse of tolerance is sensitivity.

Host genotypes that are totally free of parasites when grown under infested conditions would be termed **immune** (= totally resistant); however, this term might be somewhat misleading as it was applied instead of resistance throughout the extensive Russian literature on *Orobanche cernua/cumana*.

LITERATURE

Berner DK, JG Kling & BB Singh 1995. *Striga* research and control, a perspective from Africa. Plant Disease 79: 652-660.

Cubero JI, AH Pieterse, SA Khalil & J Sauerborn 1994. Screening techniques and sources of resistance to parasitic angiosperms. Euphytica 73: 51-58.

Doggett H 1988. Sorghum. 2nd edition. John Wiley, New York. 392-394.

Ejeta G & LG Butler 1993. Host plant resistance to *Striga*. In: International Crop Science I. 1993. Crop Science Society of America, 677 pp. Segoe Rd, Madison, WI53711, USA. 561-569.

Ejeta G, LG Butler & AG Babiker 1992. New approaches to the control of *Striga*. *Striga* Research at Purdue University. Research Bulletin RB-991. Agricultural Experiment Station. Purdue University. West Lafayette, IN.

Author: BIG Haussmann & DE Hess

6.3.1 Breeding for resistance to *Orobanche*

6.3.1.1 Screening techniques and sources of resistance

INTRODUCTION

Most screening for resistance to *Orobanche* was conducted in the field so far. However, field screening on its own can be rather unreliable, time consuming and expensive. Hence, also reliable greenhouse and laboratory techniques are required for this purpose. The understanding of the respective mechanism(s) of resistance may lead to methods for reliable, quick, cheap and out of season screenings.

OBJECTIVES

- To provide precise and reliable screening techniques in the laboratory, greenhouse and field. These are indispensable prerequisites to breeding for resistance to any biotic or abiotic stress factor,
- to identify sources of resistance to *Orobanche* spp. in various host crops.

TECHNIQUES

1. Laboratory screenings

OBJECTIVES

- To screen under controlled environment and/or for special measurements,
- to screen for individual resistance mechanisms,
- to test a large number of entries within a short period of time.

PROCEDURES

i) Germination test (→ chapter 4.4.1)

ii) Agar-gel assay (→ chapter 4.4.2)

iii) Root chamber technique (→ chapter 5.1.1)

iv) Petri dish method (→ chapter 5.1.2)

v) *In vitro* culture of parasites (→ chapter 5.1.2)

vi) Histological, cytochemical and microscopical studies on resistance mechanisms

☞ Fixation and dehydration of samples.

☞ Embedding in Histowax.

☞ Staining of paraffin sections with astra blue, safranin red, auramin yellow or toluidine blue.

☞ Near infrared spectroscopy (NIRS) may be used to quantitatively determine the lignin content of host root cells.

(for further details see → Literature).

EVALUATION

✶ A screening index has to be developed for every special purpose, e.g.,

- germination rate (%),
- distance of germination from the stimulating source,
- length of germtube,

- induction of swelling at the tip of the germtube,
- thickness/staining/expression of mechanical barriers in the host root,
- further evaluation of resistance mechanisms.

COMMENTS

- Any results of the laboratory testing will need further confirmation under field conditions. This holds true especially for the testing of root exudates as germination stimulant, as they are subject to large spectrum of factors and variation.
- The agar-gel test, although not yet widely applied to *Orobanche* or *Alectra*, may represent a reliable tool for further screening with these parasites.

LITERATURE

Antonova TS 1994. Biochemical aspects of the development of new virulent forms in the Moldavian population (race C) of *Orobanche cumana* Wallr. against the background of resistant sunflower cultivars. In: AH Pieterse, JAC Verkleij & SJ ter Borg (eds.), Biology and Management of *Orobanche*. Proceedings of the 3rd International Workshop on *Orobanche* and related *Striga* research. Amsterdam, The Netherlands, Royal Tropical Institute, 290-292.

Attawi F 1977. Morphologisch-anatomische Untersuchungen an den Haustorien einiger *Orobanche*-Arten. Berichte der Deutschen Botanischen Gesellschaft 90: 173-182.

Chang M, DH Netzly, LG Butler & DG Lynn 1986. Chemical regulation of distance: Characterisation of the first natural host germination stimulant for *Striga asiatica*. Journal of the American Chemistry Society 108: 7858-7860.

Dörr I, A Staack & R Kollmann 1994. Resistance of *Helianthus* to *Orobanche* – histological and cytological studies. In: AH Pieterse, JAC Verkleij & SJ ter Borg (eds.), Biology and Management of *Orobanche*. Proceedings of the 3rd International Workshop on *Orobanche* and related *Striga* research. Amsterdam, The Netherlands, Royal Tropical Institute, 276-289.

Hess DE, G Ejeta & LG Butler 1992. Selecting sorghum genotypes expressing a quantitative biosynthetic trait that confers resistance to *Striga*. Phytochemistry 31: 493-497.

Netzly DH, JL Riopel, G Ejeta & LG Butler 1988. Germination stimulants of witchweed (*Striga asiatica*) from hydrophobic root exudate of sorghum (*Sorghum bicolor*). Weed Science 36: 441-446.

Luft JH 1961. Improvement in epoxy resin embedding methods. Journal of Biophysical and Biochemical Cytology 9: 409-414.

Olivier A, N Benhamou & GD Leroux 1991. Cell surface interactions between sorghum root and the parasitic weed *Striga hermonthica*: cytochemical aspects of cellulose distribution in resistant and susceptible host tissues. Canadian Journal of Botany 69: 1679-1690.

Olivier A, JC Glaszmann, C Lanaud, G Salle & GD Leroux 1996. An insight into the population structure and genetic diversity of *Striga hermonthica* in West Africa. In: MT Moreno, JI Cubero, D Berner, D Joel, LJ Musselman & C Parker (eds.), Advances in Parasitic Plant Research; 6th International Parasitic Weed Symposium, Cordoba, Spain, 113-121.

Petzoldt K, Y Nemli & J Sneyd 1994. Integrated control of *Orobanche cumana* in sunflower. In: AH Pieterse, JAC Verkleij & SJ ter Borg (eds.), Biology and Management of *Orobanche*. Proceedings of the 3rd International Workshop on *Orobanche* and related *Striga* research. Amsterdam, The Netherlands, Royal Tropical Institute, 442-449.

2. Screening in pots (→ chapter 5.2)

3. Screening in the field

OBJECTIVE

- Evaluation of promising material under natural conditions and different environments.

MATERIALS

✓ Usual equipment for field experiments; good and sufficient labels.

- ✓ Seed of parasites for artificial infestation.
- ✓ Crop entries to be tested.

PROCEDURE

- ☞ A row length of 1 m is the minimum (if test seed is limited). Generally, it is recommended to use small plots but with sufficient replications.
- ☞ Use one or two rows of test entries (**T**) with susceptible (local) check (**C**) at both sides in the following way: **C**, $\mathbf{T_1}$, **C**, $\mathbf{T_2}$, **C**, $\mathbf{T_3}$, etc., or **C**, $\mathbf{T_1T_1}$, **C**, $\mathbf{T_2T_2}$, **C**, $\mathbf{T_3T_3}$, etc..
- ☞ Replicate! For preliminary evaluation a single row or one replication might be adequate in order to screen a large number of accessions rather than providing a confident indication of resistance. In verification trials an increased number of replications is required (usually with less entries to be evaluated) to confirm resistance.
- ☞ To ensure a homogenous soil infestation of the screening site, artificial infestation is recommended (→ chapter 5.3.2). Less susceptible crops such as chickpea require a substantially higher soil infestation to allow for reliable selections than do e.g., faba bean or lentil.
- ☞ To avoid cross pollination among test entries measures appropriate to the respective crop have to be considered (for faba beans this involves covering the plants during the whole period of flowering to avoid access of pollinating insect).
- ☞ Multiply selected seed at appropriate locations (greenhouse or climatically suited environment at higher elevations).
- ☞ For breeding strategies see → chapter 6.3.2.5.
- ☞ For crossing techniques with individual crops it is recommended to consult a respective breeder in charge of that particular crop.

EVALUATION

- ★ The selection should be carried out according to the number of emerged shoots of the parasite on test entries compared to the susceptible check. The most commonly used selection criteria are the number of emerged parasites per host plant or per area. Also, the crop seed yield and straw biomass are criteria for selection. The evaluation of underground stages of *Orobanche* is possible, but extremely time consuming; it might be applied more efficiently in greenhouse screenings.

COMMENTS

- Note: Due to usually uneven distribution of parasite seed in soil, field screening requires replications and adequate numbers of check rows in order to be reliable.
- The number of emerged *Orobanche* shoots usually is a reliable index for selection due to it's high correlation with underground attachments (exceptions are (1) low number of emerged parasites under extremely high infestation and (2) highly branching types of *Orobanche aegyptiaca*/*ramosa*); however, relying only on number of emerged shoots might be misleading.
- Seasonal weather conditions further influence the absolute level of attack from one season to the other. Never plant all available test seed but keep some for later multiplication or screening in case the present crop can not be evaluated or harvested for any reason.

- The crop material to be tested in the field usually originates from fast screening systems in the laboratory or in the greenhouse. In this way management of limited resources can be optimized. Hence, co-ordination of field, greenhouse and laboratory work generally should receive high priority in research planning.

4. Sources of resistance

MATERIALS

✓ Screening facilities (laboratory or greenhouse for germination tests, root chamber technique or pot studies, suitable area for field screening).

✓ Any seed of host plants from sources that might provide a basis for the selection of resistance to *Orobanche*.

Orobanche crenata

Vicia faba (faba bean): Cultivar „Giza 402" (Egypt), line VF1071 and cultivar „Baraca" selected from the Egyptian Giza 402 (Spain); lines that induce a low germination *in vitro* include ICARDA BPL2210 (Syria), VF 1071,VF 172 , L1 and L2 (Spain) and the landrace „Locale di Castellano" (Italy). In general, the smaller the mean seed size of the faba bean (*Equina*- and *Minor*-types), the greater the level of resistance (Spain, Italy).

Lens culinaris (lentil): Negligible variation of resistance was detected despite extensive screening of cultivated and wild accessions of lentils (ICARDA, Syria).

Pisum sativum (pea): Evaluation of the accessions at ICARDA not conclusive so far; low level of resistance in accessions No. 290, 21, 271 and 291 in greenhouse screenings and accessions No. 290, 291 and 38 in field tests (ICARDA, Syria).

Cicer arietinum (chickpea): Low level of resistance in ILC 3279 („Ghab 2", Syria) and FLIP 81-293 (Syria). In general, the level of attack is lower than in faba bean, lentil or pea.

Vicia sativa (common vetch): Various lines resistant (Spain), low level of resistance in accession No. 715 and 254 (ICARDA, Syria).

Vicia narbonensis (narbon vetch): resistance in accession No. 577 and 578 (ICARDA, Syria).

Lathyrus spp. (Lathyrus peas): Low level of resistance in accession No. 201 and 188 (ICARDA, Syria).

Lathyrus cicera (chickling): Low level of resistance in accession No.143 and 145 (ICARDA, Syria).

Lathyrus ochrus (ochrus pea): Immunity in the field and in laboratory tests (Petri dish method, ICARDA; Syria).

Vicia dasycarpa (winter vetch): High level of resistance in the field and laboratory (ICARDA; Syria).

Orobanche cernua, O. cumana

Helianthus annuus (sunflower): Resistance was established in sunflowers for oil production in Russia at the beginning of this century. With the spread of these population varieties into other regions this resistance was inadequate due to the occurrence of a more virulent race of the parasite. So far, some variation in resistance to *Orobanche cumana* is available (Russia, Rumania, Bulgaria, Turkey). The resistance of sunflower to the parasite however, seems to be subject to rather fast decrease of resistance due to the common occurrence of new races or biotypes of the parasite. The most resistant line in Spain is P-1380-2, a differential line carrying the Or5 gene. Resistance to *O. cumana* races in Romania is dominant monogenic with five resistance genes.

Helianthus tuberosus (topinambur): Resistant; included as interspecific crosses in breeding programs with sunflower. Other wild relatives included in breeding programs were *H. petiolaris* and *H. lenticularis* (Russia). A recent screening of 42 wild species of sunflower in Spain indicated resistance in *H. agrestis, H. anomalus* and *H. exilis.*

Nicotiana tabacum (tobacco): Some variation in resistance was detected after evaluation of > 400 cultivars and wild relatives of tobacco (Russia). The varieties Bright Caspalia, Bel 61-11 and Coker 254 showed low levels of resistance after several years of selection in India. In the future the emphasis might shift to genetic engineering, as with tobacco this appears most promising.

Solanum melongena (eggplant): Several resistant cultivars (India), e.g., Pusa Purple Long x Manjri Gota, or Pusa Purple Long x Nurki.

Orobanche aegyptiaca/O. ramosa

Nicotiana tabacum (tobacco): Low level of resistance (Russia, Italy).

Lycopersicon lycopersicum (tomato): Line UZ-1 and cultivar „Ora" displayed field resistance in former USSR, but failed in other countries. Resistance in one cultivar in India. Wild relatives *Lycopersicon hirsutum, L. pyriforme* and *L. pruniforme* seem to be excellent sources of resistance (Russia). Mostly, a variation in susceptibility was recorded instead of resistance. Genetic engineering might shift the focus to herbicide resistance in this crop.

Cannabis sativa (hemp): Low level of resistance is reported (Russia).

Cucurbitaceae (cucurbits): No resistance available despite screening program (Russia).

Brassicaceae (brassica family): Various resistant cultivars of rape and mustard (India). Various levels of resistance in rape (Spain). *Brassica juncea* (Indian mustard) seems to be less susceptible than *Brassica campestris.* (wild biotype of turnip rape).

Orobanche foetida

Vicia faba (faba bean): This dark-red flowering parasite was recognized only a few years ago as a serious pest of faba bean in Tunisia, attacking even breeding material resistant to *Orobanche crenata.* No source of resistance is available so far (Tunisia).

(Above compilation on sources of resistance is based *inter alia* on the extensive information on Russian results provided on the subject by Cubero 1991).

PROCEDURE

☞ Collect the required crop seed material for evaluation.

☞ Conduct the appropriate screening method(s) in the lab, greenhouse or field (→ chapter 5).

EVALUATION

(Screening techniques → chapter 5)

LITERATURE

Antonova TS 1994. Biochemical aspects of the development of new virulent forms in the Moldavian population (race C) of *Orobanche cumana* Wallr. against the background of resistant sunflower cultivars. In: AH Pieterse, JAC Verkleij & SJ ter Borg (eds.), Biology and Management of *Orobanche*. Proceedings of the 3rd International Workshop on *Orobanche* and related *Striga* research. Amsterdam, The Netherlands, Royal Tropical Institute, 290-292.

Cubero JI & A Martinez 1980. The genetics of the resistance of faba beans to *Orobanche crenata*. FABIS 2: 19-20.

Cubero JI 1986. Breeding for resistance to *Orobanche* and *Striga*: a review. In: SJ ter Borg (ed.), Proceedings of a workshop on biology and control of *Orobanche*. LH/VPO, Wageningen, The Netherlands, 127-139.

Cubero JI 1991. Breeding for resistance to *Orobanche* species: a review. In: K Wegmann & LJ Musselman (eds.), Progress in *Orobanche* research. Eberhard-Karls-Universität, Tübingen, Germany, 257-277.

Cubero JI, AH Pieterse, SA Khalil & J Sauerborn 1994. Screening techniques and sources of resistance to parasitic angiosperms. Euphytica 73: 51-58.

Dörr I, A Staack & R Kollmann 1994. Resistance of *Helianthus* to *Orobanche* – histological and cytological studies. In: AH Pieterse, JAC Verkleij & SJ ter Borg (eds.), Biology and Management of *Orobanche*. Proceedings of the 3rd International Workshop on *Orobanche* and related *Striga* research. Amsterdam, The Netherlands, Royal Tropical Institute, 276-289.

Gil J, LM Martin & JI Cubero 1987. Genetics of resistance in *Vicia sativa* L. to *Orobanche crenata* Forsk. Plant Breeding 99: 134-143.

Linke KH 1991. *Orobanche* studies. In: Food legume improvement program - Annual report for 1990. ICARDA, Aleppo, Syria, 222-242.

Linke KH, A Abdel-Moneim & MC Saxena 1992. Inter- and intra-specific variation in susceptibility of forage legumes to the parasitic weed *Orobanche crenata* Forsk. Field Crops Research 32: 277-285.

Linke KH, KB Singh & MC Saxena 1991. Screening technique for resistance to *Orobanche crenata* Forsk. in chickpea. Chickpea Newsletter 24: 32-34.

Nassib, AM, AA Ibrahim & SA Khalil 1982. Breeding for resistance to *Orobanche*. In: G Hawtin & C Webb (eds.), Faba Bean Improvement. Martinus Nijhoff, The Hague, 199-206.

Parker C & CR Riches 1993. Parasitic Weeds of the World - Biology and Control. CAB International, Wallingford, UK, 332 pp.

Petzoldt K, Y Nemli & J Sneyd 1994. Integrated control of *Orobanche cumana* in sunflower. In: AH Pieterse, JAC Verkleij & SJ ter Borg (eds.), Biology and Management of *Orobanche*. Proceedings of the 3rd International Workshop on *Orobanche* and related *Striga* research. Amsterdam, The Netherlands, Royal Tropical Institute, 442-449.

Raju CA 1996. Variability in tobacco germplasm towards *Orobanche* infection. In: MT Moreno, JI Cubero, D Berner, D Joel, LJ Musselman & C Parker (eds.), Advances in Parasitic Plant Research; 6th International Parasitic Weed Symposium, Cordoba, Spain, 609-613.

Sillero JC, D Rubiales & JI Cubero 1996. Risks of *Orobanche* resistance screenings based only on number of emerged shoots per plant. In: MT Moreno, JI Cubero, D Berner, D Joel, LJ Musselman & C Parker (eds.), Advances in Parasitic Plant Research; 6th International Parasitic Weed Symposium, Cordoba, Spain, 651-657.

Snelder Y, MT Moreno, A Martin & J Gil 1995. Screening for resistance to *Orobanche crenata* Forsk. in *Vicia faba* L.: a comparison between methods. In: Anonymous, 2nd European Conference of Grain Legumes. Copenhagen, 230-231.

Sukno S, J Rusao, J Melero-Varo & J Fernandez-Martinez 1996. Evaluation of wild *Helianthus* germplasm for resistance to highly virulent races of *Orobanche cernua* Loefl. and transfer to cultivated sunflower. In: MT Moreno, JI Cubero, D Berner, D Joel, LJ Musselman & C Parker (eds.), Advances in Parasitic Plant Research; 6th International Parasitic Weed Symposium, Cordoba, Spain, 625-632.

Vranceanu AV, N Pirvu, FM Stonescu & M Pacureanu 1986. Some aspects of the interaction *Helianthus annuus* L./*Orobanche cumana* Wallr. and it's implications in sunflower breeding. In: SJ ter Borg (ed.), Proceedings of a workshop on biology and control of *Orobanche*. LH/VPO, Wageningen, The Netherlands, 181-189.

Author: K-H Linke

6.3.1.2 Genetics of resistance

INTRODUCTION

Knowledge on the genetic basis of resistance to *Orobanche* is poor. In general, the expression of resistance to *Orobanche* is subject to large genotype-environment interaction. Biotypes of *Orobanche* species are able to overcome resistance. The occurrence of biotypes of *Orobanche* species is evident in sunflower. It is likely that biotypes occur in crops such as faba bean, tobacco and hemp. Pollination and hence recombination of genes in *Orobanche* species is strongly autogamous, with the exception of *Orobanche aegyptiaca* and *O. crenata* which may be outcrossers but facultatively autogamous.

The inheritance of resistance seems to be quantitative. Hernandez *et al.* (1984) suggested a no-dominance model for the resistance of faba bean to *O. crenata*. However, in some crosses effects of dominance (at least very weak) cannot be excluded (Cubero 1986). Maternal effects were not recorded. Additive effects are responsible for the resistance in vetch to *O. crenata*, whereas dominance seems to be absent (Gil 1986).

In sunflower (*O. cumana*) various systems have been suggested, e.g., quantitative genetic action (Pustovit 1966), gene for gene relationship (Pustovit in Kolte 1985), dominance of resistance (Pogorletsky & Geshele 1976) and others. According to Vranceanu *et al.* (1986) resistance to *O. cumana* races in Romania is dominant monogenic with five resistance genes. The resistance of tomato to *O. aegyptiaca* seems to show dominance, whereas in a highly homozygous line resistance was controlled by just one pair of genes (Abdeev & Sherbinin 1978 cit. in Cubero 1986).

OBJECTIVES

- To understand the system of inheritance (mode of gene action). This will substantially improve the breeding methodologies,
- to find out whether different sources of resistance actually carry different genes. Such information is most useful for selection of parent lines for a crossing program.

LITERATURE

Abdeev Y & BM Sherbinin 1978. [Investigating tomato resistance to broomrape and the inheritance of this trait]. Weed Abstracts 27: 3074.

Cubero JI 1986. Breeding for resistance to *Orobanche* and *Striga*: a review. In: SJ ter Borg (ed.), Proceedings of a workshop on biology and control of *Orobanche*. LH/VPO, Wageningen, The Netherlands, 127-139.

Gil JT 1986. Variacion intraespecifica en *Vicia sativa*. Ph. D. Thesis, University of Cordoba, Spain.

Hernandez L, MT Moreno & JI Cubero 1984. Studies on the resistance of *Vicia faba* to *Orobanche crenata* Forskall. In: Parker C, LJ Musselman, RM Polhill & AK Wilson (eds.), Proceedings 3rd International Symposium on Parasitic Weeds, Aleppo, Syria, 210-220.

Musselman LJ, C Parker & N Dixon 1982. Notes on autogamy and flower structure in agronomically important species of *Striga* (Scrophulariaceae) and *Orobanche* (Oro-

banchaceae). Beiträge zur Biologie der Pflanzen 56: 329-343.
Pogorletsky BK & EE Geshele 1976. Sunflower immunity to broomrape, downy mildew and rust. Proceedings of the 7th International Sunflower Conference, 238-243.
Sukno S, J Rusao, J Melero-Varo & J Fernandez-Martinez 1996. Evaluation of wild *Helianthus* germplasm for resistance to highly virulent races of *Orobanche cernua* Loefl. and transfer to cultivated sunflower. In: MT Moreno, JI Cubero, D Berner, D Joel, LJ Musselman & C Parker (eds.), Advances in Parasitic Plant Research; 6th International Parasitic Weed Symposium, Cordoba, Spain, 625-632.
Vranceanu AV, N Pirvu, FM Stoenescu & M Pacureanu 1986. Some aspects of the interaction *Helianthus annuus* L./*Orobanche cumana* Wallr. and its implications on sunflower breeding. In: SJ ter Borg (ed.), Proceedings of a workshop on biology and control of *Orobanche*. LH/VPO, Wageningen, The Netherlands, 181-189.

Author: K-H Linke

6.3.1.3 Breeding strategies

INTRODUCTION

A breeding strategy has to be developed for and adjusted to the respective crop plant, e.g., sunflower, faba bean, tobacco, etc. A good strategy involves diligent planning on the basis of the *status quo*, co-ordination of laboratory and field experiments and the consideration of new technical/genetic methods.

OBJECTIVES

- To develop and follow up a breeding system which is most promising in the development of resistance to *Orobanche* species,
- to improve the currently available sources of resistance for better agronomic performance.

STRATEGIES

- Characterisation of crop germplasm.
- Transfer of resistance genes into genotypes better adapted to target areas.
- Replacement of individual selection with progeny tests by recurrent selection with selfing to fix resistance genes.
- To pyramid resistance genes to obtain more durable and stable, polygenic resistance.
- Development of open-pollinated synthetic varieties by recombining resistant inbred lines.
- Development of random-mating populations with multiple sources of resistance using recurrent selection procedures.
- Inclusion of wild relatives in the breeding programme.
- Development and employment of marker-assisted selection techniques for broad-based resistance to *Orobanche.*

COMMENTS

- The occurrence of biotypes (e.g., in *Orobanche cernua*/*O. cumana*) and the absence of efficient and reliable single-plant screening techniques have been a major drawback in the past. As with *Striga*, the development of laboratory assays which allow the non-destructive, rapid and inexpensive evaluation of individual plants, would largely facilitate early generation testing and increase selection efficiency.

LITERATURE

Cubero JI 1986. Breeding for resistance to *Orobanche* and *Striga*: a review. In: SJ ter Borg (ed.), Proceedings of a workshop on Biology and Control of *Orobanche*. LH/VPO, Wageningen, The Netherlands, 127-139.

Hernandez L, MT Moreno & JI Cubero 1984. Studies on the resistance of *Vicia faba* to *Orobanche crenata* Forskall. In: Parker C, LJ Musselman, RM Polhill & AK Wilson (eds.), Proceedings 3rd International Symposium on Parasitic Weeds, Aleppo, Syria, 210-220.

Vranceanu AV, N Pirvu, FM Stoenescu & M Pacureanu 1986. Some aspects of the interaction *Helianthus annuus* L./*Orobanche cumana* Wallr. and it's implications on sunflower breeding. In: SJ ter Borg (ed.), Proceedings of a workshop on biology and control of *Orobanche*. LH/VPO, Wageningen, The Netherlands, 181-189.

Author: K-H Linke

6.3.2 Breeding of cereals for resistance to *Striga*

6.3.2.1 Screening techniques and selection traits

INTRODUCTION

Individual mechanisms conferring resistance to *Striga* may be screened in the laboratory, while complex resistance must be assessed under field conditions. Screening in pot trials falls between the two extremes and has, compared to field testing, the advantage of superior control over the environmental conditions. For definitions and mechanisms of resistance see → chapter 6.3.

OBJECTIVE
- To identify resistant genotypes in the laboratory, greenhouse or field.

PROCEDURES

1. Screening for individual resistance mechanisms

☞ The agar gel test provides a simple means for screening host genotypes for low stimulation of *Striga* seed germination (→ chapter 4.4.2).

☞ Before the development of the agar gel test, other, more laborious techniques were used to identify low stimulant producing genotypes: For example, host plant root exudates may be collected in 'double pots' (Double Pot Technique) or in Pasteur pipettes (Pasteur Pipette Technique) and applied to preconditioned *Striga* seeds to determine the percentage of germinated *Striga* seeds (→ chapter 4.4).

☞ The Eplee Bag Technique allows the determination of germination (→ chapter 4.7).

☞ *In vitro* growth systems or root chambers are techniques to identify factors concerning attack/establishment on the host root like low production of haustoria-inducing factors, or hypersensitivity reactions (→ chapter 5.1.1).

☞ Histological studies of the host roots under the microscope can elucidate mechanical barriers like endodermal cell wall thickening. The section of interest may be stained with safranin or toluidine blue.

☞ Near infrared spectroscopy (NIRS) may be used to quantitatively determine the lignin content of host root.

☞ *In vitro* cultured sorghum cells can be treated with extracts of *Striga* plants in order to screen for resistance to the *Striga* toxin. In a different approach, extracts of host roots or other tissues can be evaluated for their ability to kill *in vitro* cultures of suspended *Striga* cells (→ chapter 5.1.2). Such genotypes, if identified, should be highly resistant to *Striga*.

2. <u>Screening for resistance in pot trials or under field conditions</u>

☞ In pot experiments, host plants are usually grown in pots filled with soil infested with a known quantity of *Striga* seed (→ chapter 5.2).

☞ Field screening for *Striga* resistance is hampered by the following problems: heterogeneity of infestation in naturally infested fields, large environmental effects on *Striga* emergence, and complex interactions between host, parasite and environment affecting parasite establishment and its reproductive success (→ chapter 5.3).

☞ Artificial infestation of the field is used to improve uniformity of infestation and the accuracy of trials (→ chapter 5.3.2).

☞ Susceptible check varieties have to be included at frequent intervals so that all observations can be expressed as percentage values relative to the nearest susceptible check. Carried to an extreme, this results in the "checkerboard layout", where each test variety is surrounded by the susceptible cultivar. This layout, however, requires considerable space and, therefore, reduces the number of entries which can be evaluated or the number of replications which can be employed. It is only necessary where infestation variability in the screening field is very high.

☞ Plot size and experimental layouts for *Striga* resistance screening largely depend on trial objectives. In preliminary observation nurseries, a large number of entries may be evaluated in two-row plots, with one or two replications and frequently repeated susceptible checks. Promising entries may be tested at two or three locations in three-row plots with a minimum of three replications. Most advanced selections should be screened at more than three locations, on-station and on-farm, and across several years to prove the stability of resistance and screen for grain yield stability, drought resistance, seedling establishment and other important adaptation traits.

☞ When performing genetic studies of *Striga* resistance which require very high precision, the number of replications may be increased up to six. This normally requires a reduction in plot size, due to limited availability of uniformly infested fields. For example two-row plots, separated from the neighbouring entry by one empty row, are an acceptable compromise: for each entry traits can be assessed in both rows, no land is "lost" to border rows. Neighbour effects are reduced due to the empty row, and more light reaches the ground, reducing shading which is deleterious to *Striga*.

☞ Entries may also be evaluated under infested and non-infested conditions in adjacent strips of the same field. This technique allows quantification of yield reduction and convenient assessment of stress symptoms relative to the control treatment. After each experiment, the area is planted with a highly efficient trap crop for two seasons and then treated with ethylene to germinate any remaining *Striga*. Thereafter, the field can be newly infested and reused for *Striga* resistance trials (→ chapters 5.3.2, 5.3.3 and 6.4.1).

EVALUATION

- ✶ Traits to be assessed in pot trials are essentially the same as those evaluated in field trials.
- ✶ With regard to *Striga*, the following traits are frequently assessed: (1) days to *Striga* appearance in each plot; (2) days to the onset of *Striga* flowering in each plot; (3) total number of emerged *Striga* plants; (4) number of flowering *Striga* plants; and (5) number of *Striga* plants with capsules. Counts are made in two-week intervals during the course of the season, generally beginning one or two weeks after *Striga* emerges in the experiment. The counts may be accompanied by a visual estimation of *Striga* vigour (Table 4, Fig. 18) as five *Striga* plants of 5 cm height differ significantly from five fully developed *Striga* plants with many branches, flowers and capsules and a plant height exceeding 40 cm. *Striga* dry matter may be another trait of interest, but it is cumbersome to measure and usually has a large error variance. It is correlated to *Striga* counts and vigour score. The assessment of the number of subterranean attached *Striga* plants is very laborious and is generally restricted to small trials or selected entries.
- ✶ Data collected from host plants may include standard variables like number of plants after thinning and at harvest, days to 50% anthesis, plant height, grain yield and grain yield components, disease resistance scores and notes on the general agronomic performance.
- ✶ In addition, visual *Striga* damage or "syndrome" ratings have been recommended. The rating on a 1 to 9 scale reflects damage of the host plant, as its reaction to *Striga*, in the form of leaf chlorosis or firing ("scorching"), poor ear or panicle development, and stunting (Table 5). Evaluation is recommended about two weeks after anthesis, and the rating is reportedly useful to assess tolerance to *Striga* in maize. However, in sorghum and pearl millet, specific, *Striga*-induced symptoms are infrequent and may only occur under extremely high infestation levels

Table 4. The *Striga* vigour ratings

0 = no emerged *Striga* plants
1 = average height of *Striga* plants 5 cm, without branches
2 = average height of *Striga* plants 6-20 cm, without branches
3 = average height of *Striga* plants 6-20 cm, with branches
4 = average height of *Striga* plants 21-30 cm with 5 branches
5 = average height of *Striga* plants 21-30 cm, with > 5 branches
6 = average height of *Striga* plants 31-40 cm, with 10 branches
7 = average height of *Striga* plants 31-40 cm, with > 10 branches
8 = average height of *Striga* plants >40 cm, with 0 branches
9 = average height of *Striga* plants >40 cm, with > 10 branches

(Note: the score should be independent of the number of emerged *Striga* plants).

Fig. 18. The *Striga* vigour ratings

Table 5. The *Striga* "syndrome" ratings in maize

1 =	Normal maize growth; no visible symptoms.
2 =	Scattered small and vague, whitish leaf blotches visible. Otherwise, normal plant growth.
3 =	Blotching and streaking easily noticeable. Mild wilting. Only a trace of scorching, restricted to tips of leaves.
4 =	Extensive blotching and streaking, mild wilting. Only a trace of leaf scorching. Slight but noticeable stunting and reduction in ear and tassel size.
5 =	Extensive blotching and streaking, wilting. Leaf scorching on a small portion of the leaf area. Moderate stunting; ear and tassel size reduction.
6 =	Extensive streaking, now obscuring the blotches, turning to scorching. Leaf scorching covering about a third of the leaf area. About one-third reduction in plant height. Reduced stem diameter, reduced ear and tassel size.
7 =	Extensive streaking/scorching, turning gray and necrotic. About half of the plant's surface is scorched. Severe stunting, about 50% reduction in height. Noticeable reduction in stem diameter and in ear and tassel size. Some stalks breaking.
8 =	Scorching on most of the leaf area. Stunting reduces height by more than 50%. Stalks look thin and weak; many are broken.
9 =	Virtually all leaf area scorched; two-thirds or more reduction in height; most stems collapsing; no useful ear formed; no pollen production; plants dead or nearly dead.

(Source: Berner *et al.* 1993)

* If data from non-infested control plots are available, a *Striga* tolerance index may be calculated as the ratio of grain yield under *Striga*-infested and *Striga*-free conditions. Other selection indices may include the number of emerged *Striga* plants, assessment of host plant damage (*Striga* syndrome), yield under infested conditions and agronomic performance of non-infested plants.

LITERATURE

Berner DK, JG Kling & BB Singh 1995. *Striga* research and control, a perspective from Africa. Plant Disease 79: 652-660.

Berner DK, MD Winslow, AE Awad, KF Cardwell & DR Mohan Raj 1993. *Striga* Research Methods - A manual (Draft Edition). IITA, Ibadan, Nigeria. 71 pp.

Efron Y 1993. Screening maize for tolerance to *Striga hermonthica*. Plant Breeding 110: 192-200.

Ejeta G, LG Butler & AG Babiker 1992. New approaches to the control of *Striga*. *Striga* Research at Purdue University. Research Bulletin RB-991. Agricultural Experiment Station. Purdue University. West Lafayette, IN.

Hess DE, G Ejeta & LG Butler 1992. Selecting sorghum genotypes expressing a quantitative biosynthetic trait that confers resistance to *Striga*. Phytochemistry 31: 493-497.

Kim SK 1991. Breeding for tolerance and general resistance in maize: a novel approach to combating *Striga* in Africa. In: Improving *Striga* management in Africa, Proceedings, 2nd General Workshop of the Pan-African *Striga* Control Network (PASCON), 23-29 June 1991, Nairobi, Kenya. 168-176.

Kim SK 1994. Genetics of maize tolerance of *Striga hermonthica*. Crop Science 34: 900 - 907.

Lane JA, JA Bailey & PJ Terry 1991a. An *in vitro* growth system for studying the parasitism of cowpea (*Vigna unguiculata*) by *Striga gesnerioides*. Weed Research 31: 211-217.

Lane JA, MJ Kershaw, THM Moore, DV Child, PJ Terry & JA Bailey 1991b. An *in vitro* system to investigate the parasitism of cereals by *Striga* species and resistance of cowpea to *Alectra vogelii*. In: JK Ransom, LJ Musselman, AD Worsham & C Parker (eds.), Proceedings of the 5th International Symposium of Parasitic weeds, June 1991, Nairobi, Kenya. 37-240.

Linke KH & W Vogt 1987. A method and its application for observing germination and early development of *Striga* (Scrophulariaceae) and *Orobanche* (Orobanchaceae). In: HChr Weber & W Forstreuter (eds.), Proceedings, 4th Int. Symposium on Parasitic Flowering Plants, Marburg, Germany, 501-509.

Vasudeva Rao MJ 1985. Techniques for screening sorghums for resistance to *Striga*. ICRISAT Information Bulletin No. 20, Andhra Pradesh, India.

Authors: BIG Haussmann & DE Hess

6.3.2.2 Variability within and among *Striga* species, and stability of resistance

INTRODUCTION

Striga is a highly variable parasite and seems to have an extraordinary plasticity and capacity to adapt to new hosts. Resistance to *Striga* is partially species-specific, i.e., resistance to *S. asiatica* does not necessarily hold against *S. hermonthica* and *vice versa*. Some sorghum varieties revealed resistance only in some locations and were susceptible in others, which may be due to the presence of specific *Striga* races or biotypes with different degrees of virulence or aggressiveness. Actually, *S. hermonthica* strains specific for sorghum and millet have been reported, while others could attack both host species. Both, inter-specific variability among *Striga* spp. as well as intra-specific variation within a given species must be taken into account when breeding for *Striga* resistance. Furthermore, edaphic and climatic factors can affect the virulence/aggressiveness of *Striga*.

OBJECTIVES

- To develop stable resistance.

PROCEDURE

☞ To obtain stable, polygenic resistance, the breeding materials should be evaluated against various *Striga* morphotypes, host-specific races and at various locations/ under differing environmental conditions.

☞ In so doing, quarantine regulations must be strictly respected, and *Striga* species or strains should not be introduced into regions where they do not already occur.

EVALUATION

(→ chapter 5, 6.3.2.1)

LITERATURE

Ejeta G, LG Butler & AG Babiker 1992. New approaches to the control of *Striga*. *Striga* Research at Purdue University. Research Bulletin RB-991. Agricultural Experiment Station. Purdue University. West Lafayette, IN.

Freitag J, DE Hess, HG Welz & HH Geiger 1996. Pearl millet and sorghum specific races of *Striga hermonthica* in Niger. In: MT Moreno, JI Cubero, D Berner, D Joel, LJ Musselman & C Parker (eds.), Advances in Parasitic Plant Research. Proceedings of the 6th International Parasitic Weed Symposium, April 16-18, 1996, Cordoba, Spain, 472-478.

Kim SK, AY Akintunde & P Walker 1994. Responses of maize, sorghum and millet host plants to infestation by *Striga hermonthica*. Crop Protection 13(8): 582-589.

Ramaiah KV 1987. Breeding cereal grains for resistance to witchweed. In: LJ Musselmann (ed.), Parasitic Weeds in Agriculture, Volume I *Striga*. CRC Press, Boca Raton, Florida, 227-242.

Vasudeva Rao MJ & LJ Musselman 1987. Host specificity in *Striga* spp. and physiological "strains". In: LJ Musselmann (ed.), Parasitic weeds in Agriculture, Volume I *Striga*. CRC Press, Boca Raton, Florida, 13-25.

Authors: BIG Haussmann & DE Hess

6.3.2.3 Sources of resistance

SEED MATERIALS

✓ Several sorghum cultivars or breeding lines were reported as *Striga* resistant. Examples are 'Dobbs' from East Africa, 'Radar' and 'Framida' from South Africa, 'Seguetanas' from West Africa, '555' and 'N-13' from India, 'IS-9830' from Sudan, 'SAR (*Striga asiatica* resistant)-lines' developed by ICRISAT, and 'SRN-39', extensively investigated at Purdue University. Among wild relatives, *Sorghum versicolor* has recently shown resistance.

✓ Little work has been done with pearl millet, but "less susceptible" lines are reported to include 'TIF 23A', 'Serere 2A-9', '80S-221', 'P-2671', 'P-2950', and 'PT499'. Progress at the ICRISAT Sahelian Center (Niamey, Niger) is being made in selecting for *Striga* resistance in variety 'ICMV IS 89305'. *Pennisetum setosum* is a wild relative of pearl millet with resistance to *Striga*.

✓ Maize is a new world crop which was never subjected to *Striga* during its evolution. It is therefore unlikely that any resistance mechanism has evolved. Tolerance sources for short-term use were found among materials from the US Corn Belt, and in lines and crosses of temperate × tropical, or subtropical × tropical origin. Among the wild relatives, *Zea diploperennis* (perennial maize), *Tripsacum* spp., and *Teosinte* seem to be promising sources of resistance to both *S. hermonthica* and *S. asiatica*.

LITERATURE

Berner DK, JG Kling & BB Singh 1995. *Striga* research and control, a perspective from Africa. Plant Disease 79: 652-660.

Kim SK 1991. Breeding for tolerance and general resistance in maize: a novel approach to combating *Striga* in Africa. In: Improving *Striga* management in Africa, Proceedings, 2nd General Workshop of the Pan-African *Striga* Control Network (PASCON), 23-29 June 1991, Nairobi, Kenya, 168-176.

Kim SK 1994. Genetics of maize tolerance of *Striga hermonthica*. Crop Science 34: 900-907.

Lane JA, DV Child, THM Moore & JA Bailey 1995. Wild relatives of cereals as new sources of resistance to *Striga*. In: Parasitic Plants: Biology and Resistance. International Workshop, 30 May - 2 June 1995, IACR-Long Ashton Research Station, Bristol, UK, 13.

Parker C & CR Riches 1993. Parasitic Weeds of the World - Biology and Control. CAB International, Wallingford, UK, 332 pp.

Ramaiah KV 1987. Breeding cereal grains for resistance to witchweed. In: LJ Musselmann (ed.), Parasitic Weeds in Agriculture, Volume I *Striga*. CRC Press, Boca Raton, Florida, 227-242.

Authors: BIG Haussmann & DE Hess

6.3.2.4 Genetics of resistance

INTRODUCTION

The low stimulation of *S. asiatica* seed germination by the sorghum cultivars 'Framida', '555', and 'SRN-39' has been shown to be controlled by a single recessive gene. Differing GCA ("General combining ability") effects for the germination distance in the agar gel assay (using *S. hermonthica*) indicated that the alleles or genes for low stimulant production may differ for '555' and 'Framida'. F_1-diallel studies and line× tester analyses with sorghum clearly indicated the presence of genetic variation in sorghum with preponderance of additive effects for stimulation of *S. hermonthica* seed germination in the agar gel assay, the number of aboveground *Striga* plants supported in pots, and the number of emerged *Striga* under field conditions. Estimates of broad-sense heritabilities were 0.97 and 0.91 for germination distances of lines and hybrids, while for *Striga* counts in a pot trial, heritabilities ranged between 0.1 and 0.58.

Resistance (or lower susceptibility) in pearl millet, where found, was reported as dominant.

Additive-genetic variation has been found for the *Striga hermonthica* damage rating in maize whereas for the *Striga* emergence counts, non-additive gene action was predominant. Broad-sense heritability estimates were very low (<0.1) for the number of emerged *Striga* plants, and moderate to high (0.76) for the *Striga* damage rating.

OBJECTIVES

- To find out whether different sources of resistance actually carry different genes. Such information is most useful for selection of parent lines for a crossing program,
- to enlarge knowledge about the mode of gene action (dominant or recessive, additive or non-additive) and heritability estimates, which is essential for optimising breeding procedures.

LITERATURE

Berner DK, JG Kling & BB Singh 1995. *Striga* research and control, a perspective from Africa. Plant Disease 79: 652-660.

Haussmann BIG, DE Hess, BVS Reddy, HG Welz & HH Geiger 1996a. Quantitative-genetic parameters for resistance to *Striga hermonthica* in sorghum. In: MT Moreno, JI Cubero, D Berner, D Joel, LJ Musselman & C Parker (eds.), Advances in Parasitic Plant Research. Proceedings of the 6th International Parasitic Weed Symposium, April 16-18, 1996, Cordoba, Spain, 681-688.

Haussmann BIG, DE Hess, BVS Reddy, HG Welz & HH Geiger 1996b. Analysis of resistance to *Striga hermonthica* in diallel crosses of sorghum. Proceedings of the 4th General Workshop of the Pan-African *Striga* Control Network (PASCON), October 28 - November 1, 1996, Bamako, Mali (in press).

Hess DE & G Ejeta 1992. Inheritance of resistance to *Striga* in sorghum genotype SRN 39. Plant Breeding 109: 233-241.

Hess DE, G Ejeta & LG Butler 1992. Selecting sorghum genotypes expressing a quantitative biosynthetic trait that confers resistance to *Striga*. Phytochemistry 31: 493-497.

Kim SK 1994. Genetics of maize tolerance of *Striga hermonthica*. Crop Science 34: 900 - 907.

Ramaiah KV 1984. Patterns of *Striga* resistance in sorghum and millets with special emphasis on Africa. In: Proceedings of the International Workshop on the Biology and Control of *Striga*, held at Dakar, Senegal, 14 - 17 November 1983. ICSU Press, France, 71-92.

Ramaiah KV 1987. Breeding cereal grains for resistance to witchweed. In: Musselmann LJ (ed.), Parasitic Weeds in Agriculture, Volume I *Striga*. CRC Press, Boca Raton, Florida, 227-242.

Ramaiah KV, VL Chidley & LR House 1990. Inheritance of *Striga* seed germination stimulant in sorghum. Euphytica 45: 33-38.

Shinde VK & N Kulkarni 1982. Genetics of resistance to *Striga asiatica* in sorghum. In: Proceedings, ICRISAT-AICSIP (ICAR) Working Group Meeting on *Striga* Control. ICRISAT Center 30 Sept. - 1 Oct. 1982. ICRISAT, Patancheru PO, Andhra Pradesh 502 324, India. 134-140.

Authors: BIG Haussmann & DE Hess

6.3.2.5 Breeding strategies

OBJECTIVE
- To develop *Striga*-resistant adapted cultivars with stable performance.

STRATEGIES

- To characterise crop germplasm and improve the currently available sources of resistance for better agronomic performance;
- Transfer resistance genes into genotypes better adapted to target areas;
- To pyramid resistance genes to obtain more durable and stable, polygenic resistance;
- Develop *Striga* resistant parent lines to be used in hybrid breeding programs;
- Develop open-pollinated synthetic varieties by recombining resistant inbred lines;
- Develop random-mating populations with multiple sources of resistance using recurrent selection procedures;
- Include wild relatives in the breeding program;
- Develop and employ marker-assisted selection techniques for broad-based *Striga* resistance;
- Develop a “low-emergence pool” from traditional African cultivars, improved inbreeds and hybrids, and wild relatives for further selection in laboratory and field (pertains to IITA maize breeding program).

COMMENTS

- The absence of reliable single-plant screening techniques to distinguish resistant and susceptible plants in segregating populations has been a major drawback in the past. Selection for *Striga* resistance was therefore usually deferred until true breeding progenies were produced. This meant that large numbers of progeny had to be advanced before the trait of interest could be assessed, a time- and cost-intensive procedure. The development of laboratory assays which allow the non-destructive, rapid and inexpensive evaluation of individual plants, would largely facilitate early generation testing and increase selection efficiency. To date, the only available assay screens for the low stimulant character. However, breeders must bear in mind that selection for low germination distance in parent or early generation lines, although attractive, could result in a loss of valuable materials possessing resistance mechanisms other than low-stimulant production.

LITERATURE

Berner DK, JG Kling & BB Singh 1995. *Striga* research and control, a perspective from Africa. Plant Disease 79: 652-660.

Efron Y 1993. Screening maize for tolerance to *Striga hermonthica*. Plant Breeding 110: 192-200.

Ejeta G & LG Butler 1993. Host Plant Resistance to *Striga*. In: International Crop Science I. 1993. Crop Science Society of America, Segoe Rd, Madison, WI53711, USA, 561-569.

Ejeta G, LG Butler & AG Babiker 1992. New approaches to the control of *Striga*. *Striga* Research at Purdue University. Research Bulletin RB-991. Agricultural Experiment Station. Purdue University. West Lafayette, IN.

Hess DE, G Ejeta & LG Butler 1992. Selecting sorghum genotypes expressing a quantitative biosynthetic trait that confers resistance to *Striga*. Phytochemistry 31: 493-497.

Ramaiah KV 1987. Breeding cereal grains for resistance to witchweed. In: LJ Musselmann (ed.), Parasitic Weeds in Agriculture, Volume I *Striga*. CRC Press, Boca Raton, Florida. 227-242.

Authors: BIG Haussmann & DE Hess

6.3.3 Breeding for resistance to *Alectra vogelii* and *Striga gesnerioides*

INTRODUCTION

Breeding for resistance to *Alectra* (basically *A. vogelii*) and *S. gesnerioides* includes the same techniques as discussed above for *Orobanche* and *Striga* (→ chapter 6.3.1 and 6.3.2). This comes from the almost identical life cycle of these species, their dependence on stimulation of seed germination and the related production of large numbers of seed. Also the variation within the species and the potential occurrence of biotypes is high due to the predominant selfing coupled with a certain level of out-crossing and possibly mutations (as a result of the high seed number produced by this species). The existence of particular biotypes in West and South Africa was demonstrated.

COMMENTS

- Resistant cowpea accessions/cultivars have to be further tested for resistance to potential other biotypes of the parasites at different locations (necessity of multi-location field trials).
- Different sources of resistance to *A. vogelii/S. gesnerioides* might be required for different regions due to

the occurrence of biotypes of the parasite, and

- to ensure a high stability of resistant lines due to the narrow genetic base for resistance to this parasite.
- Use *Alectra* seed of the previous season to infest field plots, as after storing for more than 3 years seed viability decreases substantially.

Sources of resistance:

- B301, B319 and B359 are cowpea land-races resistant to *A. vogelii*; resistance was also found in several local accessions of cowpea in Botswana; B301 is also resistant to *S. gesnerioides* (dual parasite resistance). The resistance in B301 is controlled by duplicate dominant genes.
- IITA lines IT86D-472 and IT 84D-666 (cowpea) have resistance to A. vogelii.
- IITA line 58-57 (cowpea) which has resistance to *S. gesnerioides* has also a low level of resistance to *A. vogelii*.

LITERATURE

Aggarwal VD, N Muleba, I Drabo, J Souma & M Mbewe 1984. Inheritance of *Striga gesnerioides* resistance in cowpea. In: C Parker, LJ Musselman, RM Polhill & AK Wilson (eds.), Proceedings 3rd International Symposium on Parasitic Weeds, ICARDA/International Parasitic Seed Plant Research Groupe, 7-9 May 1984, Aleppo, Syria. ICARDA, P.O.Box 5466, Aleppo, Syria,, 143-147.

Lane JA, MJ Kershaw, THM Moore, DV Child, PS Terry & JA Bailey 1991b. An *in vitro* system to investigate the parasitism of cereals by *Striga* species and resistance of cowpea to *Alectra vogelii*. In: JK Ransom, LJ Musselman, AD Worsham & C Parker (eds.), Proceedings of the 5th International Symposium of Parasitic Weeds. CIMMYT, Nairobi, Kenya, 237-240.

Polniaszek TI, C Parker & CR Riches 1991. Variation in the virulence of *Alectra vogelii* populations on cowpea. Tropical Pest Management 37: 152-154.

Riches CR 1989. The biology and control of *Alectra vogelii* Benth., (Scrophulariaceae) in Botswana. Ph.D. Thesis, University of Reading, UK, 208 pp.

Riches CR, KA Hamilton & C Parker 1992. Parasitism of *Alectra* species by grain legumes. Annals of Applied Biology 121, 361-370.

Singh BB & AM Emechebe 1991. Breeding for resistance to *Striga* and *Alectra* in cowpea. In: JK Ransom, LJ Musselman, AD Worsham & C Parker (eds.), Proceedings of the 5th International Symposium of Parasitic Weeds. CIMMYT, Nairobi, Kenya, 303-305.

Author: K-H Linke

6.3.4 Resistance to *Cuscuta*

INTRODUCTION

Cuscuta spp. are parasites of numerous plants. However, *Cuscuta* was never observed on any member of the grass family. The intensity of attack on various host plants differs, indicating a preference of infestation for certain plant species. This is attributed to hypersensitive reaction of host cells in tomato, the formation of suberized layers of cells from a secondary meristem in cotton and the presence of inhibitors of cell wall degrading enzymes in sweet potato.

Differences in resistance between cultivars were reported only for green gram (*Vigna radiata*) in India, of which several varieties seem to allow less vigorous growth of the parasite. However, there are no reports of outstanding differences within any crop species.

A common type of resistance to *Cuscuta* can be observed with older host plants becoming less infected than younger plants. Therefore, contradicting reports on resistance/susceptibility of a certain host species to *Cuscuta* may be attributed to the age of the respective host plant studied.

PROCEDURES

→ chapter 5.2

LITERATURE

Al-Menoufi OA & FM Ashton 1991. Studies on the parasitism of *Cuscuta* spp. Series 8: susceptibility and resistance of some *Lycopersicon* species to *Cuscuta campestris* infection. In: JK Ransom, LJ Musselman, AD Worsham & C Parker (eds.), Proceedings of the 5th International Symposium of Parasitic Weeds. CIMMYT, Nairobi, Kenya, 293-297.

Capdepon M, A Fer & P Ozenda 1985. Sur un système inédit rejet d'un parasite: exemple de la *Cuscuta* sur cotonnier (*C. lupuliformis* Krock. sur *Gossypium hirsutum* L.). Comptes Rendues de l'Académie des Sciences III. Sciences de la vie 300(6): 227-237.

Ihl B & I Miersch 1996. Susceptibility and resistance of *Lycopersicon* to infection by *Cuscuta*. In: MT Moreno, JI Cubero, D Berner, D Joel, LJ Musselman & C Parker (eds.), Advances in Parasitic Plant Research; 6th International Parasitic Weed Symposium, Cordoba, Spain, 599-607.

Nagar N & GC Sanwal 1984. Biochemical aspects of parasitism in *Cuscuta reflexa*: Inhibition of cell wall degrading enzymes of *Cuscuta* by non-susceptible plants. In: C Parker et al (eds.), Proceedings of the 3rd International Symposium on Parasitic Weeds, ICARDA/ISPRG, Aleppo, Syria, 175-183.

Parker C & CR Riches 1993. Parasitic Weeds of the World - Biology and Control. CAB International, Wallingford, UK, 332 pp.

Rao KN & RSN Rao 1991.Studies on tolerance of black gram and green gram varieties to dodder (*Cuscuta chinensis* Lam.). In: JK Ransom, LJ Musselman, AD Worsham & C Parker (eds.), Proceedings of the 5th International Symposium of Parasitic Weeds. CIMMYT, Nairobi, Kenya, 170-175.

Tsivion Y 1979. Morphogenetic sequences in the formation of the haustorium of *Cuscuta campestris*. In: Musselman LJ, AD Worsham & RE Eplee (eds.), Proceedings of the 2nd International Symposium on Parasitic Weeds. Raleigh, North Carolina, USA, 174-181.

Author: K-H Linke

6.3.5 Use of molecular markers

INTRODUCTION

Traditionally, selection is based on phenotype. The phenotype of a plant is determined not only by its genetic composition but also by the environment in which it is grown. In many cases, the effects of the environment mask those of the genotype, so that the phenotype provides an imperfect measure of a plant's genetic potential. As early as 1923, Sax proposed identifying and selecting "minor genes" of interest by linkage with "major genes" (i.e., markers) which could be scored more easily. Generally, three types of markers can be distinguished: morphological markers, isoenzyme markers, and DNA markers.

DNA markers are preferable because they can be assessed most accurately and are practically unlimited in number. Restriction fragment length polymorphism (RFLP) is most widely used, but the method is laborious and expensive. Polymerase chain reaction (PCR)-based markers like RAPDs (random amplified polymorphic DNAs) or AFLPs (amplification fragment length polymorphisms) are becoming increasingly popular as they require less sophisticated laboratory equipment and are more cost-efficient, i.e., yielding more "data points" per time and chemical input. Isoenzyme markers may be affected by environmental conditions and hence expressed differentially.

The application of marker-assisted selection is particularly advantageous when

- resistance tests are difficult, complex, expensive and unreliable,
- the pathogen is quarantined,
- breeding materials are advanced in off-season nurseries where the pathogen does not occur,
- resistance genes are recessive, i.e., restricting the utility of back-cross schemes,

The integration of molecular marker selection techniques into plant breeding promises enhanced, quicker movement of desirable genes among varieties, and allows the transfer of novel genes from related wild species. Knowledge on the genetic variability of a parasite is basic for any resistance breeding (→ chapter 2.3). Detecting resistance genes by their linkage to DNA markers makes it possible to screen for many different resistance genes simultaneously, without the need to inoculate with the various pathogens. Pyramiding of resistance genes to provide durable resistance is therefore greatly facilitated.

OBJECTIVES

- To enhance breeding efficiency,
- to provide durable resistance through pyramiding of resistance genes,
- when resistance genes are transferred from wild relatives into a cultivated crop, molecular markers can assist in selecting against the undesired genetic background of the donor parent.

PROCEDURES

☞ In order to find genes or "quantitative trait loci" (QTL) for resistance to *Striga,* by their linkage to DNA-markers, two lines which differ widely for *Striga* resistance and marker genotype must be crossed and the progeny selfed for one or several generations. Due to recombination and segregation of the various resistance genes, the resulting "mapping population" will reveal a wide range of resistance, from extremely susceptible to highly resistant with all levels in between. This population is evaluated for *Striga* resistance, and at the same time the DNA profile of each individual plant or line is examined for marker loci in the laboratory. The data are then analyzed for co-segregation, i.e., linkage of the markers with the resistant phenotypes.

☞ In principle, the association of marker genotype and resistant phenotype could be assessed directly in the F2 generation. However, the level of quantitative resistance to *Striga* cannot be accurately measured on a single plant but must be assessed in replicated field experiments. Therefore, frequently F2:3 lines (F3 plants tracing back to the same F2 plant, also called F3 families) or further inbred generations are used. They may be referred to as recombinant inbred lines (RILs).

EVALUATION

✶ Resitance data (→ chapters 6.3.1.1 and 6.3.2.1).

✶ Marker data (→ chapters 2.3.1)

COMMENTS

- The utility of DNA markers in resistance breeding is based on finding tight linkages between these markers and the resistance genes of interest. In subsequent marker-assisted breeding programs, such linkage permits inference of the presence of a resistance gene by assessing for the DNA marker - instead of evaluating the materials for resistance traits. Selection, particularly among early segregating generations, can then be done in the absence of parasite.

- The accuracy of locating resistance genes (QTL) by their linkage to DNA markers depends on the number of markers applied and the evenness of their distribution within the genome, the size of the mapping population, the phenotypic distance between the parents, and the heritability of the trait. High number of applied markers, even distribution, large mapping population, large phenotypic distance between the parents, and high heritability of the trait of interest, all contribute to increase the accuracy of locating QTL. Multilocation testing and a large number of replications can partially compensate for a low heritability value.
- At present, routine use of DNA markers is severely limited by the significant costs and time required for the marker analyses. More efficient and simpler laboratory methods are needed and may be developed in the near future. Therefore, the prospects for cost-effective implementation of marker-based techniques in breeding for *Striga* resistance are very promising.

LITERATURE

Joel DM, V Portnoy, G Tzuri, R Greenberg & N Katzir 1996. Molecular markers for the identification of *Orobanche* species. In: MT Moreno, JI Cubero, D Berner, D Joel, LJ Musselman & C Parker (eds.), Advances in Parasitic Plant Research; 6th International Parasitic Weed Symposium, Cordoba, Spain, 151-160.

Melchinger AE 1990. Use of molecular markers in breeding for oligogenic disease resistance. Plant Breeding 104: 1-19

Paterson AH, SD Tanksley & ME Sorrells 1991. DNA markers in plant improvement. Advanced Agronomy 46: 39-90.

Sax K 1923. The association of size differences with seed coat pattern and pigmentation in *Phaseolus vulgaris*. Genetics 8: 552-560.

Stuber CW 1990. Biochemical and molecular markers in plant breeding. Plant Breeding Review 9: 37-61.

Tanksley SD, ND Young, AH Paterson & MW Bonierbale 1989. RFLP mapping in plant breeding: new tools for an old science. Biotechnology 7: 257-264.

Authors: BIG Haussmann, DE Hess & HG Welz

6.3.6 Genetic engineering

INTRODUCTION

Principally, genetic engineering permits the transfer of resistance genes from whatever organism in which they occur into chosen crops. Regarding resistance to parasitic weeds, the main limitation is the lack of well-defined resistance genes. At present, genetic engineering of herbicide resistance is fairly popular, allowing to control weeds with efficient, "environment-friendly" herbicides. Two kinds of resistance to herbicides may be genetically transferred into crops: (1) Target-site resistance and (2) metabolic resistance. The former is more convenient for parasitic plants as the herbicide remains intact and gets transported to the parasite. Hence, biotechnology may be used to confer target-site herbicide resistance to heretofore sensitive crops.

An *Orobanche* sp. was successfully controlled on transgenic chlorsulfuron-resistant tobacco (*Nicotiana tabacum*) by foliar application of chlorsulfuron at 10, 20, or 30 days after planting. *Striga hermonthica* can be controlled on biotechnologically-derived maize with resistance to acetolactate syntase (ALS) inhibitors like imazaquin or chlorsulfuron.

Appropriate gene constructs, e.g., for resistance to herbicides like chlorsulfuron, glyphosate, or asulam, are actually available and could be transferred readily. However, tobacco is much easier to transform than maize, pearl millet or sorghum.

The development of herbicide resistance in crop plants represents a rather complex operation, including sophisticated laboratory equipment. Only certain large companies were able to develop such resistance so far. Examples are maize, tobacco and oilseed rape with acetolactacte synthase target site resistance or glyphosate-resistant rape plants.

OBJECTIVES

- To test and evaluate herbicide-resistant crops,
- to introduce transgenic, herbicide-resistant crops in order to achieve immediate, cost-effective and selective control of parasitic weeds by the use of herbicides,
- (to develop herbicide-resistant crop plants).

MATERIALS & PROCEDURES

✓ __Note__: Within the framework of this manual however, the target groups are researchers studying the performance of biotechnology-derived crops in their response to parasitic weeds and *vice versa* rather than developing crops modified by biotechnology. Hence, for details on biotechnological procedures itself we refer to the relevant literature.

☞ For the evaluation of biotechnology-derived crops the techniques as mentioned above (→ chapter 5) may be applied as usual.

COMMENTS

- Herbicide resistance in crops affected by parasitic weeds has several positive properties: it (1) allows the selective control of parasitic weeds, often at very low dose rates; (2) is effective against several strains or species of the parasite; (3) replaces cultivation methods for control of other weeds. The greater efficacy and lower labour time and energy requirements of herbicides are the major reasons for their cost effectiveness. However, it should also be taken into consideration that herbicides may have negative impact on the environment. Furthermore, herbicide resistance should only be used in crops which do not interbreed with related weeds in the same locality. Even if this condition is respected, there exists the possibility of evolution of herbicide resistance in the parasitic weeds. The high natural frequency of such mutations and the huge seed output of parasitic weeds only serve to exacerbate the problem. Nonetheless, the large soil seed bank of susceptible parasitic weeds would serve to slow the rate of evolution of resistant populations.
- Acetolactate syntase (ALS) inhibitors like imazaquin or chlorsulfuron are subject to faster development of

resistant parasite populations than e.g., glyphosate; delaying the development of resistance may include e.g., changes of herbicide or herbicide mixtures.

- In the long term, selection for genetic resistance to *Striga* and *Orobanche* is a priority.

LITERATURE

Abayo GO, JK Ransom, J Gressel & GD Odhiambo 1996. *Striga hermonthica* control with acetolactate synthase inhibiting herbicides seed-dressed to maize with target site resistance. In: MT Moreno, JI Cubero, D Berner, D Joel, LJ Musselman & C Parker (eds.), Advances in Parasitic Plant Research; 6th International Parasitic Weed Symposium, Cordoba, Spain, 761-768.

Gressel J, L Segel & JK Ransom 1996. Managing the delay of evolution of herbicide resistance in parasitic weeds. International Journal of Pest Management 42(2): 113-129.

Gressel J 1995. The potential role of herbicide resistant crops in world agriculture. In: SO Duke (ed.), Herbicide Resistant crops. Lewis Publishers, Boca-Raton, USA, 231-250.

Gressel J, Y Kleifeld & DM Joel 1994. Genetic engineering can help control parasitic weeds. In: AH Pieterse, JAC Verkleij & SJ ter Borg (eds.), Biology and Management of *Orobanche*, Proceedings of the 3rd International Workshop on *Orobanche* and Related *Striga* Research, Royal Tropical Institute, Amsterdam, The Netherlands. 406-418.

Joel DM, Y Kleifeld, Y Losner-Goshen, Y Herzlinger & J Gressel 1995. Transgenic crops against parasites. Nature 374: 220-221.

Marx JL 1985. Plant gene transfer becomes a fertile field. Science 230: 1148-1150.

Authors: BIG Haussmann, DE Hess & K-H Linke

6.4 Chemical

6.4.1 Soil disinfection

INTRODUCTION

For certain studies like the comparison of treatments of infested with non-infested plots under otherwise comparable conditions, parasite free areas close to infested ones are required. For this purpose a field has to be partly or entirely disinfected. This is a rather expensive operation and should only be done if absolutely necessary.

The methods described below may not always clean a field completely (especially considering the soil depth where seeds are buried), but they are sufficient to allow comparisons and assess the actual damage on the host crop inflicted by the parasite.

OBJECTIVES

- To develop parasite free areas (soil) in the field,
- to quantify yield reduction,
- to assess stress symptoms relative to the control treatment.

PROCEDURES

1. Application of ethylene gas

☞ → chapter 6.4.3.1

2. Application of analogues of strigol

☞ → chapter 6.4.3.2

3. Soil solarization

☞ → chapter 6.5.5

4. Chemical soil sterilisation (fumigation)

☞ Chemical soil sterilisation is probably the most effective method to disinfect a soil. But it is also a rather hazardous method for the environment and the persons in charge of its application.

☞ The agents used are methylbromide (in quantities of about 350 kg/ha), also metham (sodium N-methyldithiocarbamate, in quantities of 400 to 500 l/ha), or dazomet. The latter two are less effective but also less hazardous to man and less expensive than methylbromide. A further chemical is Telone (1,3-dichlorpropene) which has to be applied at rates of about 400 kg/ha.

☞ For efficient treatment methylbromide has to remain in the ground for about two days. For this purpose the soil needs to be covered with an air-tight plastic sheet immediately after application of the chemical. Metham, dazomet and Telone do not require soil covering with plastic but incorporation into the soil by tilling operations or light irrigation is recommended.

☞ A considerable disadvantage of this method is that the entire soil flora and fauna are killed, creating rather artificial soil conditions which might influence plant growth and host root-parasite interactions.

☞ <u>Note:</u> Methylbromide is highly toxic; hence, specialized firms should be contracted for application. It has also been identified as a chemical agent involved in depleting the ozone layer. It is already banned for agricultural use in several countries and from the year 2005 it will be in countries of the tropics and subtropics.

<u>5. Physical soil sterilisation</u>

☞ Treatment of moist soil in special containers with hot water vapour (min. 90°C) for about half an hour destroys the parasite seed in the soil.

☞ Side effects are an increased amount of available nitrogen (ammonium) and manganese in the soil. The biological balance of the soil is greatly affected, the entire soil flora and fauna gets killed (same as with chemical soil sterilisation).

☞ This technique is economic only for small quantities of soil as e.g., required in certain pot studies.

EVALUATION

* Soil samples are taken before and after the treatment and checked for viable seeds of parasites; this can be supplemented with seed bags buried in the soil to be treated (→ chapters 4.3 and 4.5 and 4.7).
* Count emerged parasites and assess their dry weight.

LITERATURE

Egley GH & JE Dale 1970. Ethylene, 2-chloroethylphosphonic acid and witchweed germination. Weed Science 18: 586.

Jacobsohn R, Y Kleifeld, FP Agrawal, PJ & K Marton 1991. Soil fumigation with Telone II for broomrape (*Orobanche* spp.) control. In: K Wegmann & LJ Musselman (eds.), Progress in *Orobanche* research. Eberhard-Karls-Universität, Tübingen, Germany, 185-190.

Egley GH, Eplee RE & Norris RS 1990. Discovery and development of ethylene as a witchweed seed germination stimulant. In: PF Sand, RE Eplee & RG Westbrooks (eds.), Witchweed research and control in the United States. Weed Science Society of America, Champaign, IL, USA. 56-66.

Author: K-H Linke

6.4.2 Herbicides

INTRODUCTION

In modern agricultural production systems herbicides account for weed control on about 95% of the arable land. Parasitic plants however, predominantly occur in less developed agricultural systems, where application of herbicides due to lack of resources is low. Moreover, due to the close interrelation between host and parasite herbicidal control of one partner, the parasite, becomes particularly difficult due to phytotoxicity on the host crop. Hence, the success in controlling parasitic weeds with herbicides so far is limited.

The objectives of using herbicides for the control of parasitic weeds are (1) to save at least part of the crop yield, (2) to stop seed production of the parasite, (3) to prevent infestation of the crop and/or (4) to control the parasite selectively.

OBJECTIVES

- To screen herbicides for their efficiency in controlling parasitic weeds,
- to screen herbicides for their phytotoxicity on the host crop,
- to evaluate the dosage required for optimum control.

MATERIALS

✓ Normal equipment for field experiments.

✓ Herbicide(s), equipment for spraying (field sprayer and tractor or knapsack sprayer; size and complexity of the sprayer are determined by the demands and the resources of the situation) with nozzle types adjusted to the desired spray pattern, spare nozzles, water, graduated measure (beaker), pipette, safety clothing.

✓ If the product label recommends it, protective gloves and respirator should be worn.

✓ Time requirement ranges from 0.5 h/ha on fields >1 ha to more than 10 h/ha on experimental plots.

PROCEDURES

☞ Measure the exact size of field to be sprayed.

☞ Calculate the concentrations of herbicide required for spraying.

☞ Prepare the required amount of liquid for all replications of a treatment at once in one container. This will ensure identity of the sprayed concentrations.

☞ Pre-plant herbicides are applied prior to the planting of a host crop. This requires mixing into the top 2 to 9 cm of soil by tillage operation (e.g., disking).

☞ Pre-emergence herbicides are applied after sowing of the host. It has to be distinguished between application before host emergence and application before parasite emergence.

☞ Adjust the forward speed (3.5 km/h for hand sprayers, 6-8 km/h for tractor mounted sprayers) to the amount of liquid required per ha or m^2.

☞ Calibration is especially important when using knapsack sprayers: Determine the time taken to spray and the liquid applied to one hundredth of a hectare; then calculate the required liquid and time allocated for one plot or field.

☞ Crop susceptibility and label restrictions have to be considered carefully.

☞ Do not spray during windy weather or when rainfall is expected to occur.

☞ The method of seed-dressed crops has recently been identified as an efficient application technique: The herbicide may be either (1) applied directly to the seed in the individual planting hole as an aqueous solution or (2) the crop seed gets coated with a dressing substance to which the desired herbicide is added. Advantages beside good control effects are that considerably less herbicide is required, and that no spray equipment is needed.

1. *Orobanche*

- Spraying glyphosate for *O. crenata* control in faba bean requires proper timing: Usually an amount of 60 to 80 g a.i. has to be sprayed twice, i.e. at tubercle and at bud stage of *Orobanche* development. Only repeated digging out of crop plants and examination of the underground development during the vegetative growth of the crop will indicate the right time to spray.
- This holds as well for the application of imidazolinone compounds, but here lower rates of 20 to 40 g a.i./ha are already sufficient. The rate has to be adjusted to the respective crop and whether the application is split. This group of herbicides may become more important in *Orobanche* control than glyphosate. They are also promising for crop seed dressing.
- For the pre-emergence application of imidazolinones the timing is easier as no underground stages have to be considered. Only one application is required, but under severe infestation a post-emergence application may be recommended.
- Recently, the application of rimsulfuron was reported to selectively control *Orobanche* in tomatoes. This herbicide affects the acetolactate synthase (ALS), which is the same target as with the imidazolinones.
- Consider phytotoxicity which varies considerably from crop to crop.

2. *Striga*

- Pre-plant incorporation: Pendimethalin, benfluralin, trifluralin, fluchloralin, each at 0.5 to 0.75 kg/ha; pendimethalin gets washed into the ground by rain.
- Pre-emergence application: Dicamba, oxyflurofen, fomesafen, DCPA, dichlobenil.
- Post-emergence treatments: 2,4-D, dicamba, ametryn, oxyflurofen, fomesafen and paraquat.
- Contact herbicides such as paraquat also affect the crop; hence, direct application to *Striga* is essential except for the control of weed hosts.
- Control of weed hosts (host avoidance): Paraquat, glyphosate.
- In crop mixtures care has to be taken to choose herbicides that can be tolerated by a range of crops normally grown in such mixtures (e.g., ametryne, bromoxynil; but not 2,4-D).
- Seed dressing with imazaquin and chlorsulfuron was recently found to be efficient for the control of *S. hermonthica* and *S. gesnerioides*.
- For the use of herbicide resistant crop plants see → chapter 6.3.6

3. *Cuscuta*

- Soil applied herbicides before/at seedling stage of *Cuscuta*: Pendimethalin (1 to 4.5 kg/ha), chlorpropham (7 kg/ha), pronamide (2.5 kg/ha) or trifluralin (3 to 9 kg/ha).

- Contact herbicides for spot treatments to kill isolated patches of parasite and host: Paraquat, dinoseb and 2,4-D.
- Selective control in alfalfa: Glyphosate; this herbicide is translocated from host to parasite. The dosage required for sufficient control ranges between 0.075 and 0.150 kg a.i./ha. Other herbicides like imidazolinones or rimsulfuron for the selective control of *Cuscuta* are being tested.

4. *Alectra*

- Recently, the treatment of crop seed with imazaquin was reported to control *Alectra* in cowpea.

EVALUATION

* Design: randomized block design with 3 to 4 replications. In combination with other control methods split-plot designs may be appropriate.
* Count the number and assess the dry weight of emerged parasites (with *Cuscuta* only dry weight).
* Assess crop yield parameters like biomass and seed yield (→ chapter 5.3.5).
* Special assessment is required for symptoms of phytotoxicity on both parasite and crop. The symptoms vary especially according to the herbicide sprayed, the target plant, their developmental stage and the prevailing weather conditions. For the evaluations of herbicides special visual ratings may be helpful (e.g., EWRS-scale, → Table 6) In legume crops treated against *Orobanche* typical symptoms of phytotoxicity are, for example, various degrees of chlorosis or narrowing leaves.
* Sometimes it might be necessary to excavate the root system and to evaluate the herbicide effects on the underground stages of the parasite (e.g., with pre-plant and pre-emergence herbicides like pendimethalin and imidazolinones or with systemic herbicides like glyphosate).
* For other aspects of data collection and evaluation see → chapter 5.3.5

Table 6. EWRS-scale for visual ratings of phytotoxicity

class	Symptoms of damage
1	No damage/healthy plants
2	Very slight symptoms, weak suppression
3	Slight, but clearly visible symptoms
4	Severe symptoms (e.g., chlorosis), which do not lead to as negative effect on yield
5	Thinning, severe chlorosis or suppression; yield reduction expected
6–9	Severe damage up to complete destruction

COMMENTS

- Avoid conditions which promote drift. Consideration of drift is especially important where small plots are to be treated with a herbicide, and are surrounded by crops susceptible to the formulation selected.
- Sometimes, soil applied herbicides may not sufficiently control root parasites as their seedlings attack the crop root also at soil depths of 10 to 15 cm, which is hardly reached by the herbicide.
- Consider the waiting time after treatments with imidazolinones.
- Numerous hints for the chemical control of parasitic weeds are provided by Parker & Riches (1993).

LITERATURE

Abayo GO, JK Ransom, J Gressel & GD Odhiambo 1996. *Striga hermonthica* control with acetolactate synthase inhibiting herbicides applied to maize seed with target-site resistance. In: MT Moreno, JI Cubero, D Berner, D Joel, LJ Musselman & C Parker (eds.), Advances in Parasitic Plant Research; 6th International Parasitic Weed Symposium, Cordoba, Spain, 761-768.

Anon. Guidelines for the safe and effective use of pesticides. G.I.F.A.P. Brussels, Belgium.

Berner DK, AE Awad & EI Aigbokhan 1994. Potential of imazaquin seed treatment for controls of *Striga gesnerioides* and *Alectra vogelii* in cowpea (*Vigna unguiculata*). Plant Disease 78: 18-23.

Coutts HH 1984. The application of pesticides. - Assessment techniques for spray pattern analysis. GTZ, Eschborn, Germany.

Dawson JH, FM Ashton, WV Welker, JR Frank & GA Buchanan 1984. Dodder and its control. USDA, Farmer Bulletin No. 2276.

Eplee RE & RS Norris 1987. Chemical control of *Striga*. In: LJ Musselman (ed.), Parasitic Weeds in Agriculture, Vol. I, *Striga*. CRC Press, Boca Raton, Florida, USA, 173-182.

Garcia-Torres L, M Jurado-Esposito, M Castejon-Munoz & F Lopez-Granados 1996. Herbicide treated crop seeds for control of *Orobanche* spp. In: MT Moreno, JI Cubero, D Berner, D Joel, LJ Musselman & C Parker (eds.), Advances in Parasitic Plant Research; 6th International Parasitic Weed Symposium, Cordoba, Spain, 700-705.

Kleifeld Y, Y Goldwasser, G Herzlinger, S Golan, T Blumenfeld & H Buxbaum 1995. Selective control of *Orobanche* in tomatoes with rimsulfruron (DPXE-9636). In: AH Pieterse, JAC Verkleij & SJ ter Borg (eds.), Biology and Management of *Orobanche*. Royal Tropical Institute, Amsterdam, The Netherlands, 561-571.

Liu Z, A Fer & FM Lecoq 1991. L'imazaquine: un herbicide prometteur pour la lutte curative contre la cuscute (*Cuscuta* spp.) dans les cultures de soja (*Glycine max*). Weed Research 31: 33-40.

Nir E, B Rubin & SW Zharasov 1996. On the biology and selective control of field dodder (*Cuscuta campestris*). In: MT Moreno, JI Cubero, D Berner, D Joel, LJ Musselman & C Parker (eds.), Advances in Parasitic Plant Research; 6th International Parasitic Weed Symposium, Cordoba, Spain, 809-816.

Linke KH 1990. *Orobanche* studies. In: Food legume improvement program - Annual report for 1989. ICARDA, Aleppo, Syria, 243-266.

Linke KH 1989. *Orobanche* studies. In:. Food legume improvement program - Annual report for 1988. ICARDA, Aleppo, Syria, 214-242.

Parker C & CR Riches 1993. Parasitic Weeds of the World - Biology and Control. CAB International, Wallingford, UK, 332 pp.

Rohrmoser, K 1985. Handbook for field trials in technical cooperation. GTZ, TZ-Verlagsgesellschaft, Roßdorf, Germany.

Sauerborn J, H Masri, MC Saxena & KH Linke 1987. Control of *Orobanche* with Scepter. FABIS Newsletter 19: 14-17.

Sauerborn J, MC Saxena & A Mayer 1989. Broomrape control in faba bean (*Vicia faba* L.) with glyphosate and imazaquin. Weed Research 29: 97-102.

Author: K-H Linke

6.4.3 Germination stimulants

6.4.3.1 Ethylene gas

INTRODUCTION

Ethylene gas (C_2H_4) has been successfully applied under field conditions in the USA to induce germination of *Striga asiatica* seed in the soil. Only about 2 kg of ethylene gas/ha injected into the soil caused up to 90% of the *Striga* seed to germinate (Table 7). With the absence of a suitable host root in the immediate vicinity (<4 mm) the germinated seedlings will die. This procedure was named "suicidal germination". Besides *S. asiatica* also *S. hermonthica* is triggered to germinate after exposure to ethylene gas. Germination rate however, seems to be much less. A low effectiveness has been recorded also for *Orobanche* spp.

Table 7. Biological activity of ethylene in the soil, demonstrated by the germination induction of *Striga asiatica* seed.

	Distance from the point of injection (cm)						
Seeds at various soil depth (cm)	30	45	60	75	90	105	120
	Striga seeds germinated (%)						
30	81	93	89	91	95	95	97
45	88	92	91	68	92	98	92
60	83	96	89	90	93	81	92
75	75	94	85	88	90	97	93
90	76	88	87	90	94	97	91

Ethylene rate = 1.7 kg ha^{-1} (Source: Eplee 1975)

OBJECTIVES

- To provide parasite free areas for crop production and for experimental purpose (→ chapter 5.3.3),
- to evaluate the effectiveness of ethylene gas for germination of other parasitic weeds.

MATERIALS

✓ Hand-operated ethylene injector consisting of gas cylinder mounted on a backpack frame, regulator and injection probe, ethylene gas.

✓ In the USA also a hand-carried unit, a larger, hand-operated unit mounted on wheels and a tractor-mounted ethylene injector for the treatment of a large acreage were developed. These systems are not considered here.

PROCEDURE

☞ *Striga* seeds have to be conditioned prior to the ethylene application; therefore, the soil has to be wet for several weeks before the treatment is carried out. Ethylene gas is usually applied 3 to 6 weeks after the main rains have begun to ensure the *Striga* seeds are conditioned.

☞ No host plants should grow in the field to assure the absence of any host roots.

☞ Ethylene application by a backpack system: the entire unit, except the injector, is carried on the back by the operator; with this equipment, a further burst of gas every m^2 is sufficient.

☞ A rate of 1.5 to 3 kg/ha has been determined to be suited for most soil types.

EVALUATION

* Estimation of parasite seed (number/kg soil) in soil samples taken before and 10 days after the ethylene application (→ chapter 4.5).
* In addition seed bags may be buried in the soil and evaluated in the same way (→ chapter 4.7).
* Comparison of infestation on subsequently sown host plants in treated and untreated plots.

COMMENTS

- The ethylene gas in the gas cylinder is under high pressure. For safety reasons, special application equipment which sustains the pressure and which is safe for the application is required for its use in the field. The gas is explosive. Different regulations may have to be observed in various countries for transport and application.
- Light and sandy soils are most convenient for the treatment.
- The application of ethephon (2-chloroethylphosphonic acid) as an ethylene releasing chemical has been tested without promising results.

LITERATURE

Babiker AGT & AM Hamdoun 1983. Factors affecting the activity of ethephon in stimulating seed germination of *Striga hermonthica* (Del.) Benth. Weed Research 23: 125-131.

Bebawi FF & RE Eplee 1986. Efficacy of ethylene gas as a germination stimulant for *Striga hermonthica* seed. Weed Science 34: 694-698.

Egley GH & JE Dale 1970. Ethylene, 2-chloroethylphosphonic acid and witchweed germination. Weed Science 18: 586.

Egley GH, Eplee RE & Norris RS 1990. Discovery and development of ethylene as a witchweed seed germination stimulant. In: PF Sand, RE Eplee & RG Westbrooks (eds.), Witchweed research and control in the United States. Weed Science Society of America, Champaign, IL, USA, 56-66.

Eplee RE & RS Norris 1987. Chemical control of *Striga*. In: LJ Musselman (ed.), Parasitic Weeds in Agriculture. Vol. I, *Striga*. CRC Press, Boca Raton, Florida, USA, 173-182.

Eplee RE 1975. Ethylene: a witchweed seed germination stimulant. Weed Science 23: 433.

Author: K-H Linke

6.4.3.2 Strigol and analogues

INTRODUCTION

Seeds of *Striga*, *Orobanche* and *Alectra* germinate only under the influence of root exudates of certain plants. Stimulation of germination in the absence of a host will kill the seeds and is called "suicidal germination". Such a substance was isolated in 1966 from cotton root exudate and named "strigol" as it was very effective in stimulating germination of *Striga*. Also the analogues GR7 and GR24 were found to trigger germination, even on *Orobanche*. However, the use in the field proved to be not satisfying so far due to the instability of the compounds in the soil and due to their production costs. Stimulants chemically similar to strigol were isolated from sorghum and cowpea root exudate and named "sorgolactone" and "alectrol", respectively. So far, germination stimulants produced by hosts of *Orobanche* have not been identified.

OBJECTIVES
- To improve the efficacy/stability of strigol and analogues in field applications,
- to find optimum conditions for their application,
- to find appropriate formulations of the stimulant in order to overcome problems of instability.

MATERIALS

✓ Usual equipment for field trials.

✓ Strigol or analogues, knapsack sprayer, pipette.

✓ Equipment for soil analysis.

PROCEDURE

☞ Test soil for pH-value. High pH-values require higher dosage of the stimulant. Assess soil water content.

☞ The seeds of the parasite have to be conditioned prior to the application; therefore, the soil has to be wet for several weeks before the treatment is carried out.

☞ Application by spraying (knapsack sprayer) at rates according to the material under investigation (e.g., Saghir 1986 found 0.3 to 1.5 kg/ha to be sufficient in *Orobanche crenata*).

EVALUATION

★ Estimation of parasite seed (number/kg soil) in soil samples taken before and 20 days after the application (→ chapter 4.5).

★ In addition seed bags may be buried in the soil and evaluated in the same way (→ chapter 4.7).

★ Comparison of infestation on subsequently sown host plants in treated and untreated plots.

COMMENTS

- The application technology required for these allelochemicals is by far simpler (in terms of cost, safety, technical needs) than that for ethylene gas.
- Soil water content and pH-value seem to be a crucial point concerning the efficacy of these allelochemicals.
- The development and production of further analogues that are cheaper to fabricate and more stable under field conditions would be desired.

LITERATURE

Babiker AGT, AM Hamdoun, A Rudwan, NG Mansi & HH Faki 1987. Influence of soil moisture on activity and persistence of the strigol analogue GR24. Weed Research 27: 173-178.

Babiker AGT, NE Ibrahim & WG Edward 1988. Persistence of GR7 and *Striga* germination stimulant(s) from *Euphorbia aegyptiaca* Boiss. in soils and in solutions. Weed Research 28: 1-6.

Mangnus E & B Zwanenburg 1991. Design and synthesis of germination stimulants for seeds of *Striga* and *Orobanche* spp. In: K Wegmann & LJ Musselman (eds.), Progress in *Orobanche* research. Eberhard-Karls-Universität, Tübingen, Germany, 157-166.

Müller S, C Hauck & H Schildknecht 1992. Germination stimulants produced by *Vigna unguiculata* Walp cv. Saunders Upright. Journal of Plant Growth Regulation 11: 77-84.

Saghir AB 1986. Dormancy and germination of *Orobanche* seeds in relation to control methods. In: SJ ter Borg (ed.), Proceedings of a workshop on biology and control of *Orobanche*. LH/VPO, Wageningen, The Netherlands, 25-34.

Thuring JWJF 1996. Synthesis and bioactivity of germination stimulants for the seeds of parasitic weeds - a molecular study. PhD-Thesis, University of Nijmegen, The Netherlands.

Author: K-H Linke

6.5 Physical

6.5.1 Cleaning of crop seed

INTRODUCTION

Seed of *Striga*, *Orobanche*, *Cuscuta* and *Alectra* may be disseminated among others by contaminated crop seed.

OBJECTIVES

- To assess a potential infestation of crop seed with seed of parasitic weeds,
- to avoid seed of parasitic plants invading non-infested fields, regions or countries.

PROCEDURES

☞ List all possible crop seed of a given company that are potential host crops of parasitic weeds.

☞ Identify the source of the crop seed. If it originates from infested areas there is an increased probability of carrying parasitic weed seed.

☞ Most of the dust-like parasite seed of a contaminated batch of crop seed in a seed bag or box will aggregate on the bottom of the bag or the box; these places deserve special attention.

☞ Special sieving equipment has been developed by the seed industry for cleaning crop seed from any other undesired accompanying material. A thorough cleaning of seed after the usual cleaning during harvest (e.g., by combine, by hand) may include for example the following units:

- pre-aspiration
- bottom screen
- tail aspiration
- gravity separator
- top screen.

☞ The cleaning effect of such units depend on various factors such as (1) the type of crop seed (surface, size), (2) the contamination of the harvested material, (3) the flow rate of seed through the system, (4) the intensity of aspiration, and others (Table 8). In this regard, operating instructions or experienced personnel can be of further valuable help.

☞ For highly contaminated crop batches several runs through the cleaning units might be required.

☞ For the further evaluation of the cleaning effect it is necessary to take samples at various steps of the cleaning process, label the samples and subject 100 g each to an extraction in the lab (→ chapter 4.6).

☞ The cleaning is repeated until no more seed are recovered in the lab test. This final lab test includes 5 samples of each 100 g.

EVALUATION

✶ Laboratory test of crop seed for contamination (→ chapter 4.6).

✶ Expression as parasite seed/kg crop seed.

COMMENTS

- In order to know after a prophylactic cleaning whether a crop batch was infested at all, the waste of the various sieving/separation units might be analyzed. Usually, only at one unit most of the parasite seed are detected. With *Orobanche crenata* seed in a combine harvested batch of lentil seed, most of the parasite seed (91%) were detected in the waste of the pre-aspiration unit.

- With the consideration of origin and type of seed only some seed batches might require cleaning at all.
- A thorough cleaning of crop seed (mainly the effect of wind) is able to reduce the contamination of crop seed with seed of parasitic weeds almost to zero even in a highly infested batch.

Table 8. Example of data on crop seed (lentil) heavily contaminated with parasite seed at various stages of cleaning

Description	Kg of crop seed in respective stage	Number of *Orobanche* seed/kg crop seed	Number of *Orobanche* seed in respective stage
Seed as harvested by combine	903.5	136.610	123.358.380
Waste from pre-aspiration	48.0	235.296	112.942.080
Waste from bottom screen	60.0	54.000	3.240.000
Waste from tail aspiration	9.0	260.400	2.343.000
Waste from gravity separator*	8.0	50	400
Waste from top screen	1.5	44.960	67.440
Waste from gravity separator**	27.0	5.250	141.750
Cleaned crop seed	740.0	6	4.440

* heavy material; ** light material

Source: Linke (1999)

LITERATURE

Berner DK, KF Cardwell, BO Faturoti, FO Ikie & OA Williams 1994. Relative roles of wind, crop seeds, and cattle in the dispersal of *Striga* species. Plant Disease 78: 402-406.

Linke KH 1999. Status quo of *Orobanche* management: Preventive, cultural, and physical control. In: J Kroschel, M Abderabihi, H Betz (eds.), Advances in Parasitic Weed Control at On-farm Level. Vol. II. Joint Action to Control *Orobanche* in the WANA Region. Margraf Verlag, Weikersheim, Germany, 107-133.

Author: K-H Linke

6.5.2 Hand weeding

INTRODUCTION

Hand weeding and hoe weeding are the most common practices for the control of parasites (especially *Striga*) used so far by small-scale farmers. Hand weeding of parasitic plants represents a labour-intensive method. Although too late to prevent damage to the crop it will stop a further increase of the parasite's seed bank. Too early hand weeding brings about the parasite to re-establish from underground stages. Too late hand weeding bears the potential of already mature seed on the parasite to be dispersed.

OBJECTIVES

- To find out the optimum period for hand weeding,
- to estimate the cost (time, labour requirement) of hand weeding at various densities of parasites in the field,

- to estimate the effect of hand weeding on crop yield.
- to estimate the effect of hand weeding on the dynamics of a parasite population.

MATERIALS

✓ Bags for the collected parasite plants.

✓ High labour requirement with increased infestation.

PROCEDURE

☞ Manual removal of emerged, flowering and mature plants of the respective parasite.

☞ Collection of plants at the field border.

☞ Burning of collected plants to ensure all potential seed is destroyed (after harvest dry weather conditions with low relative humidity is the rule; only in exceptional conditions may material for burning be necessary).

EVALUATION

∗ Assessment of the time required for weeding of a given area (e.g., h/ha) and under low and high infestation.

∗ Comparison of hand weeding carried out at different stages of the parasite development.

∗ Comparison of hand weeding carried out in different host crops of a parasite.

∗ Assessment of repeated hand weeding in the same crop.

∗ Comparison of treated and untreated areas on the basis of soil samples analyzed for parasite seeds.

∗ Comparison of treated and untreated areas on the basis of emerged parasites in a following susceptible crop.

∗ For detailed studies also the estimation of the stage of maturity of the parasite might be required. In this case the plants have to be analyzed for mature seed immediately after weeding.

COMMENTS

- Hand weeding in high infested areas will never be an economic method (exception: for experimental purpose). In very low infested areas hand weeding may be the most economic means of control.
- *Striga* and *Cuscuta* require more labour input for hand weeding than *Orobanche* or *Alectra*. Labour requirement is also greatly influenced by the crop itself, i.e. tall and bushy crops impede the identification of parasite plants. Detection for example of *Cuscuta* plants in onion is much easier, thus faster, than in alfalfa; *Orobanche* in faba beans are more difficult to find than in lentils or chickpea.

LITERATURE

Carson AG 1986. Research and development strategies for the control of *Striga hermonthica* in the Gambia. In: Proceedings of the FAO/OAU All-African Government Consultation on *Striga* control, 20-24 October 1986, Maroua, Cameroon, 100-117.

Geipert S 1997. Potentiale und Grenzen der Bekämpfung von *Orobanche crenata* Forsk. im Ackerbohnenanbau (*Vicia faba* L.) Marokkos. PLITS 15(5).

Linke KH 1990. *Orobanche* studies. In: Food legume improvement program - annual report for 1989. ICARDA, Aleppo, Syria, 243-266.

Ogborn JEA 1984. Research priorities in agronomy. In: ES Ayensu, H Doggett, RD Keynes, J. Marton-Lefevre, LJ Musselman, C Parker & A Pickering (eds.), *Striga*, Biology and Control. IRL Press Ltd Oxford, UK, 195-212.

Ramaiah KV 1985. Hand-pulling of *Striga hermonthica* on pearl millet. Tropical Pest Management 31(4): 326-327.

Ransom JK & GD Odhiambo 1994. Long-term effects of fertility and hand-weeding on *Striga* in maize. In: AH Pieterse, JAC Verkleij & SJ ter Borg (eds.), Biology and Management of *Orobanche*. Proceedings of the 3rd International Workshop on *Orobanche* and related *Striga* research. Amsterdam, The Netherlands, Royal Tropical Institute, 513-519.

Vyas BP 1966. Control of *Orobanche* (on tobacco). Indian Farming 16(2): 51.

Author: K-H Linke

6.5.3 Burning

INTRODUCTION

Two types of burning have to be considered: (1) Complete burning of an entire field or large infested area and (2) flaming of infested spots.

OBJECTIVE

- To assess the effect of burning/temperature on the seed of parasitic weeds,
- to assess the effect of burning on the entire seed bank of parasitic weeds,
- to destroy plants and seeds of parasitic weeds.

MATERIAL

✓ Dry residues of crop and parasite, propane or butane fuelled equipment for flaming.

✓ Soil tillage equipment (harrow, tractor).

PROCEDURE

☞ Determine the area to be burnt according to infestation level, availability of dry residues, distance to other fields.

☞ Use a convenient soil tillage technique to delimit the border of the area to be burnt and to avoid the fire to expand beyond that area (not required for spot treatments).

☞ Burn the infested, delimited area.

☞ For spot treatments (especially for the control of *Cuscuta*, e.g., in alfalfa) the flaming has to be carried out long enough to ensure the destruction of mature seed already dropped to the ground. Another technique is to kill the topgrowth with a light flaming, and then several days later when the vegetation has dried, entirely burn the patches.

EVALUATION

∗ Assess the amount/number of new produced parasite seed/m^2 (→ chapter 4.1) prior to the burning and/or

∗ Assess the seed bank before and after burning (assess the amount of seed on the soil surface only, i.e. the top 2 cm; this can be achieved by taking 20 subsamples/m^2 at different places); eventually assessment of vertical distribution and respective viability of seed (→ chapter 4.5).

∗ Check viability of seed, as the heat treatment is expected to damage seed (→ chapter 4.3).

COMMENTS

- For security reasons it has to be assured that no other dry crop is present in the neighbourhood of the area to be burnt. The direct availability of a fire brigade is recommended as well as the information of adjacent landowners.
- The wind direction has to be considered, also it's potentially rapid change.

- Only economic where no crop yield is to be expected due to an extremely high infestation.
- **Note:** For ecological reasons burning of larger field has to be considered very critically.

LITERATURE

Dawson JH, FM Ashton, WV Welker, JR Frank & GA Buchanan 1984. Dodder and its control. USDA, Farmer Bulletin No. 2276.

Linke KH, H Schnell & MC Saxena 1991. Factors affecting the seed bank of *Orobanche crenata* in fields under lentil based cropping systems in Northern Syria. In: JK Ransom, LJ Musselman, AD Worsham & C Parker (eds.), Proceedings of the 5th International Symposium of Parasitic Weeds. CIMMYT, Nairobi, Kenya, 321-327.

Sprich H 1994. Bedeutung der Fruchtfolge zur Ertragssicherung in getreidebetonten Produktionssystemen der Guinea-Savanne unter besonderer Berücksichtigung des parasitischen Unkrauts *Striga hermonthica* (Del.) Benth. PLITS 12(3).

Author: K-H Linke

6.5.4 Deep ploughing

INTRODUCTION

Tillage operations influence the distribution of parasite seed in the soil. In undisturbed soil the largest amount of seed occurs in the very top layer of soil. Conventional tillage was found to mix the seed with the soil and hence increases the infestation compared to minimum or zero tillage. Deep ploughing buries the seed in a deep soil layer from where the parasite is unable to infect or emerge.

OBJECTIVES

- To evaluate the effect of deep ploughing (other tillage systems) on the incidence of parasitic weeds on host crops,
- to assess an efficient and economic ploughing depth.

MATERIALS

✓ Tractor powered with at least 100 horse power; heavy plough with a type of share(s) that ensure optimum turning effect of the soil.

✓ Whether one, two or even three shares are used depends on tractor and soil conditions.

PROCEDURE

☞ Assess the infestation level of that particular area in order to avoid the deep ploughing on non-infested areas.

☞ The ploughing is carried out at the regionally usual time. However, wait for optimum soil conditions.

☞ Then plough the area; check the depth of ploughing to ensure the required working depth (e.g., 45 cm). The evaluation of the vertical distribution of parasite seed in the soil is the basis for the decision on how deep to plough.

☞ After the ploughing any disturbance of the soil has to be minimized. In order not to bring up parasite seed from deeper soil layers any minimum tillage system is recommended for the following years.

EVALUATION

∗ Take soil samples before and after the ploughing and estimate the number of seed/m^2.

* The soil samples have to be taken from the top 10 cm of the soil at both samplings.
* Assess the vertical distribution of the parasite seed in the soil up to 100 cm depth before and after the ploughing (→ chapter 4.5).
* Assessment of emerged parasites after the treatment (unploughed checks are required) and yield parameters of the crop (→ chapter 3.1 and 5.3.5).
* Assess the soil depth from which the parasites emerge.

COMMENTS

- For preliminary studies and for orientation on the efficacy of the technique the soil turning effect of a plough may be simulated by digging the soil by hand: From a plot of 1 x 2 m the top soil layer (0 to 25 cm) is dug out and placed at one side of the plot, the lower soil layer (25 to 45 cm) is placed on the other side; then the hole is filled again with the lower layer on top and subsequently planted to a host crop for evaluation of the effects.

LITERATURE

Aleksiev A 1967. Trench ploughing as a control measure against broomrape on tobacco. B'lgarski Tyutyun (Bulgaria), 12(3): 13-16.

Konate A 1986. *Striga* in Mali. In: Proceedings of the FAO/OAU All-African Government Consultation on *Striga* control, 20-24 October 1986, Maroua, Cameroon, 58-61.

Kukula ST & H Masri 1984. Integrated control practices and chemical control of *Orobanche crenata* in faba bean. In: C Parker, LJ Musselman, RM Polhill & AK Wilson (eds.), Proceedings of the 3rd International Symposium on Parasitic Weeds, ICARDA, Aleppo, Syria, 1984. 256-261.

Mathews LJ 1989. Improved package for *Striga* control for the small-scale farmer. In: TO Robinson & HR Broad (eds.), *Striga*-Improved Management in Africa. Proceedings FAO/OAU All-Africa Government Consultation on *Striga* control, Maroua, 1986. FAO Plant Production and Protection Paper 96, 183-186.

Author: K-H Linke

6.5.5 Solarization

INTRODUCTION

Solarization is a method of disinfecting wet soil by heating using transparent polyethylene sheet and natural radiation (Fig. 19).

OBJECTIVES

- To destroy parasite seed in the soil (field),
- to avoid the use of chemicals,
- to provide parasite-free plots for experimental purposes (check).

MATERIALS

✓ Transparent polyethylene (PE) sheet (0.15 to 0.25 mm thickness), sun light irradiation, irrigation water (about 50 l/m^2), capacity for field work, tractor for harrowing the area to be covered, thermometer or equipment for permanent recording of temperature in the soil.

✓ Time requirement depends on the size of the area to be covered and the techniques applied for field preparation and covering the ground with the plastic. To cover an experimental area of 2000 m^2 may need between 20 and 100 hours of fieldwork until the plastic is fixed on the soil.

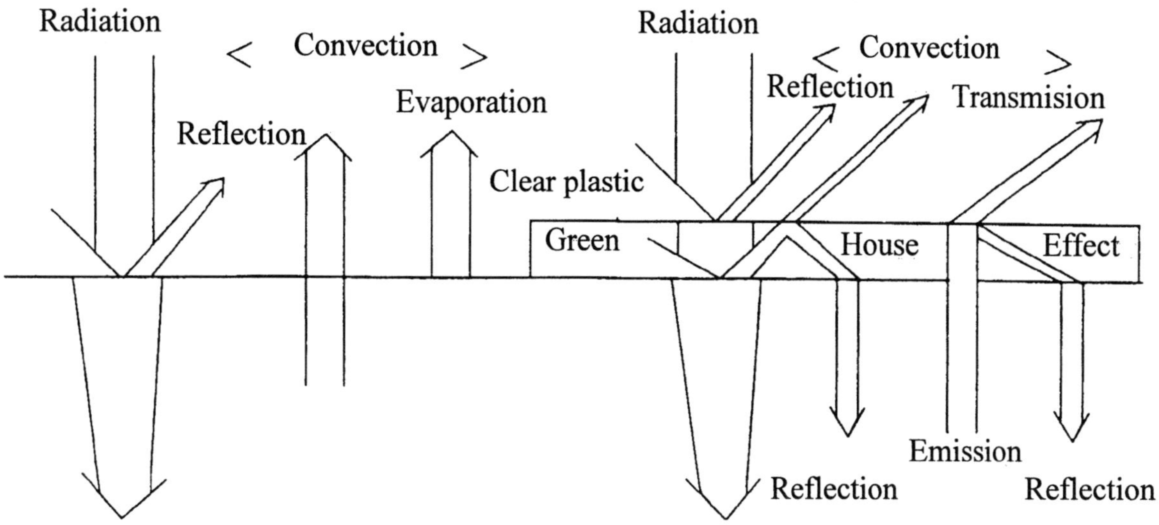

Fig. 19. Concept of solarization (Braun *et al.* 1987)

PROCEDURE

- ☞ The soil has to be moist (near field capacity) so the heat spreads uniformly and into deeper soil layers. The water will ensure higher susceptibility of seeds to temperature. Therefore, the field has to be irrigated (sprinkler irrigation) or the PE sheet is spread after sufficient rainfall.
- ☞ Wait about one day after irrigation to avoid excess moisture on the surface of the ground.
- ☞ Then the soil gets covered with the PE-foil. This has to be done very carefully to avoid any damage of the plastic (tears have to be sealed using adhesive tape). The borders of the plastic sheets have to be covered with soil to fix the PE-sheet (→Fig. 20).
- ☞ The PE sheet has to lay as close as possible to the soil surface to avoid air buffers which impede the penetration of radiation into the ground. The field should therefore be harrowed prior to the irrigation to obtain a fine textured surface.
- ☞ The timing and duration of solarization are critical to its disinfecting effect. It is best done during the hottest and driest period of the year for not less than 20 days.
- ☞ Care has to be taken that the PE-sheet withstands the wind, the radiation and that it does not disintegrate before time. It is recommended to check the plastic cover regularly.

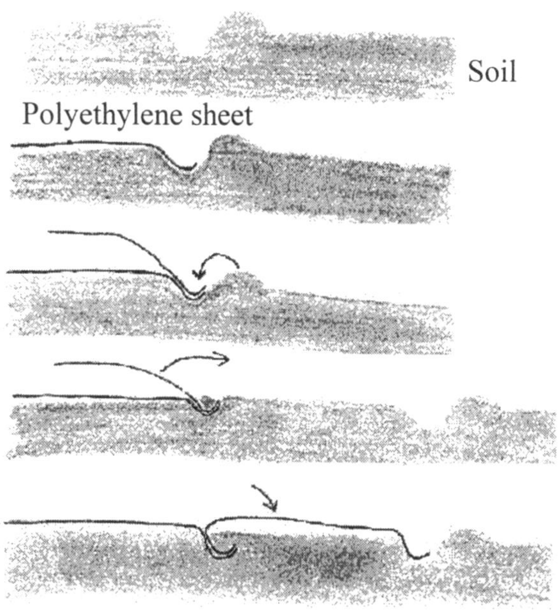

Fig. 20. How to cover the soil with the PE-foil

EVALUATION

- ⋆ Take soil samples before and after covering the soil and estimate the number of parasite seed/m^2 or kg soil.
- ⋆ Thermometers are placed in the field at different soil depths to monitor and record the temperature. The temperature down to about 15 cm soil depth should be monitored. Graphically plotted data of the course of the temperatures during a day provides good comparison of the temperature regimes in covered and uncovered soil.
- ⋆ Record air temperature at 5 cm above ground.
- ⋆ Efficiency of the treatment is assessed by counting the underground and/or emerged stages of the parasites during the growth period of the crop planted subsequently (→ chapter 5.3.5).
- ⋆ For detailed observations seed bags are placed in the soil at various depths and evaluated after the solarization (including checks) for seed viability (→ chapters 4.3 and 4.7).
- ⋆ Assessment of crop development like height, weight and yield (→ chapter 5.3.5).
- ⋆ In addition, occurrence and biomass of weeds other than parasites and other pests in the treated and untreated area might be recorded.
- ⋆ Calculate the LD_{100} and LD_{50} for the parasite species (LD = lethal dose of temperature, LD_{100} (LD_{50}) = duration of exposure to a permanent temperature after which 100% (50%) of the seed have died).

COMMENTS

- This method produces side effects such as destruction of most weed seed and nematodes. *Rhizobium* bacteria get affected too but re-establish fast. Availability of nitrogen and phosphorous in the soil after solarization is increased (Linke *et al.* 1991; Sauerborn *et al.* 1989).
- Avoid any deep tillage of the soil after solarization to prevent undamaged seed from lower soil layers being brought to the surface of the soil and infesting the crop. No or low tillage after solarization provides the maintenance of a long term effect.
- The plastic sheets can be re-used up to two times, but only once if the material is thin and torn.
- An important fact of solarization is the cost of the method which ranges from about 400-1200 US$ per ha (depending on quality of plastic and labour cost). Hence, for most field crops solarization will not be an economic technique, whereas for some valuable crops and for field experiments it represents an efficient, profitable and non-toxic tool.
- Occasionally, black PE is recommended for soil solarization. It provides better control effect against already emerged weeds due to its shading effect, but the green-house effect is much less pronounced compared with clear PE.
- On large plots or fields it is helpful to use tractor mounted equipment for preparing the furrows and fixing the plastic sheets on the ground.
- The development of sprayable foils for solarization could reduce some limitations of plastic foils. However, the evaluation of several materials was not satisfying so far due to their failure to produce a stable coat on the soil surface.

LITERATURE

Braun M, W Koch & M Stiefvater 1987. Solarisation for soil sanitation. - Possibilities and limitations. Gesunde Pflanzen 7: 301-309.

Jacobsohn R, A Greenberger, J Katan, M Levi & H Alon 1980. Control of Egyptian broomrape (*Orobanche aegyptiaca*) and other weeds by means of solar heating of the soil by polyethylene mulching. Weed Science 28: 312-316

Linke KH, MC Saxena, J Sauerborn & H Masri 1991. Effect of soil solarization on the yield of food legumes and on pest control. In: J DeVay, JJ Stapleton & CE Elmore (eds.), Proceedings of the 1st International Conference on Soil Solarization, Amman, Jordan. FAO Plant Production and Protection Paper 109: 139-154.

Sauerborn J 1991. Parasitic flowering plants - Ecology and Management. Margraf Verlag, Weikersheim, Germany, 127 pp.

Sauerborn J, KH Linke, MC Saxena & W Koch 1989. Solarization; a physical control method for weeds and parasitic plants (*Orobanche* spp.) in Mediterranean agriculture. Weed Research, Volume 29: 391-397.

Linke KH 1990. *Orobanche* studies. In: Food legume improvement program - Annual report for 1989. ICARDA, Aleppo, Syria, 243-266.

Author: K-H Linke

6.6 Biological

6.6.1 Phytophages

INTRODUCTION

The use of natural antagonists, especially of insect herbivores, is an additional tool for the control of parasitic weeds in an integrated control programme. Two approaches of biological control of weeds can be distinguished: (1) The classical or inoculative approach which consists in the introduction and release of potential exotic antagonists for durable naturalisation; (2) The **inundative approach**, where endemic as well as exotic but non-persistent antagonists are mass-reared and released periodically in high numbers. A wide diversity of phytophagous insects have been reported for *Striga* and *Orobanche*. Most of them are polyphagous without any host specificity and therefore not for use in biocontrol. Only some monophagous and stenophagous species are of interest: *Eulocastra agentisparsa* Hampson and *E. undulata* (Lepidoptera, Noctuidae), *Ophiomya strigalis* Spencer (Diptera, Agromyzidae), *Smicronyx umbrinus* Hustache and *S. albovariegatus* Faust (Coleoptera: Curculionidae) for *Striga* spp. and *Phytomyza orobanchia* Kalt. (Diptera: Agromyzidae) and *Smicronyx cyaneus* Gyll. (Coleoptera: Curculionidae) for *Orobanche* spp.

Efficient mass-rearing methods have not been developed yet for use of insects in an inundative approach for *Striga*. Since the occurrence of potential herbivores of *Striga* spp. is limited to certain regions (e.g., within Africa) or continents (India or Africa) releases into new areas seem to be the most promising method. The use of inundative releases of *P. orobanchia* has been successfully applied to control *O. cumana* and *O. cernua* in the former Soviet Union on a large scale. First results indicate that this method could be applied in other regions where *P. orobanchia* has been co-evolved with *Orobanche* spp. In countries, where *Orobanche* spp. has been unintenionally introduced (e.g., in Chile), the classical approach could be considered.

The following chapters are concerned with methods applied to the study of *Smicronyx* spp. and *Junonia orythia* L. (Lepidoptera: Nymphalidae) for *S. hermonthica* in West Africa, to *P. orobanchia* and *S. cyaneus* and their parasitoids for *O. crenata* in Morocco as well as to *Smicronyx* spp. as antagonists of *Cuscuta* spp.

LITERATURE

Anderson DM 1974. Some species of *Smicronyx* (Coleoptera: Curculionidae) associated with *Cuscuta* species (Convolvulaceae) in Pakistan. In: Proceedings of the Entomological Society Washington 76, 359-374.

Anderson DM & ML Cox 1997. *Smicronyx* species (Coleoptera: Curculionidae), economically important seed predators of witchweeds (*Striga* spp.) (Scrophulariceae) in sub-saharan Africa. Bulletin of Entomological Research 87, 3-17.

Bashir MO & LJ Musselman 1984. Some natural enemies of *Striga hermonthica* in the Sudan. Tropical Pest Management 30, 211.

Davidson A 1963. Insects attacking *Striga* in Kenya. Nature 197, 923.

Greathead DJ & JED Milner 1971. A survey of *Striga* spp. (Scrophulariacaea) and their natural enemies in East Africa with a discussion on the possibilities of biological control. Tropical Agriculture (Trinidad) 48, 111-124.

Greathead DJ 1984. The natural enemies of *Striga* spp. and their utlization as biological control agents. In: ES Ayensu, H Doggett, RD Keynes, J Marton-Lefèvre, LJ Musselman, C Parker & A Pickering (eds.), *Striga* Biology and Control. IRL Press Ltd Oxford, UK, 133-160.

Kapralov SI 1974. [*Phytomyza* against broomrape]. Zernovoe Chozjajstvo, Moskva 13(7), 43-44.

Kapralov SI 1975. [*Phytomyza* Einsatz im Gebiet Rostov]. Zascita Rastenij 20(6), 22-23.

Khan MQ & DV Murthy 1955. *Smicronyx albovariegatus* Faust (Curculionidae: Coleoptera) on *Striga* spp. Indian Journal of Entomology 17, 362.

Kroschel J, AA Abbasher & J Sauerborn 1995. Herbivores of *Striga hermonthica* in northern Ghana and approaches to their use as biocontrol agents. Biocontrol Science and Technology 5, 163-164.

Kroschel J, A Jost & J Sauerborn 1999. Insects for *Striga* control – Possibilities and constraints. In: J Kroschel, H Mercer-Quarshie & J Sauerborn (eds.), Advances in Parasitic Weed Control at On-farm Level. Vol. I. Joint Action to Control *Striga* in Africa. Margraf Verlag, Weikersheim, Germany, 117-132.

Kroschel J & O Klein 1999. Biological control of *Orobanche* spp. with *Phytomyza orobanchia* Kalt., a review. In: J Kroschel, M Abderabihi & H Betz (eds.), Advances in Parasitic Weed Control at On-farm Level. Vol. II. Joint Action to Control *Orobanche* in the WANA Region. Margraf Verlag, Weikersheim, Germany, 135-159.

Lekic MB 1970c. The role of the dipteran *Phytomyza orobanchia* Kalt. (Agromyzidae) in reducing parasitic phanerogam populations of the *Orobanche* genus in Vojvodina. Contemporary Agriculture 18(7/8), 59-68.

Lekic MB 1974. [Investigations of the dipteran *Phytomyza orobanchia* Kalt. as a controler of parasitic phanerogams of the genus *Orobanche*]. Savremena Poljoprivreda, Novi Sad 22(1-2), 93-99.

Linke KH, C Vorländer, MC Saxena 1990. Occurence and impact of *Phytomyza orobanchia* (Diptera: Agromyzidae) on *Orobanche crenata* (Orobanchaceae) in Syria. Entomophaga 35(4), 633-639.

Murthy DV & AS Rao 1949. *Precis orithya* feeding on *Striga* spp. Current Science 9 (18), 342.

Murthy DV 1960. Insects attacking *Striga*. Nature 187, 959.

Musselman LJ, PC Matteson & S Fortune 1983. Potential pollen vectors of *Striga hermonthica* (Scrophulariaceae) in West Africa. Annals of Botany 51, 859 - 862.

Norambuena H, J Kroschel, O Klein 1998. Introduction of *Phytomyza orobanchia* for biocontrol of *Orobanche* spp. in Chile. In: J Kroschel, H Betz, M Abderahibi (eds.), Advances in Parasitic Weed Control at On-farm level. Joint Action to control *Orobanche* in the WANA region. Margraf Verlag, Weikersheim, Germany, 197-204.

Sankaran T & VP Rao 1966. Insects attacking witchweed (*Striga*) in India. Common-

wealth Institute Biological Control, Technical Bulletin 7, 63-73.
Sankaran T, MK Rajendran & Y Ranganath Bhat 1969. The biology of *Eulocastra argentisparsa* Hampson (Lep.: Noctuidae) with special reference to the occurrence of dimorphism. Commonwealth Instiute of Biological Control, Technical Bulletin 11, 31-43.
Smith MC, J Holt & M Webb 1993. Population model of the parasitic weed *Striga hermonthica* (Srophulariaceae) to investigate the potential of *Smicronyx umbrinus* (Coleoptera: Curculionidae) for biological control in Mali. Crop Protection 12 (6), 470-476.
Traoré D 1995. Bionomics of *Smicronyx guineanus* Voss and *S. umbrinus* Hustache (Coleoptera: Curculionidae), potential agents of *Striga hermonthica* (Del.) Benth. PhD thesis, Macdonald Campus, McGill University, Quebec, Canada, (unpublished).
Traoré D, C Vincent & RK Stewart 1995. Life history of *Smicronyx guineanus* and *S. umbrinus* (Col.: Curculionidae) on *Striga hermonthica* (Scrophulariceae). Entomophaga 40 (2), 211-221.
Traoré D, C Vincent & RK Stewart 1996. Association and synchrony of *Smicronyx guineanus* Voss, *S. umbrinus* Hustache (Coleoptera: Curculionidae), and the parasitic weed *Striga hermonthica* (Del.) Benth. (Scrophulariceae). Biological Control 7, 307-315.
Williams CN & GH Caswell 1959. An insect attacking *Striga*. Nature 184, 1668.
Zermane N, G Sallé, J Kroschel & Z Bouznad 1998. Prospects for biological control of the parasitic weed *Orobanche* in Algeria. In: J Kroschel, M Abderabihi & H Betz (eds.), Advances in parasitic weed control at on-farm level. Volume II. Joint action to control *Orobanche* in the WANA region. Margraf Verlag, Weikersheim, Germany, 173-184.

Author: J Kroschel & O Klein

6.6.1.1 Collection of insect species

INTRODUCTION

For the identification of potential natural antagonists of parasitic weeds, insects damaging the vegetative or generative parts can be observed, sampled and identified.

OBJECTIVES

- To survey the occurrence of herbivores and their antagonists (e.g., parasitoids) of parasitic weeds,
- to collect the herbivores at different developmental stages,
- to preserve them for identification purposes, or
- for further entomological studies, and releases.

MATERIALS

✓ Glass tubes and different containers of suitable sizes for storing the insect samples (Eppendorf capsules, Petri dishes, gelatine capsules etc.), paper envelopes for butterflies.

✓ Exhaustors (→Fig. 21), killing jar (killing substances are cyanide or ethyl acetate), different tweezers, sweeping and beating net.

✓ Preservation liquid, i.e. ethanol 70% or Scherpel-solution (55% deionized water, 30% ethanol, 10% acetic acid, 5% ethyl acetate).

✓ Scalpel, razor-blades, sieves of 1 and 2 mm mesh width.

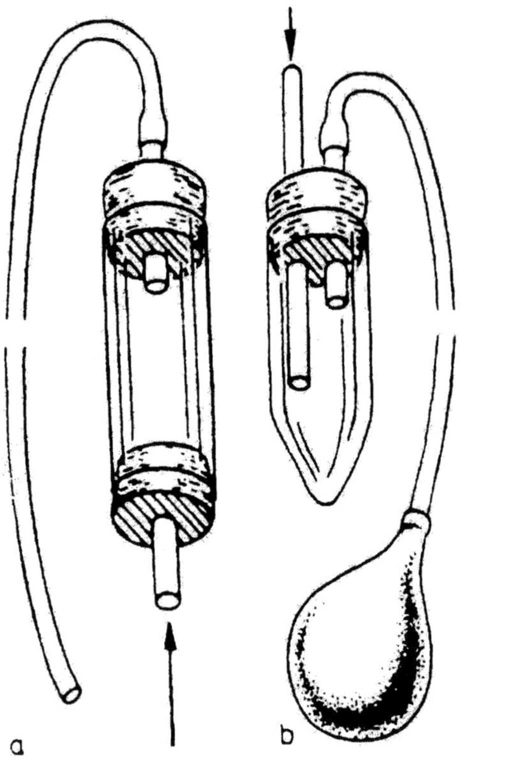

Fig. 21. Two different types of exhaustors

PROCEDURES

***Striga* spp.**

☞ *Junonia* adults can be caught with the sweeping net and transferred carefully into paper envelopes. It is advisable, if they need to be killed, to put them in the killing jar within the paper envelope. Preservation of *Junonia* adults will be as dried specimen in the paper envelopes, or they are pinned out and labelled (date, location etc.).

☞ *Junonia* larvae can be collected by hand or using a beating net. They can be preserved in glass tubes with 70% ethanol.

☞ *Smicronyx* adults are removed from the *Striga* plants with an exhaustor and transferred to the killing jar or to a plastic tube. For a later pinning on labelled cards it is advisable to preserve them in glass tubes with Scherpel-solution to maintain them softer.

☞ For the collection of *Smicronyx* larvae, galls of *Striga*-plants have to be removed and carefully opened with a scalpel.

☞ For the collection of *Smicronyx* pupae the galls of *Striga* plants have to be removed and distributed in containers filled with a 5 to 10 cm soil layer. The *Smicronyx* larvae will leave the galls and pupate in the soil. *Smicronyx* pupae can be sieved out of the soil using a 2 mm and a 1 mm sieve. The content of the 1 mm sieve has to be searched for the pupae.

☞ *Smicronyx* larvae and pupae can be preserved in glass tubes with 70% ethanol.

☞ For the determination of parasitoids, pupae and larvae have to be kept singly in tubes, until parasitoids have been developed.

***Orobanche* spp.**

☞ *P. orobanchia* adults can be caught directly from *Orobanche* plants using an exhaustor. They can be used for further studies alive, preserved in 70% ethanol or killed with ethyl acetate and preserved dried in small containers (e.g., gelatine capsules).

☞ *P. orobanchia* larvae can be dissected from *Orobanche* seed capsules and shoots. The capsules and shoots have to be opened carefully. The first two stages of the larvae are very sensitive and are difficult to keep alive after preparation. The third stage can be kept with fresh *Orobanche* tissues to pupate. Larvae can be preserved in 70% ethanol.

☞ Pupae of *P. orobanchia* can be prepared from *Orobanche* seed capsules and shoots. Whereas in the beginning of the development of *Orobanche* the larvae pupate in the capsules, they pupate later in the shoots, especially in the lower part of the shoots. Pupae should be stored singly in gelatine capsules to prevent them from parasitisation when parasitoids emerge.

☞ For the determination of parasitoids, at least 100 pupae should be kept singly in gelatine capsules. Emerged parasitoids can be detected regularly and determined. To monitor the development of parasitisation during the vegetation period, samples of 100 pupae each have to be taken regularly in the field.

☞ *Smicronyx* larvae can be found in seed capsules and in the lower parts of the shoots.

***Cuscuta* spp.**

☞ *Smicronyx* adults are removed from *Cuscuta* spp. using an exhaustor (→ *Striga*).

☞ *Smicronyx* larvae can be found in seed capsules only.

☞ For the collection of *Smicronyx* larvae → *Striga*.

EVALUATION

* The insect specimen collected can be used for preservation and identification or for further investigations.
* Parasitisation rates can be calculated.

COMMENTS

- Parasitisation of *J. orythia* larvae or pupae was very rare and had only a minor impact on its population dynamics in northern Ghana.

LITERATURE

(see end of chapter 6.6.1.4)

Author: O Klein & A Jost

6.6.1.2 Field observation of insects

INTRODUCTION

Field observations can be made either through surveys or through insect trapping and the collection of soil and plant samples.

OBJECTIVES

- To collect data on the population dynamics of herbivores during the vegetation period and the natural infestation of parasitic weeds in different fields and regions.

<u>**Field surveys**</u>

MATERIALS

✓ Vehicle, data sheets, pen, paper bags.

✓ Scalpel, razor-blades.

✓ The time requirement depends on the number of data to be collected and on the distance of the different fields to be observed. It includes the preparation of the survey (several days), the survey itself (15 to 30 min. per field plus transport from field to field) and the time for the evaluation (days or several weeks according to number of samples and extent of analysis).

✓ Agaragar solution 0.3%, 20 ml glass containers, 10 ml pipette, 0.5 ml Eppendorf pipette, Petri dishes, binocular.

✓ identification material.

PROCEDURE

Striga spp.

☞ A survey on the infestation of *Striga* by *Smicronyx* spp. and *Junonia orythia* should include at least 15 fields selected at random per region surveyed.

☞ Each field is walked through on a diagonal course and after a determined distance (e.g., every tenth step) one crop plant is selected at random and the following data are collected as required: (1) Number of *Striga* plants per stand, (2) number of *Striga* plants infested with an antagonist, (3) number of *J. orythia* larvae and/or *Smicronyx* adults, (4) *Smicronyx* galls and (5) *Striga* seed capsules per infested plant.

☞ The plant samples are also used to assess the number of larvae per gall and the remaining *Striga* seeds per gall.

Orobanche spp. and *Cuscuta* spp.

☞ The infestation of *Orobanche* spp. by *Phytomyza orobanchia* and *Smicronyx cyaneus*, and the *Cuscuta* infestation with *Smicronyx* spp. should include at least 15 fields selected at random per region surveyed.

☞ Each field is walked through on a zigzag course and after a determined distance (e.g., every tenth step) one *Orobanche* plant is selected by random, in total at least 10 plants per field. For the sampling of *Cuscuta*, 10 seed capsules should be taken per stand.

☞ To study the population dynamics of herbivores, these field surveys should be carried out regularly throughout the vegetation period.

☞ For the quantification of infestation by parasites see → chapter 3.1.

☞ The number of plants and seed capsules infested and the number of *P. orobanchia* and *Smicronyx* larvae are counted by dissection of plants and capsules with a scalpel or razorblades.

☞ The number of *Orobanche* seeds destroyed by an infestation is determined by the method described in chapter 4.1 and 4.3. Seeds of infested seed capsules can not be counted by their weight. The number of seeds of each capsule and their viability have to be tested. The viability is tested according to the method described in chapter 4.3. For the calculation of the seed number, the seeds are put in a small glass container (20 ml) containing 10 ml of a 0.3% agar agar solution. The seeds have to be distributed evenly by a slight movement. The seeds are counted from 5 samples of 0.5 ml. The method has to be tested with a known number of seeds to calculate a correction factor. The found number of seeds has to be multiplied by four and the calculated correction factor.

☞ The rate of destroyed *Cuscuta* seeds can be counted directly.

EVALUATION

* Choose the number of replications according to the variation of data.
* The mean of each parameter is determined for each field and for all fields of the surveyed area.

Striga spp.

* The infestation intensity of *Striga* is measured as the mean number of *Striga* plants per stand.

* The infestation frequency is the percentage of *Striga* plants infested with *Smicronyx* adults, *Smicronyx* galls or *Junonia* larvae.
* The infestation intensity with *Smicronyx* is expressed as the mean number of galls per *Striga* plant.

***Orobanche* spp.**

* For *Orobanche* infestation rate and the infection intensity see chapter 3.1.
* The infestation frequency is the percentage of *Orobanche* plants infested with *Phytomyza* or *Smicronyx.*
* The infestation intensity of *Orobanche* is expressed as the percentage of seed capsules infested.

***Cuscuta* spp.**

* The infestation frequency is the percentage of *Cuscuta* stands infested with *Smicronyx*.
* The infestation intensity of *Cuscuta* is expressed as the percentage of seeds infested.

<u>**Insect trapping and collection of soil and plant samples**</u>

MATERIALS

Smicronyx* spp. parasitizing *Striga*, *Orobanche* and *Cuscuta

✓ As pit-fall traps glass or plastic containers filled with formalin can be used. In this kind of trap, insects which walk on the ground (e.g., ground beetles, weevils) can be caught and monitored.

✓ Emergence cages are boxes of a certain size, where a glass or transparent plastic container with a funnel is fixed to collect insects which emerge from the soil (→Fig. 22).

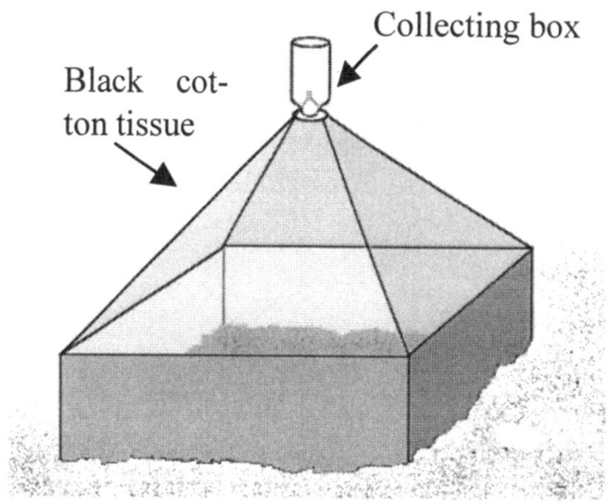

Fig. 22. Emergence cage

✓ For taking soil samples a frame of a certain square and depth is necessary, a shovel and containers for the soil, and sieves of 2 and 1 mm mesh width.

✓ For trapping *P. orobanchia* Malaise (Fig. 23) or window traps (glass barrier) (Fig. 24) are used.

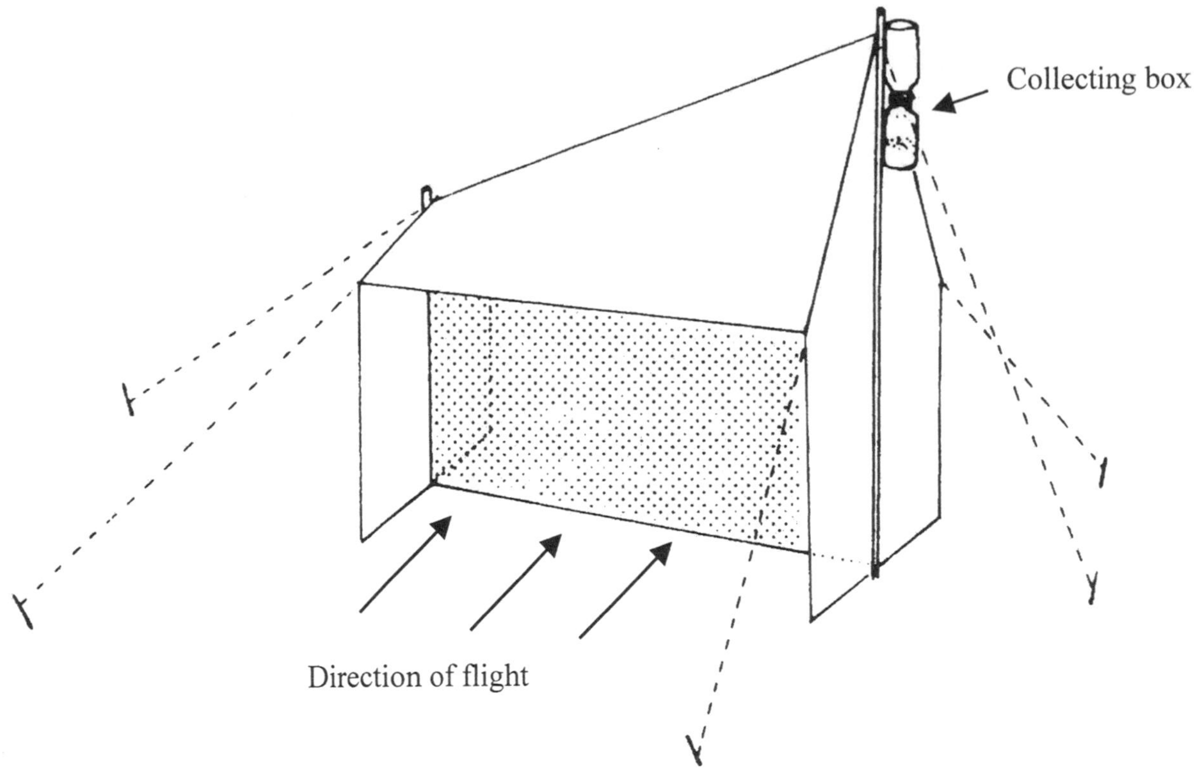

Fig. 23. Malaise trap (Model Townes) (Behre)

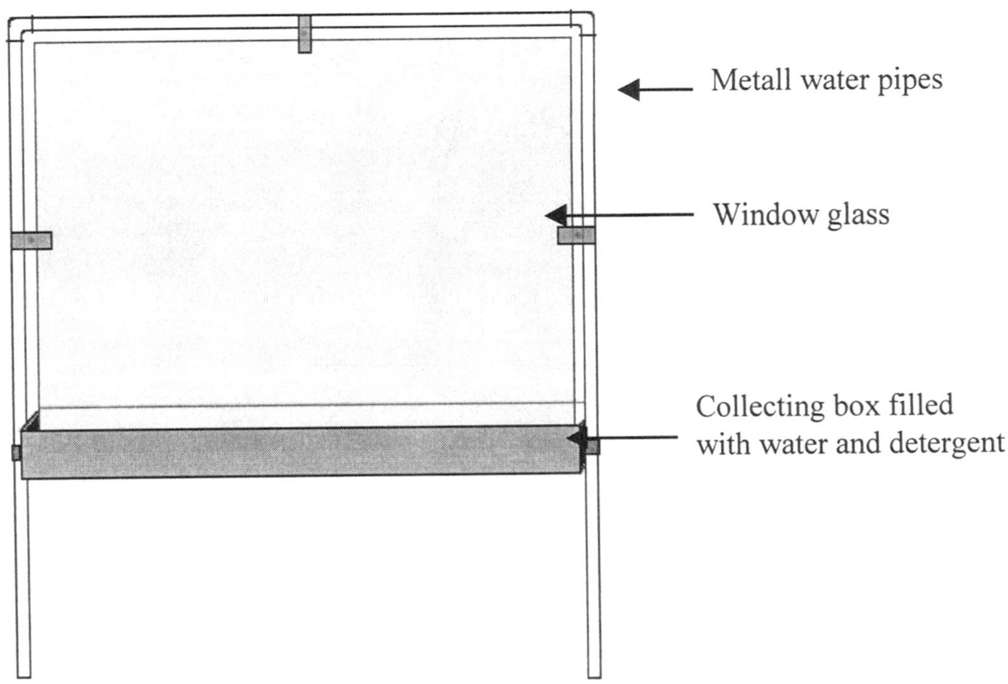

Fig. 24. Window trap (glass barrier) according to Kühn (in Mühlenberg 1993)

PROCEDURE

☞ The emergence cages are distributed randomly in the field to collect and monitor the adults of *P. orobanchia* or *Smicronyx* spp., which emerge from the soil where they pupate. Within the cage they fly to the light source, where they are collected in a transparent container.

☞ To trap adults of *Smicronyx* spp. which move on the ground, pit-fall traps are dug into the soil up to surface level. Formalin in the container will preserve the insects, if the containers are not emptied daily.

☞ For taking soil samples in a field the frame is pushed into the ground at random, and the soil is taken out. The soil samples have to be sieved through a 2 mm sieve to remove the gravel. The content of the 1 mm sieve is assessed for *Smicronyx* pupae and adults in dormancy.

☞ In order to monitor the flying activity of a population of *P. orobanchia*, two methods proved to be effective: the Malaise trap and the window trap. The traps have to be installed at random in the field. The Malaise trap works such that adults of *P. orobanchia* fly against a fine, invisible mesh and try to escape by moving up to the highest part of the tent, where they are collected in a container filled with formalin solution. Insects flying against the window trap fall down into a basin filled with formalin solution. Traps should be emptied daily and insects should be preserved in alcohol.

☞ The effectiveness of other trapping methods (sticky traps, coloured basins, sweeping net) for the monitoring of the flight of *P. orobanchia* is very low.

EVALUATION

* The number of emerged adults trapped with the emergence cages or collected pupae (*Smicronyx*) from the soil give indications about the total number of flies, which have survived from one season to another. In addition, the time of emergence of the flies and their coincidence to the development of their host plants (*Orobanche, Striga*) is determined.
* The catches in Malaise traps and window traps give only information on a relative quantity or population density. Regular trapping allows the monitoring of the development of a population and its population dynamics over a longer period of time.

COMMENTS

- A high number of replications is needed in the case of emergence cages and the soil samples to get representative results.
- Strong wind can negatively influence the effectiveness of window traps.

LITERATURE

(see end of chapter 6.6.1.4)

Author: O Klein, A Jost &J Kroschel

6.6.1.3 Rearing of insects

INTRODUCTION

Prerequisites to the use of herbivores in inundative biological control are rearing methods, which allow a sufficient production of individuals for field releases. In order to be able to rear insects or to optimise rearing techniques, a sound knowledge of their biology is needed.

OBJECTIVE

- To study the biology of herbivores (life cycle, reproduction capacity) in order to establish and optimise mass rearing techniques.

MATERIALS

✓ Seeds of host plants either to grow *Striga* (sorghum, maize or millet seed) or *Orobanche* (broad bean seed), *Striga* and *Orobanche* seed, pots, transparent stiff plastic foil of 1.50 m width, very fine gauze material, adhesive tape, insectaries made out of wood and mosquito wire, sugar or honey solution, cotton wool, wood or plastic containers of at least 15 cm depth, soil.

✓ Binocular

✓ Time requirement: Installation and maintenance of pots, daily observation of the insects.

PROCEDURES

Striga spp.

☞ As the weevil *Smicronyx* is closely related to its host *Striga*, which is a parasite itself, the rearing of *Smicronyx* depends on a subsequent culture of maize, sorghum or millet infested with *Striga*. For this purpose host plants are planted in pots with *Striga* infested soil (→ chapter 5.2). As soon as the *Striga* plants develop, the pots are surrounded with a transparent stiff plastic foil. The top is covered with fine gauze material. In this equipment *Smicronyx* adults can be observed without disturbance. The rearing should be done in a good ventilated environment in the shade.

The bottom of the pot is covered with gauze to collect the larvae dropped to the soil. They are collected into containers with a 5 to 10 cm wet soil layer, where pupation takes place. Adults develop and emerge after watering the containers with an equivalent of 30 to 40 mm of rainfall.

☞ The larvae of *Junonia* are kept in plastic containers covered with nylon gauze. They have to be fed with *Striga* material. The pupae are hanging upside down on the gauze cover. They are collected and transferred into insectaries. The butterflies are fed with sugar or honey water to prolong their life span.

Orobanche spp.

☞ As the development of *P. orobanchia* and *Smicronyx* spp. is closely related to its host, *Orobanche* spp. have to be grown in pots. For this purpose, host crops (e.g., faba bean) are planted in pots with *O. crenata* infested soil. As soon as *O. crenata* plants start to emerge, the pots are covered with fine gauze material. *Smicronyx* and *P. orobanchia* adults can be observed and reared within this "living system".

☞ Artificial rearing diets for *P. orobanchia* have been tested, but without success. Mass rearing for the biological control of *Orobanche* spp. is therefore not possible. Methods to store and rear pupae collected in the field have to be developed to obtain a high number of adults for field releases (→ chapter 6.6.1.4).

EVALUATION

★ By daily observation of the cultures data on length of the different developmental stages of the insect species can be recorded as well as mortality rates.

★ For the detection of eggs and the first larval stages of *Phytomyza* and *Smi-*

cronyx on and in *Orobanche,* a binocular microscope is necessary.

COMMENTS

- The rearing of *Junonia* seems difficult, as no egg-laying was observed so far in the insectaries.
- Insects which seem to be parasitized are to be removed from the culture and kept separately.

LITERATURE

(see end of chapter 6.6.1.4)

Author: O Klein, A Jost & J Kroschel

6.6.1.4 Biological control of *Orobanche*

INTRODUCTION

Since the rearing of *Phytomyza orobanchia* on semi-artificial or artificial diets is not successful, other methods have been developed to supply adults for inundative releases. The most appropriate method is the use of a "phytomyzarium", where collected *Orobanche* shoots containing pupae of *P. orobanchia* are stored until the next season. This method ensures a controlled rearing and release of adults with promising results. Additional releases of 500 to 1000 adults/ha led to a reduction of the *Orobanche* seed production of up to 96%. However, due to the longevity of *Orobanche* seeds of 10 to 15 years further infestations in the following cropping seasons will ocur. Consequently, releases have to be repeated continously over several years to reduce the infestation to a minimum.

<u>1. Storage of *Orobanche* plants</u>

OBJECTIVES

- To obtain a high number of adults of *Phytomyza* for field releases.

MATERIALS

✓ Sacks (50 x 100 cm or bigger) to collect and store *Orobanche* plants.

✓ Phytomyzarium (store room) (Fig. 25).

✓ Separator to separate parasitoids from *Phytomyza* using a mesh with 0.8 mm width (Fig. 25).

PROCEDURE

☞ The simplest way to store *Orobanche* plants is to dig a ditch of 20 cm depth at the border of the field, where no cultivation will take place. The *Orobanche* plants are covered with a 10 cm thick layer of soil which prevents dissemination of seeds and eventually the penetration of parasitoids.

☞ They can also be stored simply in the sacks and kept in the shelter of a roof or a shed which are exposed in the field when *P. orobanchia* begins to emerge. By this method parasitoids can not be separated and emerge together with adults of *P. orobanchia.*

☞ The best way to store the *Orobanche* plants is a store room called "Phytomyzarium" with a separator mechanism for the parasitoids. Light can penetrate through windows during the stocking period of the *Orobanche* plants to guarantee photoperiodic influence on diapause. With the first emergence of *P. orobanchia*, the windows are closed so that the adults fly to the separator, where they are separated from their parasitoids. *P. orobanchia* can fly directly to the field or can be transported in a container to the field.

☞ The number of healthy pupae kept in the "Phytomyzarium" can be determined by samples of 5 times 100 g

Orobanche plants, which are dissected to count them. The total number of pupae of a sample can be calculated (number of pupae per 100 g x 10 x kg of the sample). The rate of parasitized and diapausing pupae can be determined by pupae, which are kept single in gelatine capsules.

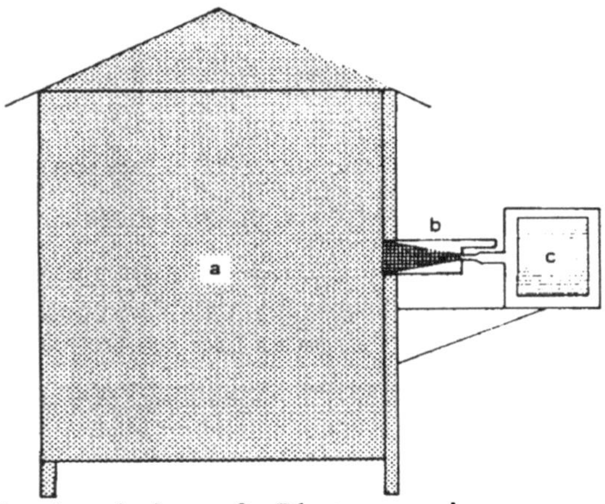

Sectional view of a Phytomyzarium

a Storage of *Orobanche* shoots
b Separator
c *Phytomyza* collecting-box

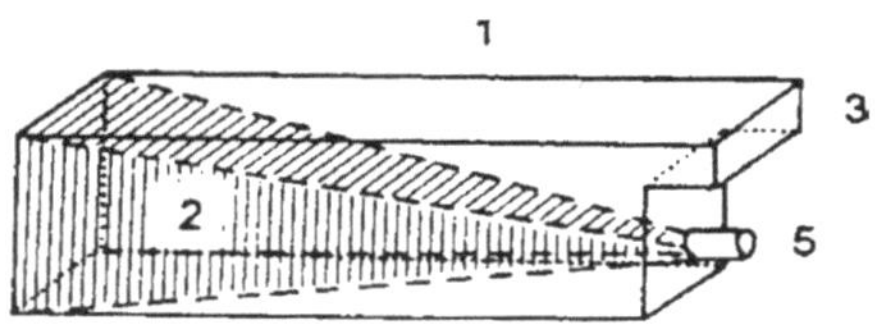

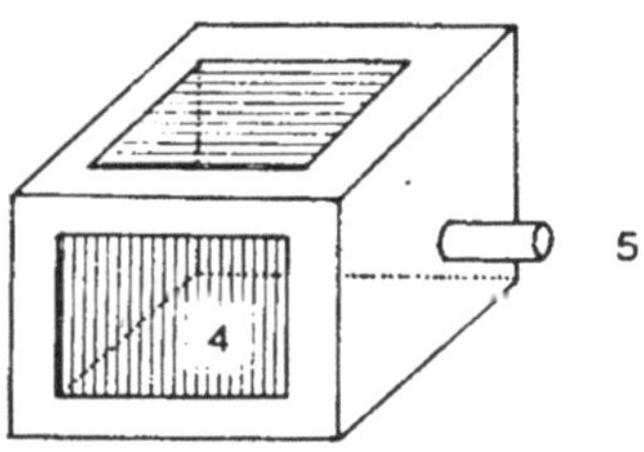

Separator

1 transparent casing
2 gauze funnel
3 killing sector (parasitoids)

Collecting-box

4 gauze window
5 connection tube

Fig. 25. Construction of a phytomyzarium (Bondarenko 1986)

COMMENTS

- If the collected *Orobanche* plants are not dry, they have to be dried before storage.
- The best time to collect *Orobanche* plants for storage is the period when pupae rest in diapause but are not highly parasitized (monitoring → chapter 6.6.1.2).
- Try to avoid storing spiders with the *Orobanche* plants in the "Phytomyzarium". They will prey on *P. orobanchia* and can reduce the number of *P. orobanchia* adults.

2. Inundative releases

OBJECTIVES
- To determine the release number of *Phytomyza* adults,
- to evaluate the effectiveness of inundative releases.

PROCEDURE

☞ The number of *Phytomyza* to be released per hectare has to be calculated using the following formula:

$$N = \frac{SR^2 \times 4 \times SC}{(SR^2 \times 2 + DS) \times IC}$$

N Number of *P. orobanchia* to be released
SR Sex-ratio of *Phytomyza* (male/female)
SC Number of expected seed capsules of *Orobanche* (where SC = *Orobanche* plants per m² x seed capsules per plant x 10 000)
DS Number of descendants per female *P. orobanchia*
IC Mean number of infested seed capsules per female *P. orobanchia*

☞ *P. orobanchia* adults which are collected in the collecting-box of the "Phytomyzarium" can be counted or their number estimated using the specific mean weight of 1000 adults of 0.74 g.

☞ The easiest way to release adults permanently is to open directly the outlet of the separator of a Phytomyzarium near by the field. For targeted releases adults are collected in the collecting-box and transported to the fields, where calculated numbers are released appropriate to the *Orobanche* infestation.

EVALUATION

* In bi-weekly intervals samples of *Orobanche* plants should be investigated to determine the rate of infested/destroyed seed capsules. Untreated naturally infested fields serve as control fields. For the evaluation of the efficacy at least 5 to 10 fields should be selected for field releases. The results should be analysed statistically using analysis of variance to detect significances.

LITERATURE

Bondarenko NV 1986. [Biologische Bekämpfung von *Orobanche*]. Biologiceskaja Zascita Rasrenij, Moskva, 226-230.

Bruckner G, F Flindt & T Kunsch 1979. Biologische Feldmethoden. Quelle & Meyer, Heidelberg, Germany.

Chapman JA & JM Kinghorn 1955. Window flight traps for insects. Canadian Entomologist 87: 46-47.

Hammad S, H El-Arosi & O Al-Menoufi 1967. *Phytomyza orobanchia* Kalt (Diptera: Agromyzidae) feeding on *Orobanche crenata* Forsk. in Egypt. Bulletin de la Société Entomologique d'Egypte 51: 141-144.

Janetschek H 1982. Ökologische Feldmethoden. Ulmer Verlag Stuttgart, Germany.

Jost A, J Kroschel & J Sauerborn 1996. Studies on *Smicronyx* Spp. and *Junonia orithya* and their potential for biological control of *Striga hermonthica* in northern Ghana. In: MT Moreno, JI Cubero, D Berner, D Joel, LJ Musselman & C Parker (eds.), Advances in Parasitic Plant Research. Proceedings 6th Parasitic Weed Symposium. Cordoba, Spain, 888-897.

Jost A 1997. Integrierter Getreideanbau in Nord-Ghana unter besonderer Berücksichtigung der *Striga*-Problematik. PLITS 15(4).

Klein O, J Kroschel & J Sauerborn 1999. Éfficacité de lachers supplementaires de *Phytomyza orobanchia* Kalt. [Diptera: Agromyzidae] pour la lutte biologique contre l'*Orobanche* au Maroc. In: J Kroschel, M Abderabihi & H Betz (eds.), Advances in Parasitic Weed Control at On-farm Level. Vol. II. Joint Action to Control *Orobanche* in the WANA Region. Margraf Verlag, Weikersheim, Germany, 161-171.

Klyueva MP & GV Pamikchi 1982. [Technologie des *Phytomyza* Einsatzes]. Zascita Rastenij 27(1): 33-34.

Kroschel J & O Klein 1999. Biological control of *Orobanche* spp. with *Phytomyza orobanchia* Kalt., a review. In: J Kroschel, M Abderabihi & H Betz (eds.), Advances in Parasitic Weed Control at On-farm Level. Vol. II. Joint Action to Control *Orobanche* in the WANA Region. Margraf Verlag, Weikersheim, Germany, 135-159.

Mühlenberg M 1993. Freilandökologie. UTB Quelle & Meyer Heidelberg, Germany.

Nishida T & T Torii 1970. A handbook of field methods for research on rice stemborers and their natural enemies. Davis Company, Philadelphia, UK.

Olroyd H 1958. Collecting, preserving and studying insects. Hutchinson & Co. Ltd London, UK.

Singh P & RF Moore 1985. Handbook of insect rearing. Vol. I. Elsevier, Amsterdam, Oxford, New York, Tokyo.

Southwood TRE 1989. Ecological methods with particular reference to the study of insect populations. University Press, Cambridge, UK.

Townes H 1972. A light-weight Malaise trap. Entomology News 83: 239-247.

Wetzel T 1984. Diagnosemethoden. Springer Verlag, Berlin, Heidelberg, New York, Tokyo.

Wyniger R 1974. Insektenzucht. Ulmer Verlag Stuttgart, Germany.

Author: O Klein & J Kroschel

6.6.2 Pathogens

INTRODUCTION

Research on biological control of weeds with fungi has been extensive enough in the past twenty-five years to identify three methods for their utilisation: the inoculative method, the augmentative method and the inundative or mycoherbicidal method. For effective parasitic weed management, the inundative or mycoherbicidal approach is a promising control strategy. It is based on the periodical application of massive doses of inoculum of phytopathogenic fungi to create a high infection level when conditions are favourable for disease development and weed control. Soil-borne pathogens are advantageous for the control of root parasites because they can infect underground developmental stages when they are incorporated preplanting into the soil. Additionally, soil-borne fungi are buffered from environmental extremes and may also persist in the soil providing residual weed control. Examples of effective control of *Orobanche* as well as *Striga* by strains of *Fusarium* spp. under field conditions are reported.

The following chapter will not give an overview of mycological work. In order to carry out the procedures described, basic knowledge on aseptic and microscopical working as well as culturing of fungi is required. It was the aim to select simple and cheap methods and to work with a minimum of technical inputs. Nevertheless, the work has to be conducted under sterile conditions (e.g., sterile bench). To sterilise media and tools, an autoclave is needed and to identify the fungi associated with diseased parasitic plants, a light-microscope has to be available.

LITERATURE

Bedi JS & N Donchev 1991. Results on mycoherbicides control of sunflower broomrape (*Orobanche cumana* Wall.) under glasshouse and field conditions. In: JK Ransom, LJ Musselman, AD Worsham & C Parker (eds.), Proc. 5th Int. Symp. on Parasitic Weeds, CIMMYT, Nairobi, Kenya, 76-82.

Charudattan R 1991. The mycoherbicide approach with plant pathogens. In: OD TeBeest (ed.), Microbial control of weeds. Chapman and Hall, New York, USA, London; UK, 24-57.

Dhingra OD & JB Sinclair 1986. Basic plant pathology methods. CRC Press, Boca Baton, USA, 355pp.

Sauerborn J, AA Abbasher, J Kroschel, DW Cornes, A Zoschke & KT Hine 1996. *Striga hermonthica* control with *Fusarium nygamai* in maize. In: VC Moran & JH Hoffmann (eds.), Proc 9th Int. Symp. on Biological Control of Weeds, Stellenbosch, South Africa, 461-466.

Sauerborn J, I Dörr, AA Abasher, H Thomas & J Kroschel 1996. Electron microscopic analysis of the penetration process of *Fusarium nygamai*, a hyperparasite of *Striga hermonthica*. Biological Control 7: 53-59.

Thomas H, A Heller, J Sauerborn & D Müller-Stöver 1999. *Fusarium oxysporum* f.sp. *orthoceras*, a potential mycoherbicide, parasitizes seeds of *Orobanche cumana* (sunflower broomrape): a cytological study. Annals of Botany, 83, 453-458.

6.6.2.1 Survey

INTRODUCTION

Fungal diseases on weeds are usually common in areas of weed origin while often less or absent in areas where weeds have been introduced and became a problem. On the other hand, new host-pathogen associations caused by newly encountered weeds and native pathogens may yield a higher number of mycoherbicidal agents explained by the lack of evolved host-pathogen homeostasis. Because of quarantine restrictions it is generally advisable to survey in the same country where the pathogen is planned to be used.

OBJECTIVES
- To survey for occurrence and incidence of fungi on parasitic plants,
- to collect fungi which attack parasitic plants.

MATERIALS

✓ Vehicle, map for the area to be surveyed, paper bags, scissors, spate.

PROCEDURE

☞ Select only fields which are not treated with herbicides. The host crop has to be in a good physiological condition and should not be affected by natural constraints or about to be harvested.

☞ Survey the fields along their peripheries, along the rows or by walking diagonally across the area.

☞ If no larger patches of diseased plants can be found within the field, investigate single plants at random. Especially in the case of *Orobanche* examine the underground stages, too.

☞ Collect plants showing disease symptoms only if they have not finished their flowering stage and are not injured by mechanical damages caused by management measures or feeding insects.

☞ Pull out the whole plants carefully.

☞ Pack the diseased plant material in paper bags and transfer it to the laboratory for isolation of the fungal organisms.

EVALUATION

✶ Evaluate parasitic plants for symptoms of a fungal disease which can include yellowing, stunting, wilting, browning, lesions, rots or dead plants.

COMMENTS

- If the isolation of fungi is conducted over a long period, dry the diseased parasitic plant tissue at room temperature to conserve potential pathogens and to prevent secondary infections.

LITERATURE

Kirk AA 1993. A fungal pathogen with potential for biocontrol of *Striga hermonthica (Scrophulariaceae).* Entomophaga 38: 459-460.

6.6.2.2 Isolation

INTRODUCTION

The isolation of parasitic fungi from infected plant material normally requires a surface sterilisation of the diseased plant piece. The concentration of the disinfectant and the duration of the sterilisation process is strongly dependent on the susceptibility of the respective fungi.

OBJECTIVES
- To obtain cultures of the fungi associated with the parasitic weeds.

MATERIALS

✓ Sodium hypochlorite (1% active chloride) plus Tween 80 (0.02%, v/v) or ethanol (70%), water (sterilized and if possible deionized), sterile glass-vessels (e.g., 40 ml), sterile Petri dishes, sterile filter paper, solid complete medium [e.g., Potato Dextrose Agar (PDA)] supplemented with antibiotics.

PROCEDURE

☞ Cut the infected plant parts at the edge of the lesion and beyond into small pieces of 0.5 to 2 cm in diam.

☞ Wash the plant parts with tap water.

☞ Soak the infected material in a 1% sodium hypochlorite solution or ethanol for 3-5 min. in glass vessels.

☞ Wash the infected plant tissues in sterile water (e.g., for 10 min. in sterile glass vessels).

☞ Use sterile filter paper to dry the material.

☞ Plant the pieces onto PDA-medium supplemented with antibiotics.

☞ Prepare pure cultures by mass transfer of mycelia when fungal growth can be observed.

EVALUATION

★ Regular visual observations of fungal growth.

LITERATURE

Abbasher AA, J Kroschel & J Sauerborn 1995. Microorganisms of *Striga hermonthica* in Northern Ghana with potential as biocontrol agents. Biocontrol Science and Technology 5: 157-161.

6.6.2.3 Classification

INTRODUCTION

Taxonomic identification of fungi is time consuming and has to be confirmed by an expert. Therefore, it might be advisable to start screening of the isolates directly and to restrict taxonomic classification on selected pathogens.

OBJECTIVES

- To identify the type(s) of fungi involved.

MATERIALS

✓ Agar plates with complete medium (e.g., PDA) and nutrient-poor medium (e.g., SNA; composition: 1 g KH_2PO_4, 1 g KNO_3, 0.5 g $MgSO_4$ x 7 H_2O, 0.5 g KCl, 0.2 g glucose, 0.2 g sucrose, and 20 g agar per litre).

✓ Staining agent.

✓ General equipment for working with a light-microscope.

PROCEDURES

☞ Culture each isolate on PDA and SNA plates.

☞ Evaluate the shape and colour of the aerial mycelium macroscopically.

☞ For light microscopy studies and preparation of slides, use distilled water and add a staining agent in order to simplify the recognition of morphological structures.

☞ Use the shape, size, colour, and manner of arrangement of spores or fruiting bodies as well as shape and colour of the sporophores for distinguishing the isolates and for their identification.

EVALUATION

★ Fungal genera belonging to *Hyphomycetes* can be identified according to Ellis (1971, 1976), *Ascomycetes* are described by Sivanesan (1987) as well as Hanlin (1990) and *Fusarium* spp. are well documented by Booth

(1971); Nirenberg (1981) and Gerlach & Nirenberg (1982).

COMMENTS

- The majority of *Orobanche* pathogens are soil-borne and belong to the genus *Fusarium*.
- A nutrient-poor medium can induce sporulation and/or the formation of fruiting bodies when their production on a complete medium failed. The type and shape of spores formed on PDA and SNA medium may also differ, which helps to classify the isolates or to group them together.
- Generally, sporulation can be enhanced by culturing the isolates under Near-Ultraviolet (NUV)-Light.

LITERATURE

Booth C 1971. The genus *Fusarium*. CMI, Kew, Surrey, England, 237pp.

Ellis MB 1971. Dematiaceous Hyphomycetes. CMI, Kew, Surrey, England, 608pp.

Ellis MB 1976. More Dematiaceous Hyphomycetes. CMI, Kew, Surrey, England, 507pp.

Gerlach W & H Nirenberg 1982. The genus *Fusarium* - a pictoral atlas. Biologische Bundesanstalt für Land- und Forstwirtschaft, Berlin-Dahlem, Kommisionsverlag Paul Parey, Berlin, Germany.

Hanlin RT 1990. Illustrated genera of Ascomycetes. APS Press, St. Paul, Minn., USA, 236 pp.

Nirenberg H 1981. A simplified method for identifying *Fusarium* spp. occurring on wheat. Canadian Journal of Botany 59: 1599-1609.

Sivanesan A 1987. Graminicolous species of *Bipolaris, Curvularia, Drechslera, Exerohilum*, and their teleomorphs. Mycological Papers 158, 216 pp.

6.6.2.4 Storage

INTRODUCTION

There is no universal method for long-term storage of plant pathogens. Selection of a method depends on the properties of the pathogen, the size of the collection, and the advantages and limitations of each method.

OBJECTIVES

- To preserve the virulence of fungal isolates until further evaluation.

MATERIALS

✓ Agar plates with nutrient-poor medium (e.g., SNA), refrigerator.

PROCEDURE

☞ Subculture the fungi on SNA medium and store the plates at 5°C in the refrigerator.

COMMENTS

- There is no universal method for long-term storage of plant pathogens. Selection of a method depends on the properties of the pathogen and the advantages and limitations of each method.
- Fungi may be stored at 5°C on a nutrient-poor medium up to half a year without losing their virulence if subculturing is avoided. If it is not possible to start the screening process within this time and no deep freezing equipment is available, many fungi can be preserved in the dried plant material (as collected) at room-temperature or at 5°C.

LITERATURE

Singleton LL, J D Mihail & CM Rush 1993. Methods for research on soil borne phytopathogenic fungi. APS Press, St. Paul, Minn., USA.

6.6.2.5 Inoculum production

INTRODUCTION

Generally, the infective units of fungi are their spores. Fungi produce sexual and asexual spores. In some cases, fragments of mycelia are used instead of spores for some fungi which do not or insufficiently produce spores.

OBJECTIVES

- To produce a sufficient amount of inoculum required for screening procedures,
- to provide different types of fungal inoculum as required for various screening methods.

MATERIALS

✓ Water (tap and if possible deionized), liquid medium, Petri dishes with solid medium, Erlenmeyer flasks (e.g., 250 ml), cotton plugs, aluminium foil, shaker, funnels, cheese cloth, micropore filter paper, Waring blender, counting chamber (hemacytometer), organic substrate (e.g., straw, or grains), wide-mouth glass flasks (e.g., 2000 ml).

PROCEDURE

1. Liquid inoculum (suspensions of spores and/or mycelia):

☞ Prepare the medium which is to be used for the production of spores.

☞ Fill 100 ml medium in 250 ml Erlenmeyer flasks.

☞ Close the flasks with cotton plugs and cover them with aluminium foil.

☞ Sterilise the flasks in the autoclave for 20 min. at 15 bar pressure and 120°C.

☞ Use agar plugs from a growing culture or a spore suspension to inoculate each flask after cooling. 1 cm diam. plug(s) can be used per flask.

☞ Place the flasks on a shaker and adjust temperature and rpm to be optimal for growth and sporulation of the fungus.

☞ After the growth period, blend the content of the flasks and separate mycelial fragments from spores by filtration through a cheese cloth.

☞ Separate the spores from the culture medium by filtration through micropore filter paper and suspend them in tap water.

☞ Determine the spore density with a counting chamber and adjust it to the desired concentration (e.g., 10^6 per ml).

☞ If mycelium is used, homogenise it in a Waring blender for 20 sec. at high speed.

2. Solid inoculum (organic substrate colonized by the fungus)

☞ Cut the organic substrate (e.g., straw) into small pieces of about 1-2 cm^2, (cereal grains can be used directly without being cut) and fill 2000 ml wide-mouth glass flasks with them.

☞ Soak the substrate in water overnight and pour-off the excess water.

☞ Sterilise the medium in the autoclave for 60 min. at 15 bar pressure and 120°C.

☞ After 24 hours, inoculate each flask with a spore suspension.

☞ Incubate the flasks at room temperature for 1-2 weeks before their use as inoculum.

☞ Observe the growth of the fungus inoculated on the grain or the straw and add sterilized water when required.

EVALUATION

✶ Observe the growth of the fungi in the liquid culture or on the solid substrate.

COMMENTS

- Optimal temperature for the growth of most fungi is about 25°C.
- Separation of spores from the liquid medium can be done by centrifugation if available.
- Isolates which did not sporulate in submerged systems can be cultured on solid medium in Petri dishes. Sporulating cultures are flooded with 10-20 ml tap water and the spores are detached with a drygalski spatula or a glass rod. This procedure is also useful when only small spore quantities are required.

6.6.2.6 Screening

INTRODUCTION

Testing for the pathogenicity of a fungal isolate consists of inoculation, incubation and the evaluation of possible disease symptoms. Optimum inoculation and incubation conditions should be provided, but they should not be too severe so that the potential of the pathogen to cause disease is not overestimated.

OBJECTIVES

- To screen fungal isolates for their pathogenicity towards parasitic weeds.

1. Root chambers

MATERIALS

✓ General equipment for working with root chambers, sterilized soil, atomiser or sprayer, prepared fungal liquid inoculum (preferable: spore suspension).

PROCEDURE

☞ Grow the parasite on its host root in root chambers (→ chapter 5.1.1) using autoclaved soil.

☞ Apply the fungal spores to *Orobanche* tubercles or juvenile *Striga* shoots on their host root. For this purpose, remove the Plexiglas of the root chamber and spray 10 ml of the spore suspension onto the glass fibre strip.

☞ Close the root chambers again and incubate them under the same conditions as before.

EVALUATION

✶ Determine the percentage of dead tubercles/shoots based on the total amount under a dissecting microscope after 10 days.

✶ Include a water-inoculated control treatment to assess other effects causing death of tubercles/shoots.

COMMENTS

- Inoculation of tubercles in root-chambers is recommended as the first step in a pathogenicity screening for *Orobanche.*
- *Orobanche* plants and their hosts can also be grown in polyethylene bags (→ chapter 5.1.1).
- Fungal isolates causing browning or death of underground stages of parasitic weeds have to be re-isolated following the procedure described in chapter 6.6.2.2. If more than one fungal species is obtained, purify and compare them with the original iso-

late used for inoculation. Use the re-isolated strain which is identical with the original strain and inoculate a second time. If the same disease symptoms occur, the organism meets the demands to be a primary pathogen according to Koch's postulates. This procedure must be followed for all screening approaches.

2. Pot trials

MATERIALS

✓ General equipment for pot trials (→ chapter 5.2).

✓ Liquid fungal inoculum (preferably a spore suspension), solid inoculum (→ chapter 6.6.2.5), sprayer or atomiser, polyethylene bags.

i) Pre-planting inoculation

PROCEDURE

☞ Incorporate solid fungal inoculum pre-planting into the soil at a rate of 10 g per kg soil or liquid inoculum at a rate higher than 10^8 spores per kg soil (→ chapter 5.2).

EVALUATION

✶ Evaluate emerged parasitic plants for symptoms of fungal infection. Additionally, *Orobanche* shoots are washed out before the end of the flowering stage to examine underground stages for diseased tissue.

✶ The impact of the fungus can also be determined by counting the number of emerged shoots of the parasitic weed and measuring the dry matter of the host plants (shoot and/or yield) as well as the parasite (shoots above- and underground). When working with solid inoculum, five controls have to be laid out:

- host plant alone
- host plant + parasite
- host plant + non-inoculated substrate;
- host plant, parasite + non-inoculated substrate
- host plant + inoculated substrate

When working with liquid inoculum, three controls have to be laid out:

- host plant alone
- host-plant + parasite
- host-plant + inoculum.

COMMENTS

- These screening procedures are especially recommended for soil-borne fungi.
- Emergence rate of parasitic plants alone is not an appropriate indicator for control efficiency.

ii) Post-emergence inoculation

PROCEDURE

☞ Spray spore suspensions with at least 10^6 spores/ml until runoff to emerged *Striga* or *Orobanche* plants of similar size before the parasites have reached their flowering stage.

☞ Keep the inoculated shoots as well as water-inoculated control shoots under a polyethylene bag for 48 h to maintain humidity after spraying.

EVALUATION

✶ Plants are examined until 10 days after inoculation for infection symptoms.

COMMENTS

- This screening procedure is especially recommended for air-borne fungi.
- Agar (0.2%, w/v) as adhesive agent and nutrient source for the fungus can be added to the suspension before spraying.

LITERATURE

Abbasher AA & J Sauerborn 1992. *Fusarium nygamai*, a potential bioherbicide for *Striga hermonthica* control in sorghum. Biological Control 2: 291-296.

Abbasher AA, J Sauerborn 1995. Pathogens attacking *Striga hermonthica* and their potential as biological control agents. In: ES Delfosse & RR Scott (eds.), Proc. 8th Int. Symp. On Biological Control of Weeds, Lincoln University, Canterbury, New Zealand, 527-533.

Ciotola M, AK Watson & SG Hallett 1995. Discovery of an isolate of *Fusarium oxysporum* with potential to control *Striga hermonthica* in Africa. Weed Research 35, 303-309.

Kroschel J, A Hundt, AA Abbasher & J Sauerborn 1996. Pathogenicity of fungi collected in Northern Ghana to *Striga hermonthica*. Weed Research, 36, 515-520.

Thomas H, J Sauerborn, D Müller-Stöver, A Ziegler, JS Bedi & J Kroschel 1998. The potential of *Fusarium oxysporum* f.sp. *orthoceras* as a biological control agent for *Orobanche cumana* in sunflower. Biological Control 13, 41-48.

Authors:
D Müller-Stöver, H Thomas & AA Abbasher

6.7 Toxins

6.7.1 Biological assays to test toxins

INTRODUCTION

Many plant pathogens produce a wide spectrum of toxins, having different chemical structures and roles in the course of the disease they cause on plants. A broad range of biological activities on host plants are expressed by these toxins: e.g., growth inhibition, necrosis, chlorosis, wilting, or inhibition of seed germination. In this chapter we will give few guidelines on the production, purification and biological characterisation of toxic metabolites produced by parasitic weed pathogens with the following objectives:

OBJECTIVES

- To evaluate the production of toxins for direct application as natural herbicides in addition or alternative to pathogens,
- to use culture filtrates, crude extracts or pure compounds to screen more promising isolates to be applied as mycoherbicides.

MATERIALS

✓ Petri dishes, flasks, pipettes, deionized water, funnels, Clorox, filter papers, dissecting microscope, corn kernels, mineral salts, multiwell plates, TLC plates, organic solvents......

PROCEDURE

Considering the large number of pathogens and their specific cultural conditions, the wide availability of biological assays and of purification protocols, the procedures described are confined to toxins produced by *Fusarium nygamai*, or in general by *Fusarium* species, which are highly effective pathogens of *Striga hermonthica.*

Production of toxins

Both, solid or liquid cultures can be used (the latter in static or shaken conditions) in combination with different media, temperature regimes and growth periods.

Liquid culture

☞ Put 200 ml of a defined medium (M-1 D) in Roux bottles and sterilise at 118°C for 20 min.

☞ Inoculate the flask with 1 ml of a conidial suspension obtained from an actively growing fungus

☞ Incubate in static conditions for 4 weeks at 25°C.

☞ Remove mycelia by paper filtration with reduced pressure.

☞ Use the culture filtrate for biological assays or extraction procedures.

Solid culture

☞ Put 100 g of corn kernels in 500 ml flasks.

☞ Keep overnight with 45 ml distilled water and

☞ sterilise in autoclave for 30 min. at 120°C.

☞ Inoculate the flask with 1 ml of a rich conidial suspension obtained from an actively growing fungus.

☞ Incubate at 25°C for 4 weeks.

☞ Dry and grind culture.

☞ Extract with solvent.

Purification of toxins

The first step of the chemical purification is based on the extraction using organic solvents. The further purification and the chemical characterisation, both of known or new compounds, can be determined only by the use of separation techniques (column chromatography, thin layer chromatography, HPLC, gel filtration) and analytical methods (mass spectrometry, nuclear magnetic resonance, ultraviolet and infrared spectroscopy, X-ray crystallography), which are available only in well equipped and specialized laboratories.

Testing toxins in biological assays

Conventional germination assay

☞ Prepare the test as described in chapter 4.4.1 (preconditioning of parasite seeds).

☞ Add your test solution (containing germination stimulant and toxin).

☞ Continue the assay as described in chapter 4.4.1.

Leaf puncture assay

☞ Detach fully-expanded young leaves (e.g., of *S. hermonthica*).

☞ Puncture leaf with a needle in one or more sites.

☞ Overlay the puncture with a droplet (10-15 µl) of assay solution or water (control).

☞ Keep leaves in a moist chamber and observe daily for the appearance of symptoms.

Seedling assay

☞ Cut young plants (4-6 cm high).

☞ Immerse the stem in test tubes containing the assay solution for 48 hours.

In vitro assay

For *Striga*:

☞ Prepare *in vitro* assays as described in → chapter 5.1.2.1.

☞ Toxins can be added to the media at different developmental stages (Fig. 26).

For *Orobanche*:

☞ Prepare *in vitro* culture as described in → chapter 5.1.2.2.

☞ For the bioassay, we must obtain massive and simultaneous seed germination. Choose only callii differentiated into stunted shoots, similar size and morphogenetic stage, each transferred into wells of ELISA plates containing MS medium with or without the tested substances. Fig. 27 demonstrates the influence of a toxin on the development of *Orobanche*.

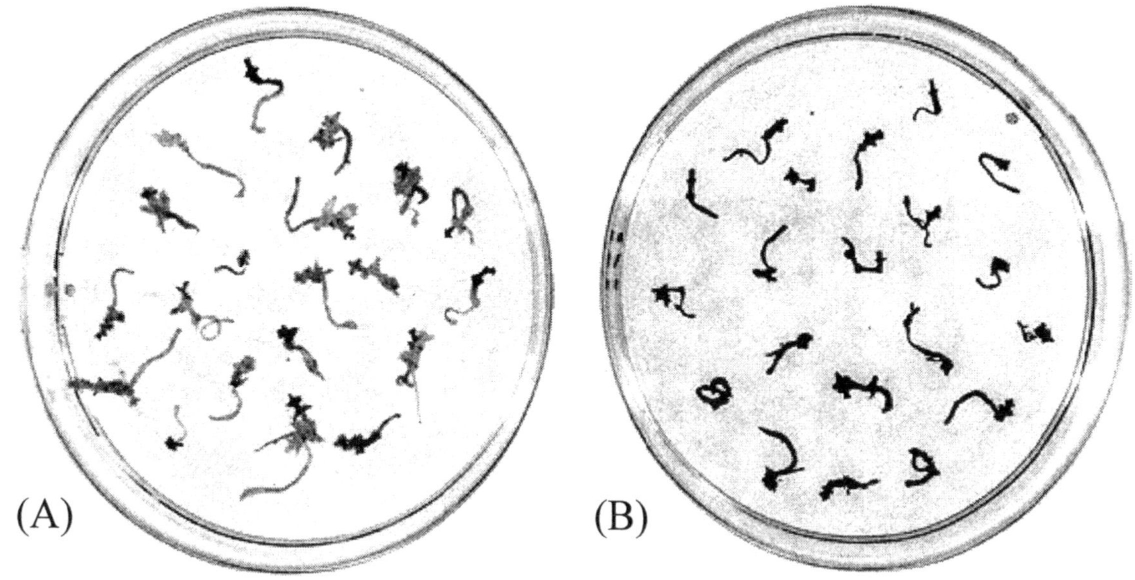

Fig. 26. ***In vitro*** **culture of** ***Striga hermonthica*** **to test toxins, (A) untreated control, (B) application of fumonisin B_1 into the growth media**

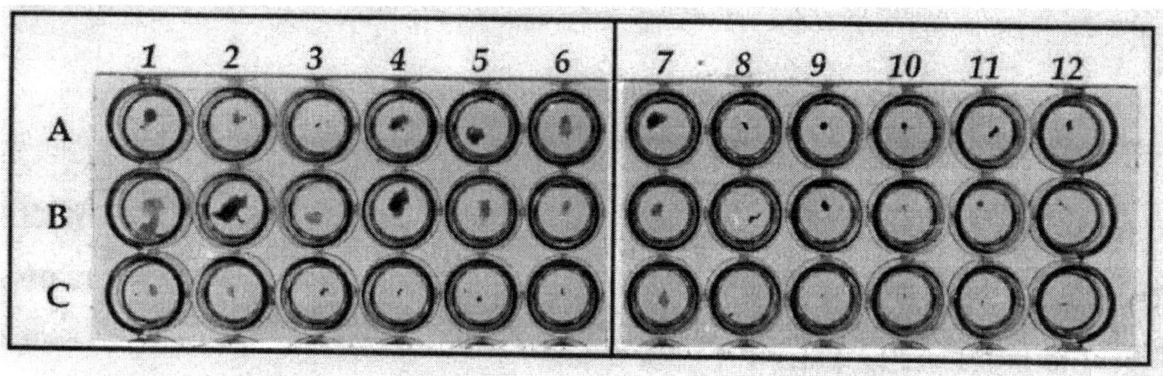

Fig. 27. Bioassay of herbicide toxicity on ***Orobanche*** **grown with and without herbicides. 2-6 different concentrations of glyphosate were added to the medium. No appreciable inhibition was found. 8-12 with the addition of different concentrations of rimsulfuron callus growth was significantly inhibited, some callii die. Note the difference between row C where growth was slow even without herbicides because slow growing** ***Orobanche*** **callii were used, and rows A and B in which the white and fast growing callii were used, and where** ***Orobanche*** **significantly developed in the non-treated wells.**

EVALUATION

Convential germination assay

- ✶ → chapter 4.4.1.
- ✶ Compare seed germination in presence of toxins against the control, and calculate the percentage of seed germination inhibition.

Leaf puncture assay

- ✶ Observe the appearance of necrotic and/or chlorotic spots, measuring their diameter (usually after 2 days).

Seedling assay

- ✶ After 2 days of immersion in water, observe the appearance of necrosis/chlorosis which usually starts from the edges of the leaves, and evaluate the toxic effects using a scale from 0 = no symptoms to 4 = 75-100% necrosis (sometimes it is possible to observe wilting symptoms).

In vitro assays

- ★ The results can be assessed either with the naked eye or under the dissecting microscope. Another possibility for observing results is under the automatic ELISA reader.
- ★ → chapter 5.1.2.

COMMENTS

- The given procedures have to be changed and adapted to the parasitic weed species investigated (mainly for the seed germination assay), the biology of the pathogen (growth condition for the production of toxins) and the chemical characteristics of the toxins (for toxin extraction procedures).
- Because not all *Orobanche* seedlings reach the final stage of shoot morphology at the same time, it is necessary to select similar shoots for each experiment.
- Using the ELISA plate is very useful. One plate is sufficient for the examination of a series of different substances. Each well is isolated from others and contains an economic quantity of substrate that needs only minute quantities of tested substances.
- If toxins have shown a herbicidal effect on parasitic weeds, before any application in the field their ecotoxicology as well as their effects on vertebrates has to be assessed in further studies, i.e. further steps to be undertaken are comparable to the registration process for pesticides.

LITERATURE

Abbas HK, CD Boyette, RE Hoagland & RF Vesonder 1991. Bioherbicidal potential of *Fusarium moniliforme* and its phytotoxin, fumonisin. Weed Science 39: 673-677.

Abbas HK & CD Boyette 1992. Phytotoxicity of fumonisin B_1 on weed and crop species. Weed Technology 6: 548-552.

Altomare C, A Lorgieco, A Bottalico, G Mulè, A Moretti & A Evidente (1995). Production of type A trichothecenes and enniatin B by *Fusarium sambucinum* Fuckel *sensu lato*. Mycopathologia 129: 177-181.

Bottalico A, A Logrieco, A Ritieni, A Moretti, G Randazzo & P Corda 1995. Beauvericin and fumonisin B_1 in preharvest *Fusarium moniliforme* maize ear rot in Sardinia. Food Additives and Contaminants 12: 599-607.

Bottalico A, A Lerario & A Visconti 1983. Production of mycotoxins (zearalenone, trichothecenes and moniliformin) by *Fusarium* species in Italy. Microbiologie-Aliments-Nutrition 1: 133-142.

Bottalico A, A Logrieco & A Visconti 1989. *Fusarium* species and their mycotoxin in infected cereals in the field and in stored grains. In: J Chelkowsky (ed.), *Fusarium*, mycotoxins, taxonomy and pathogenicity, Elsevier, Amsterdam, 85-119.

Capasso R, A Evidente, A Cutignano, M Vurro MC Zonno & A Bottalico 1996. Fusaric and 9,10-dehydrofusaric acids and their methyl esters from *Fusarium nygamai*. Phytochemistry, 41(4): 1035-1039.

Duke SO 1986. Microbially produced phytotoxins as herbicides - a perspective. In A Putnam & CS Tang (eds.), The Science of Allelopathy. John Wiley & Sons, Inc., 287-304.

Gelderblom WCA, K Jaskiewicz, WFO Marasas, PG Thiel, RM Horak, R Vleggaar & NPJ Kriek 1988. Fumonisins - Novel mycotoxins with cancer-promoting activity produced by *Fusarium moniliforme*. Applied and Environmental Microbiology 54: 1806-1811.

Hoagland RE 1990. Microbes and microbial products as herbicides, an overview. In: RE Hoagland (ed.), Microbes and microbial products as herbicides. American Chemical Society Washington DC, 2-52.

Kroschel J 1996. Bioherbicidal effect of fumonisin B_1 on parasitic weeds of the genus *Striga*. Internal Report Supra-regional

GTZ-project "Ecology and Management of Parasitic Weeds", University of Hohenheim, Germany.
Logrieco A, A Moretti, A Ritieni, J Chelkowsky, C Altomare, A Bottalico & G Randazzo 1993. Natural occurrence of Beauvericin in preharvest *Fusarium subglutinans* infected corn ears in Poland. Journal of Agricultural Food Chemistry 41: 2149-2152.
Pinkerton F & G Strobel 1976. Serinol as an activator of toxin production in attenuated cultures of *Helminthosporium sacchari*. Proc. Natl. Acad. Sci.USA, 73, 4007-4011.
Stierle A, G Strobel, D Stierle & F Sugawara 1992. Analytical methods for phytotoxins. In: Modern methods of plant analysis. New Series, Vol. 13. Plant Toxin Analysis. Springer-Verlag, Berlin, Germany, 1-32.
Strobel G, D Kenfield, G Bunkers, F Sugawara & J Clardy 1991. Phytotoxins as potential herbicides. Experientia 47: 819-826.
Thiel PG, WFO Marasas, EW Sydenham, GS Shephard & WCA Gelderblom 1992. The implications of naturally occurring levels of fumonisins in corn for human and animal health. Mycopathologie 117: 3-9.
Zonno MC & M Vurro 1999. Effect of fungal toxins on germination of *Striga hermonthica* seeds. Weed Research, 39,15-20.
Zonno MC, M Vurro, R Capasso, A Evidente, A Cutignano, J Sauerborn & H Thomas 1996. Phytotoxic metabolites produced by *Fusarium nygamai* from *Striga hermonthica*. In: VC Moran & JH Hoffmann (eds.), Proceedings of the IX International Symposium on Biological Control of Weeds, 19-26 January 1996, Stellenbosch, South Africa, 223-226.

Authors: M Vurro, J Kroschel, V Portnoy & DM Joel

6.8 Integrated

INTRODUCTION

Definition: Integrated control of parasitic weeds represents a combination of methods aiming at a restriction of chemical control to a necessary level and by forcing the inclusion of biological, bio-technical, cultural and breeding methods.

Due to the adaptation of the biology of parasitic weeds to a host plant their control is difficult and available means of control tend to be either of low efficiency or costly. With parasitic weeds a reduction of the amount of pesticides deserves low priority, e.g., due to (1) their low availability and high cost in *Striga* or (2) low amounts of active ingredient required (e.g., glyphosate) for the control of *Orobanche* and *Cuscuta*. The implementation of integrated control to parasitic weeds is focused primarily on the increase in efficiency of the control.

Generally, integrated pest management is less suitable for annual crops, where pest populations are subjected to extreme annual changes. With parasitic weeds, however, the actual host crop infestation in a given year is the product of the present parasite seed bank, the environmental conditions, and the crop susceptibility. Thus, apart from the fact of weather conditions, parasitic weeds become rather predictable pests.

In combination, even less effective (but cheap or simple) control techniques may be applied profitably as they reduce the initial infestation pressure (e.g., the inclusion of hand weeding in very low infested areas). The large number of above listed control techniques (→ chapter 6) should encourage researchers to further investigate the circumstances under which the techniques can be combined efficiently and sometimes even with synergistic effects.

OBJECTIVES

- To assess the efficiency of integrating various techniques on (1) infestation with parasitic weeds, (2) crop yield and (3) economic return,
- to develop a management system for parasitic weeds that considers economic, ecological and social aspects within an agro-ecosystem.

MATERIALS

✓ Usual equipment for field experiments.

✓ Further items according to the respective techniques to be combined.

✓ The efficiency of the respective single techniques to be combined should be already known prior to the investigation on integrated control.

PROCEDURE

☞ Characterisation of the agro-ecosystem and evaluation of efficiency and acceptability of various control techniques (→ chapter 6).

☞ Selection and combination of adequate techniques by taking into account above findings.

☞ Field preparation and planting as usual (→ chapter 5.3)

☞ Implementation of the respective control methods.

EVALUATION

★ Field layout according to the number of factors and the objectives of the study. Frequently, a split-split-plot layout may represent the adequate experimental design.

★ Assessment of crop yield and infestation as required.

★ Investigate interactions of the factors included, especially with regard to the mutual increase in efficiency (synergistic effects).

COMMENTS

- Prior to any studies on integrated control aspects detailed data on farming systems are required in order to evaluate the possibilities and constraints of the implementation of integrated control measures to be tested or applied in field trials. This includes also the socio-economic aspects of integrated control measures and the evaluation of the acceptability of integrated control schemes (for further details on this topic see → chapter 7).
- The area required for adequate evaluation of integrated systems in the field is usually much larger than for the testing of single treatments.
- The development of control packages or combinations of techniques (Singh *et al.* 1991; Linke & Saxena 1991; Berner *et al.* 1996) or control strategies (Eplee & Langston 1991) has to be considered as a step towards integrated control of parasitic weeds, but does not yet represent an integrated control as per definition.

LITERATURE

Akobundu IO 1991. Integrated weed management for *Striga* control in cropping systems in Africa. In: SK Kim (ed.), Combating *Striga* in Africa. Proceedings, International Workshop organized by IITA, ICRISAT and ICRC, 1988 IITA, Ibadan, Nigeria, 122-125.

Berner DK, MO Alabi, U Di-Umba & FO Ikie 1996. Proposed integrated control program for *Striga hermonthica* in Africa. In: MT Moreno, JI Cubero, D Berner, D Joel, LJ Musselman & C Parker (eds.), Advances in Parasitic Plant Research; 6th International Parasitic Weed Symposium, Cordoba, Spain, 817-825.

Eplee R & M Langston 1991. Integrated control methodology for *Striga* in the USA. In: JR Ransom, LJ Musselman, AD Worsham & C Parker (eds.), Proceedings of the 5th International Symposium of Parasitic

Weeds; CIMMYT, Nairobi, Kenya, 397-399.
Geipert S 1997. Potentiale und Grenzen der Bekämpfung von *Orobanche crenata* Forsk. im Ackerbohnenanbau (*Vicia faba* L.) Marokkos. PLITS 15(5).
Jost A 1997. Integrierter Getreideanbau in Nord-Ghana unter besonderer Berücksichtigung der *Striga*-Problematik. PLITS 15(4).
Linke KH & MC Saxena 1991. Towards an integrated control of *Orobanche* spp. in some legume crops. In: K Wegmann & LJ Musselman (eds.), Progress in *Orobanche* research. Eberhard-Karls-Universität, Tübingen, Germany, 248-256.
Pieterse AH, L Garcia-Torres, OA Al-Menoufi, KH Linke & SJ ter Borg 1992. Integrated control of the parasitic angiosperm *Orobanche* (broomrape). 2nd International Food Legume Research Conference, Cairo, Egypt.
Singh L, R Ndikawa & MR Rao 1991. Integrated approach to *Striga* management on sorghum in north Cameroon. In: JR Ransom, LJ Musselman, AD Worsham & C Parker (eds.), Proceedings of the 5th International Symposium of Parasitic Weeds CIMMYT, Nairobi, Kenya, 223-231.

Author: K-H Linke

7 Putting it into practise

Emphasis has been made on various control options and their combination in an integrated approach in the previous chapter. In this section, the aspect of how to make use of research findings and on how to assist farmers in reducing the threat by parasitic weeds on their livelihood will be the main emphasis.

The following sub-chapters may differ from the more technical context of previous sections of this handbook. There are no laboratory standard methods or controlled conditions. Neither will the procedures described produce uniformly the same findings since you will rarely to never have "all other factors constant". Farmer experimentation is unlike on-station agronomy field trials. Here, the "equipment" of the researcher includes social, communication and organisational skills, the right attitude and an open mind as it involves other people and unpredictable circumstances. Methods described here are not simply reproducible without adaptation to the prevailing conditions where one would like to use them.

Extension and the use of extension material like picture series could pave the path for participatory technology development and / or adaptation of innovations as a following step.

The process of generating and dissemination of agricultural innovations involves several actors amongst the most important are:

- Farmers;
- Extension staff; and
- Researchers.

The following sub-chapters will try to illuminate some relevant aspects in this process. Examples will be given and experiences shared with the reader.

7.1 Training of researchers, extension staff and farmers

Training of all three parties will help to improve their capacities in their respective field of action. Researchers need information and training on scientific methods necessary for their respective research. Additionally, researchers should acquire and train their skills to understand farmers' rationale to assist them in the development of appropriate technologies. Extension staff usually have the important task to disseminating innovations but also act as a link between farmers and researchers. Extension staff are often barely equipped with specific, in-depth knowledge about parasitic weeds and their control and therefore require targeted training on such topics. Apart from technical knowledge the extension staff may also require training on the appropriate use of extension material and to improve their communication skills.

Training is organised to learn about new things, acquire new information and skills and to be able to do things better than before. The exchange of information and experiences with others is equally important. Therefore, the training environment needs to be conducive to mutual exchange. Whenever possible and appropriate farmers, researchers and extension staff should have the chance to interact during training activities.

As training should be targeted to the specific needs there will be activities more relevant to researchers, to extension staff or to farmers.

Training of research and extension staff is organised in various ways (Table 9). Much experience can be gained while doing the job. The quality of experience and number of beneficiaries is difficult to name, since this will depend on the working environment. If it is encouraging, staff members can experiment, develop new skills and increase their knowledge and capacity to act. Often access to and offer of continuous training or supervision is little to nil.

Training courses of several days duration are organised to impart new information or to refresh the knowledge of participants. The number of courses organised is restricted by limited resources (financial and personnel). Therefore, not all staff will have the chance to attend such training. Depending on the level of staff invited to such training courses they could act as multiplicators (trainer of trainers). In-service training should ideally combine theory inputs and practice.

Table 9. Different forms of training for researchers and extension personnel

Type of training	Duration and way of training	Number of beneficiaries
On-the-job	continuous *theory and practice*	?
In-service	Courses of 1 to 14 days *theory and practice*	moderate
Short-term	Courses of up to 3 month, outside the employing organisation, often abroad *theory and practice*	little
Long-term, (formal training)	B.Sc., M.Sc., Ph.D., outside the employing organisation, frequently abroad *theory and practice*	little

Adapted from Gwinner (1997)

More profound knowledge in research methodologies, extension approaches, etc. may be obtained through short- and long-term training. Only some specialists may profit from such training since it is expensive. In the long run it should increase the capacity of an "organisation" to find appropriate solutions when gained knowledge is put into practice.

Finally a sort of training, which is often neglected, should be mentioned. Pre-service training in schools, colleges and universities could include lessons about parasitic weeds in the curriculum. Also the following sub-chapters concentrate on in-service training, which might again be an indicator that little has been done. The training of "future farmers" could already start in primary school.

Farmers have a more universal knowledge about agriculture. This knowledge as well as their potential and constraints in adopting technologies is a key factor which should be taken into consideration at all stages of development and dissemination of innovations concerning parasitic weeds. Farmers may require support, advice and training when developing, experimenting with or adopting innovations.

The examples given should be regarded as a guide line and not be used as a blue print. Adaptations to specific needs will always be necessary.

LITERATURE

Gwinner J 1997. Curriculumentwicklung - Aufgabe von Pflanzenschutzprojekten? GTZ-423 Info Nr. 1/97. GTZ, Eschborn, Germany.

Sene D 1995. The role of research and training in agricultural extension. In: Agricultural Extension in Africa, Proceedings of an international workshop, Yaoundé, Cameroon, January 1994. Technical Centre for Agricultural and Rural Co-operation (CTA), Wageningen, 21-28.

Author: S Kachelriess

7.1.1 International training courses

INTRODUCTION

The main target group for such courses are researchers and higher level extension staff from national organisations.

OBJECTIVES

- The participants will be trained in the biology and ecology of parasitic weeds, their control measures, and they will learn to apply different research techniques.
- The participants exchange and discuss experiences amongst themselves and with different resource persons.
- The participants discuss their role and the relations between research extension and farmers.

MATERIALS

✓ It is not possible to give a comprehensive list of materials and time required to organise and conduct such a course. The manual which you hold in your hands should already serve you for this purpose. The programme of the course will depend on available facilities but also on which topics the course may focus. To learn and practise some of the research methods in the laboratory, the greenhouse or the field sufficient capacity of personnel and quantities of material and equipment needs to be available.

The following information is based on experiences with a two-week training course organised and conducted by the GTZ-UH project.

PROCEDURE

☞ The focus is on practising ('learning by doing') and not on demonstration. All participants have the chance to practice research methodologies and gain first-hand experience with the help of experienced supervisors.

☞ International exchange of information and experiences is another focus of the workshop. Hence, experienced scientists and resource persons from various institutions are invited to present and discuss a range of topics with the participants. The participants also present aspects of their work to the plenary and discuss with their colleagues. Only the programme of the course will be presented here together with comments received during the evaluation through the participants and some additional remarks and comments from resource persons and the organisers.

Programme of the course:

- Ecology, biology and distribution of parasitic weeds.
- Agricultural systems with parasitic weed problems.
- Economic aspects of parasitic weeds.
- Control methods (Cultural control with aspects of hand weeding, sowing date, trap and catch crops, resistant varieties, chemical control, physical control, biological control with insects and fungi).
- Socio-economic aspects and limitations of introducing control methods to small scale farmers
- Research methodologies:
  - Quantification of infestation and yield loss.
  - Screening techniques for selection of host plant resistance
  - Methods to evaluate effectiveness of control under laboratory, greenhouse and field conditions
  - Viability tests of parasite seeds, germination tests and root chamber techniques
  - Separation of *Orobanche/Striga* seeds from the soil
- Introduction to participatory technology development
- Communication and agricultural extension

EVALUATION AND COMMENTS

* To improve the quality of the course an evaluation is done by the participants at the end. Mainly four aspects are taken into consideration, the programme, the organisation, the laboratory and field work and the accommodation. In general it can be stated that the more practical exercises the better, which has been expressed throughout all evaluations.
* The structure of the programme has been appreciated but frequently it was wished to have more time and to extend the course.
* Literature and copies of selected Literature and access to library facilities have been greatly appreciated by the participants. All demonstrated and practised methods should have been compiled in a training manual and distributed to the participants.
* To see whether expectations and objectives have been met a scoring done by the participants may help. An example is given below.

Where did I learn most?

Topic	Score	
Biology	•••••••	7
Physiology	•	1
Field methods & demonstrations	••••	4
Lab methods & demonstrations	•••••••	8
Principles of control methods	••••••••••	10
Participatory on-farm research	••••	5
Basics of agricultural extension	•••	3
Breeding techniques	••	2
Working groups	••	2
Others		

Source: Course evaluation Kenya '96

COSTS

A major share of the total costs can be attributed to the costs for air-tickets and allowances.

7.1.2 National training of extension staff

INTRODUCTION

Training of extension staff is an important component if they are to assist farmers as advisors and facilitators in the implementation of an integrated control approach. It is indispensable that extension staff know about the ecology and biology of parasitic weeds. It is equally important to be able to recognise and discuss the relation of parasitic weed control with other ecological, economical, social and cultural factors relevant to farming. The extension staff should have and be able to create the necessary awareness that only the long-term application of (combinations of) control methods can substantially reduce the threat by parasitic weeds to crop production.

Although the training course described here is more of a technical training, there is also an emphasis on communication and interaction with farmers.

OBJECTIVES

- Participants should know and be able to explain the biology and ecology of parasitic weeds (to farmers).
- Participants should know different control methods and be able to select and combine adapted control methods.
- Participants should know about appropriate extension materials and their use.
- Participants should be able to stimulate a discussion with and amongst farmers.
- Participants should discuss on-farm trials with farmers and researchers in the field.
- Participants should have the opportunity to exchange experiences with colleagues.
- Participants should be able to train other staff as trainers.

PREPARATION AND MATERIALS

✓ Prior to the organisation of a training course national resource persons should be identified and contacted to help with the adaptation of the course content and training material to local needs and conditions. The preparation and adaptation of the course content and the material together with the national resource persons may require some months depending on the communication facilities.

✓ During the course, the use of the following didactic materials is proposed:

- transparencies,
- slides,
- brochures/leaflets.

✓ The transparencies as well as additional information are documented in a training manual. Other printed material can supplement the manual e.g., technical leaflets, field guides (with photos) or selected publications.

✓ The information should be presented as much as possible in a visual form. Sketches and graphics of all relevant aspects related to the life cycle and the relation between the parasite and its host should figure on the transparencies. Other important information like seed dispersal and explanations of control measures should be presented in the form of short paragraphs of text. The presentation of transparencies should be supplemented by the use of real samples as well as slides, e.g., showing the different species and their hosts and views from the microscope.

- ✓ Trainers should not teach in a top-down approach but animate reflection and discussions. A dialogue and mutual listening should be practised and a learner-oriented way of training should be employed. It is advantageous to have more than one resource person or an additional moderator/chairman during the workshop. The team of trainers can give feed-back to each other and discuss and adapt the course when need arises.
- ✓ Generally it should be preferred to engage trainers from a similar background as the participants, e.g., coming from the same country or region. Often this may not be possible. Then, staff from outside the region or the country will need to be invited for initial training courses.

Researchers actively involved in applied research could serve as resource person. They could take such part of the training course related to their work. Preparatory talks by the organisers of a course have to be held with the researcher to ensure that the presented information is relevant to extension staff and understandable to non-academic personnel.

Previously trained staff or **trainers of extension staff** could serve as resource persons and would have the advantage of knowing the course also from a participant point of view. It would be important that the trainer has been actively involved in extension activities or field work to be able to contribute to the adaptation of the course content and programme to the specific requirements.

Apart from the didactic material (transparencies, booklets etc.) it is necessary to get or prepare the following materials for an effective training:

- ✓ Slide projector and screen.
- ✓ Over-head projector, transparency markers.
- ✓ At least 3 pinboards.
- ✓ Brown paper for pinboards, adhesive tape, glue, colour cards, pins and felt marker in 3 colours.
- ✓ Writing pads or note books and pens for the participants.
- ✓ Demonstration material:
  seeds, dried and fresh plants, herbicide containers.
- ✓ Prepared root chambers.
- ✓ Microscope, magnifying lenses.
- ✓ Sufficient copies of benchmark tests (2x) and course evaluation questionnaire.
- ✓ Samples of extension material.
- ✓ Name badges.
- ✓ Certificates of participation.
- ✓ Computer for documentation.
- ✓ Knapsack sprayer, bucket, measuring cylinder, stop watch (for the demonstration of glyphosate application for broomrape control).

PROGRAMME

Generally the length of a training course should not exceed three full days (→ annex 11.10). Whether the course is conducted before or during the growing season may depend on the pLiterature of the partner institutions and the availability of time during specific periods of the year, e.g., during the growing season staff and farmers may be too busy with their normal duties and field work, but during other times, e.g., witchweed or broomrape may not be found in the field and the presentation of the parasites or the observation of effects of applied control measures will not be possible. If the training is held early enough in the growing season staff can immediately try to use newly gained knowledge during the on-going campaign.

Several aspects of a training course could be handled as

- practical demonstration at the workshop site (e.g., as a role play) or
- at farm-level like the discussion and evaluation of control options or extension material with farmers.

In the case of herbicide application (glyphosate) a full day might be dedicated to the correct use and adjustment of spraying equipment and the discussion of possible ways to explain and give relevant information to illiterate farmers. An example of a structure of a training course is given in Table 10.

Table 10. Example modules of a training course for extension staff

Session	Day 1	Day 2	Day 3
A morning 1.5 -2 hours	Opening Introduction Expectations Benchmark test	Relation between the farming system and parasitic weeds and their impact on agricultural production	Practical demonstration (e.g., correct use of a knapsack sprayer)
		Pause/Break	
B morning 1.5 -2 hours	Parasite species and their host plants Distribution Life cycle Reproduction and dissemination	Control methods and their combinations	Field trip Visit of farmers fields, demonstration plots, research station or preparation of a role play
		Lunch break	
C afternoon 1.5 - 2 hours	Question and answers Practicals Preparation of a role play on "how to explain the biology and spread of the parasite(s) to farmers"	Questions and answers Working groups on "integrated control"	Evaluation of field trip Extension material or role play
		Pause/Break	
D afternoon 1.5 - 2 hours	Role play	Participant's presentation of the proposed integrated control strategy	Benchmark test Evaluation Closing (Certificates)

The contribution of working groups, presentations by participants and role plays can not be underestimated. Working in groups gives more room for reflections and exchange and helps to deepen the knowledge. Participants' presentations gives room to express, to share and to discuss different views and experiences. The preparation of role plays as well as the role play as such tell many problems extension staff have to face when translating "scientific" knowledge into a local, the farmers, language. This may not only be a semantic problem but also reveals different views and perceptions of the environment (even of the world) as expressed in a language.

EVALUATION

* During the workshop (e.g., each day before closing) the expectations of the participants, as expressed on the first day, could be looked at together and asked whether they have been met so far. This allows to adapt the programme if necessary.
* The results of the first benchmark test (example → annex 11.11) already gives hints where the knowledge of the participants is limited or is already quite good. This should help the trainers to specially pay attention to the weak points revealed by the test.
* The evaluation of the workshop has two aspects:
    - the evaluation of the content, the programme and the organisation done by the participants and,
    - an evaluation of the improved knowledge on parasitic weeds through a second benchmark test.
* The evaluation of the course can be done in two ways:
    1. Using cards and visualisation on pinboards;
        - What did I like at the workshop?
        - What should be improved?
    2. By a questionnaire distributed to the participants. Such a questionnaire is included in the annexe → chapter 11.12.
* To know whether the presented information has been assimilated by the participants the comparison of the benchmark tests can be used.

A brief report including

- a list of participants,
- a group photo,
- the programme,
- some visualised results of group work results supplemented by comments,
- the evaluation,
- and the test results

should be prepared by the organisers and distributed to the participants as well as to other relevant personnel.

COMMENTS

Generally the training described above was much appreciated by the participants of all workshops conducted. The question arises on how to sustain or incorporate such training into existing training programmes of extension services. In the long run in-service training might be less necessary if information on the biology and control of parasitic weeds becomes part of the pre-service training for extension staff.

LITERATURE

Reports of the training courses exist for all training courses held in co-operation with the GTZ-UH supra-regional project.

Albrecht H, H Bergmann, G Diederich, E Grosser, U Hoffmann, P Keller, G Payr & R Sülzer 1989. Agricultural Extension, Volume 1, Basic concepts and methods. Rural Development Series, Sonderpublikation der GTZ, No. 212.

Adams ME 1988. Agricultural extension in developing countries. Longman Scientific & Technical, ELBS edition, Singapore.

UNICEF 1993. VIPP Visualisation in participatory programmes. A manual for facilitators and trainers involved in participatory group events. UNICEF Bangladesh. P.O.B. 58, Dhaka 1000, Bangladesh.

ADDRESSES

List of institutions which collaborated in conducting training courses for extension staff.

INRA, Centre Régional de la Recherche Agronomique Saïs-Moyen Atlas, B.P. 578, Meknès, Maroc.

INRAT, Rue Hédi Karray, 2080 Ariana, Tunisie.

Ministère de L'Agriculture et du Développement Rural, DPV, B.P. 1042, 101 Antananarivo, Madagascar.

Department of Agricultural Research Technical Services, P.O.B. 30145, Lilongwe, Malawi.

Department of Agricultural Extension and Training, P.O.B. 30779, Lilongwe, Malawi

MoFA, P.O.B. 14, Tamale N/R, Ghana.

SARI Nyankpala, P.O.B. 52, Tamale N/R, Ghana.

ACDEP-AIS, P.O.B. 1188, Tamale N/R, Ghana.

Inst. Agronomique INA, Dept. de Botanique, El Harrache 16200, Alger, Algérie.

Service Protection des Végétaux et Controle Phytosanitaire, B.P. 58, Porto-Novo, Benin.

Author: S Kachelriess & J Kroschel

7.1.3 Training of farmers

INTRODUCTION

"Training" of farmers is done under various circumstances and in several ways, e.g.,:

- During extension activities in the field.
- As formal group training.
- By other farmers.
- In the field through demonstrations or "learning by doing".
- On topics related to agricultural production, crop protection.
- On organisational topics like group formation.
- Others.

The training referred to here is a subject matter training specially organised for young farmers at a centralised place like a training centre. Such training may also be part of the curriculum for apprentices in agriculture. Aspects of farmer to farmer training are presented in chapter 7.3.

The following chapter is mainly based on experiences with the training of apprentices (sons of farmers) organised in Morocco. It is a shortened and condensed version of the training course for extension staff concentrating on the practical aspect of broomrape control in broad beans by the use of glyphosate.

OBJECTIVES

- The apprentices have a general idea on the biology and ecology of the parasite.
- The apprentices know of different control methods.
- The apprentices know about the correct use of a knapsack sprayer.
- The apprentices are able to use glyphosate on their farms for *Orobanche* control in broad beans.

MATERIALS

The course is held in the local language (Arabic) and all didactic material is prepared in Arabic.

✓ Different types of functional knapsack sprayers and spraying nozzles.

✓ Demonstration knapsack sprayer (cut open).

✓ Bucket, measuring cylinder, stop watch.

PROCEDURES

☞ A sample programme of a day's training session is given in the annex → chapter 11.10. General information about the parasite is presented and discussed with the help of drawings. Samples of faba beans with different developmental stages of *Orobanche* should be used to discuss the appropriate time of intervention.

☞ A choice of knapsack sprayer (if possible brought by the farmers) is used to demonstrate the appropriate use and necessary adjustments of the spraying equipment. This also opens the opportunity to discuss maintenance and security issues when handling spraying equipment and pesticides (Fig. 28).

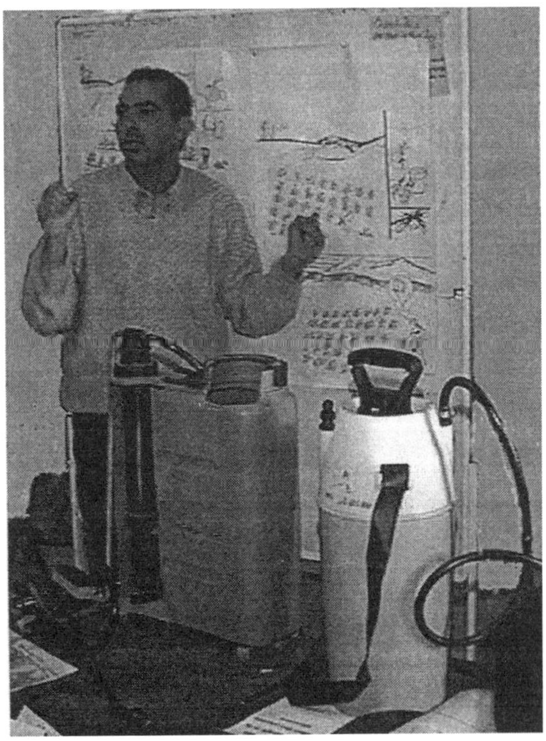

Fig. 28. Demonstration of the function and use of different knapsack sprayers

☞ The calculation necessary to know the correct amount of pesticide to be added for one filling of the sprayer and the number of sprayers needed for one hectare usually attracts lots of discussion. This point is very crucial and it needs to be verified by the animator that all participants have understood and accepted the necessity and procedure to determine the right quantity of glyphosate to be used.

☞ The theory is then put into practice. The participants measure, adjust and verify the spraying width, the walking speed and quantity of water to be utilised themselves.

☞ At a closing session remaining question can be answered. A brief evaluation of the training then closes the workshop.

EVALUATION

∗ It is not yet possible to evaluate the impact of the training courses held in Morocco. The evaluation of the workshops by the participants indicate that they appreciated the practical training and that they have gained useful knowledge.

COMMENTS

- The training described here is a special case since the apprentices were able to read and write at least in Arabic. This pre-condition will not always be fulfilled.
- Such specific training may also be done for other topics and on-site (e.g., at a field or in the village) depending on the facilities needed and available for such training.
- The costs for the training were shared amongst different parties. Therefore it is not possible to total them. Costs incurred were the trans-

port to and from the workshop venue and lunch. The animator(s) did not receive a honorarium for their assistance, as they were employed by either one of the organising institutes.

- On the long-term it is envisaged to include the topic of broomrape control into the curriculum of agricultural apprentices. This might be appropriate if agriculture is more "professionalised", meaning that farming is a profession and not a way of life to the farmer.

LITERATURE

Matthews GA & EW Thornhill 1994. Pesticide application equipment for use in agriculture, Vol. 1, Manually carried equipment. FAO Agricultural Services Bulletin 112/1. FAO, Rome.

Schwab A 1989. Pestizideinsatz in Entwicklungsländern. Ein Leitfaden für Entwicklungshelfer. Tropical Agroecology 3. Margraf, Weikersheim.

Author: S Kachelriess

7.2 Extension aids

Extension material can consist of a variety of choices. There are printed material like technical leaflets, manuals, brochures etc., demonstration material and samples, role plays, picture series, mass communication through radio or audio visuals like video tapes and more. An overview of their comparative advantages and inconveniences is given by Albrecht *et al.* (1990).

The potential users are extension staff. The material could be intended to be used with and for farmers during extension activities (e.g., pictures, role plays), but also as information material for the extension staff (e.g., technical leaflets).

The following will concentrate on the development of visual extension material to be used on-site and information material mainly targeted for the use of extension staff. However, in the annex (→ annex 11.13) we have included an example of a radio play which was developed and used in Northern Ghana in *Striga* extension).

7.2.1 Picture series

INTRODUCTION

The awareness about and the understanding of the biology of parasitic weeds is often a bottleneck for the introduction of control methods. Farmers frequently do not know much about the reproduction, number and longevity of the seeds and about the difference between host plants and trap crops. They lack technical knowledge about the appropriate use of control methods. This makes the adaptation of control methods to individual on-farm situations a problem. Farmers are therefore rarely in the position to take sound decisions on the control of parasitic weeds although the parasites can be a serious threat to their livelihood.

Information about the biology of parasitic weeds and possible control methods against them is available through research reports and publications written by scientists. This form of information is not suitable to farmers. It is essential to transform this material for extension workers and farmers to be able to make use of it. The method of communicating this information should suit the local socio-cultural conditions, e.g., taking a high rate of illiteracy into consideration.

Appropriate pictures help to stimulate a discussion with and amongst farmers. The development of visual extension material as described here is based on experiences with the GRAAP (french: *Groupe de Recherche et d'Appui pour l'Autopromotion Paysanne*) methodology and the CFSME (french: *Conscientisation, Formation, Stimulation, Moyens, Evaluation*) approach. Although the pictures series presented in this section focuses to a larger extent on the transfer of information e.g., regarding the biology of the parasite, it contains elements of an open dialogue. This leaves room for discussions and to incorporate farmers' knowledge.

OBJECTIVES

- The information given to farmers should enable them to take decisions on their own and find individual solutions.
- The information given should take the limited capacities (resources like land, labour...) into consideration that are available to farmers.
- The pictures should stimulate a discussion between farmers and the extension personnel.
- The extension series should assist a problem solving dialogue and mutual listening instead of teaching in a top-down approach.
- The materials should be easy to use and easy to be produced and transported.

PREPARATION

Materials needed

✓ The need for materials varies during the different steps of development, testing and adaptation and refining of the materials. To be able to conduct small workshops with extension agents facilities like flip charts, boards and stationary articles are needed. The multiplication of pictures requires photocopying or printing facilities. Colour can be added manually by pencils or water colours or colour prints can be made depending on available facilities.

✓ Before the picture series can be used in the field the reverse of the copies must be prepared with sawdust patches. The construction of an appropriate cloth board (→ Fig. 35) in an adequate size (1 m x 1.5 m) will be required.

Skills needed

✓ The "researcher" needs animation (moderation) skills, patience and sensibility in (cross-cultural) communication to be able to conduct working sessions with extension staff.

✓ It proves to be very useful if the researcher has some basic knowledge of the local language to be able to follow village sessions without an interpreter and to pose questions directly to the farmers.

✓ For visual materials like drawings, paintings, etc. a person with drawing skills is needed. The artist should be familiar with the agricultural practices in the region and drawing rural life.

PROCEDURES

☞ Before starting anything new it is advisable to see what others have tried in this field, some NGO or projects of other donors or organisations in other countries may have developed something useful as a base for your own work.

☞ The existing knowledge (written and oral information) with regard to the biology and control of parasitic weeds with special reference to the local situation should be analysed, discussed and evaluated with and by extension staff. This can be done in the form of team meetings. Key information (messages) could then be developed.

☞ It is advisable to split the overall information into logical components, e.g., introduction, background information, awareness creation and biology and propagation of the parasite, preventive methods and control methods with immediate or long-term effect and meaningful combinations. These units should then be dealt with separately.

☞ Team members should make their proposals for the visualisation of specific issues.

☞ Existing picture materials should be critically analysed together with the extension staff with regard to comprehensibility and their content of information.

☞ The message(s) each picture of the series should contain need to be transformed into written instructions and sketches for an artist. Possibly the artist should be present during the discussions about the design and content of the pictures.

☞ In the case of the "*Striga* Control Programme" developed in Ghana two teams of extension staff of two different districts were working in separate teams and their results were later compared and discussed at a plenary session (Table 11).

Table 11. Steps and activities to design the first version of the "*Striga* control programme"

Step	Activity / Task
1. team meeting	Introduction on the development of extension series Development of part 1: Awareness creation and long term approach
2. and 3. team meeting	Development of part 2: Biology
4. - 6. team meeting	Development of part 3: Control methods
1. plenary session	Discussion of team proposals and decision on contents and visualisation of part 1
2. plenary session	Discussion of team proposals and decision on contents and visualisation of parts 2 and 3

☞ The following questions will help to structure the team meetings:

- What issues should be discussed when presenting the picture?
- What should be its "message"? (Which information should be transmitted to the viewer?)
- What should the picture show?

☞ If possible the artist should participate at the group sessions when ideas for the pictures are discussed. Written instructions and possibly some sketches should be given to an artist and the preliminary black and white drawings discussed with extension staff. The minimum size for the draft version of pictures should be at least A4. The pictures should be drawn by using a marker (line width >2 mm). Fine pencil drawings with grey shades and many details are later on more difficult to modify, less readable from a distance and very difficult to colour.

☞ The first draft of drawings is then tested at extension staff level and the sequence of pictures to be shown needs to be determined. Farmers also may already be involved in the evaluation of the first drafts to get a better idea about their perception of the pictures and their understanding of the "message" contained (as the case for Morocco).

☞ By revisiting the two previous questions the pictures should be checked for whether they

- express the initial intention and/or
- will correctly transmit the desired information.

These questions should be clearly answered with yes, otherwise the series has to be redrawn.

☞ The following should be considered when designing, drawing and evaluating the pictures:

- The target population, those farmers or people who should be reached by the pictures, should easily recognise themselves and their situation in the pictures.
- All details shown in the pictures should be real and easy to understand.
- The picture should not show too many details that confuse or distract the viewer.
- The background shown in the pictures should be identical to the local situation with which the recipients are familiar.
- The proportion of size has to be considered correctly.
- If colours should be used they have to be close to the natural colours of the object shown.
- The message or information contained in the drawing must be easily accessible!

☞ If any of the above seems problematic or not clear the picture should be worked over by the artist and checked again.

☞ Parallel to the design of the pictures a manual containing all relevant information for the extension worker and the use of the pictures should be developed. It will serve as a reference for the staff for preparation of the extension sessions and as background information during the village visits.

☞ Only showing the pictures is not sufficient. The manual should contain necessary information to be given while discussing the picture. To stimulate a discussion between and with the farmers some questions which could be asked during the presentation need to be developed. This information and possible questions and hints for moderating a group discussion should figure in the manual.

☞ Now the extension series will be tested in the working areas of the staff who participated in the making of the materials. At first the extension agent using the material should be assisted by a colleague. This serves two purposes:

- The extension staff is trained on the use of the picture series.
- The series itself can be validated under "real" conditions.

☞ The problems encountered during the performance of the programme in the villages and the comments of the farmers serve as a new source for necessary modifications. During and after further tests the picture series and the handbook will be continuously refined.

EVALUATION

✶ Evaluation is a continuous process in the design and basis for all refinements of the information contained, the method applied and the used materials.

✶ The evaluation in the villages is very important, even if it reveals shortcomings with regard to the approach and material used. It may not always be easy to differentiate between the underlying reasons why a message is not understood, not favoured or if other rationales hinder the application of a certain proposal. A continuous dialogue and critical assessment are necessary.

EXAMPLE

The development of one picture will be illustrated step by step in the following. The example chosen is "Difference between a 'normal' weed and *Striga*". It illustrates its evolution from the first idea to its actual form. Only the black and white versions are shown.

The message the picture should contain is what makes *Striga*, being a parasite different from other plants. Its effect on the host plant (maize) should also be shown.

Proposals on how to visualise this information were made by one team at their meeting (**Fig. 29**).

During the pre-test at extension staff level the first sketch was discussed by the plenary. Instructions together with an improved sketch were given to an artist (Fig. 30).

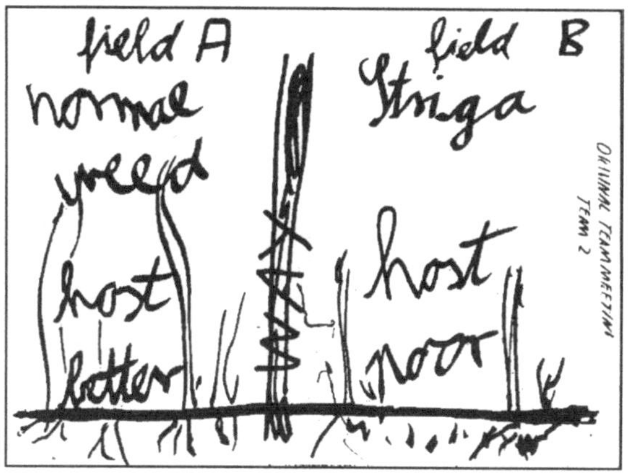

Fig. 29. Draft proposal for the design of a picture to discuss the difference between *Striga* and normal weeds.

"The picture is divided in field A with maize and a normal weed and field B with maize and Striga. *In the foreground we see the roots in the soil. A path is the border between the fields"*

Fig. 30. Sketch with drawing instructions for the artist.

An artist drafted the first set of pictures.

The first draft was not satisfactory and needed to be altered (Fig. 31). A new artist was found which improved the picture (Fig. 32).

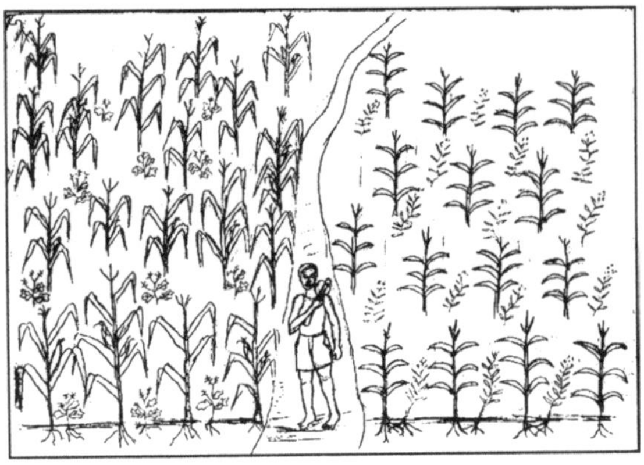

Fig. 31. First drawing of the artist

Fig. 32. Improved version of the picture

At the same time the content of the accompanying manual was discussed. The picture should be shown to the farmers while asking the question: *"What are differences?"*. The following should be mentioned by the audience:

Left:

- *Attachment between the roots of* Striga *and the maize.*
- *The maize looks very poor.*
- *The leaves are pale and have patches.*
- *Smaller cobs.*

Right:

- *No attachment between the roots of the normal weed and the maize.*
- *Maize looks better.*
- *Leaves show no changes in colour.*
- *Bigger cobs.*

The extension staff should then mention that: "Striga *takes not only water and minerals from the soil like other plants. Through the attachment to the roots of the host plant* Striga *takes water, minerals and energy from the maize. This causes the damage and the typical symptoms."*

Then, the effect of *Striga* on its host should be discussed by asking: *"What are the symptoms of* Striga *infestation?"*. The symptoms described by the farmers may include *wilting (fading), growth reduction, less yield and lower seed quality.*

The extension staff may then close the discussion and leading to a new subject: *"At the flowering stage (as shown on the image)* Striga *is very striking and everyone can recognise it. With the formation of capsules and the production of seeds a new generation of* Striga *begins".*

During the field test in the villages it became clear that the drawing needed to be further simplified to ensure a more easy transmission of the information contained (Fig. 33).

Fig. 33. Reducing the details enhanced the readability of the picture

The simplified version was then also produced in colour. Colouring of this copy was possible but very time consuming since it was done manually with colour pencil. The coloration increased the readability from greater distances and made the pictures more attractive.

For easy and faster coloration the image was further simplified (Fig. 34).

The use of marker instead of pencil forced to omit (unnecessary) details. The readability was subsequently increased.

Fig. 34. Further simplified black and white version now suitable for easier colouring

COMMENTS

- The whole process from the first discussions with the extension staff and the version shown in the last picture above took more than three years. It should not be expected to obtain immediate results. The process can not be rushed.
- Extension approaches and visual aids cannot simply be copied, but have to be adapted through a creative process, by participatory material development with researchers, extension agents and farmers. The pictures shown give only an example of such extension material. Such programmes are rarely "ideal", but, necessarily, emerge through pragmatic compromise between conflicting objectives and negotiation among different actors of differing interests and needs. The constraints of a real

working situation then make the participatory material development process more worthy of replication than the material itself.

ADDITIONAL MATERIALS REQUIRED

- ✓ All pictures need to prepared with sawdust at the reverse side. Sawdust should be applied on glue patches.
- ✓ The pictures should then adhere to a suitable cloth board. On this board they can be arranged and combined as required during the extension session (Fig. 35).
- ✓ Cloth material like felt or flannel will be used for the board. A textile store or an upholsterer may provide such material. If no felt or flannel is available you have to look for a similar cloth which the prepared pictures will stick to.
- ✓ The colour should be uni-coloured, if possible black. A multicoloured cloth is not suitable as it diverts attention too much from the pictures.
- ✓ On both edges of the cloth small wood rods will be fixed. The rods have to be a bit longer than the width of the cloth. The wood rods will be wrapped in the cloth and then the cloth will be attached to the rod with a few small nails.
- ✓ At one rod a solid string will be fixed in order to hang up the board.
- ✓ For transport the board will be rolled up.

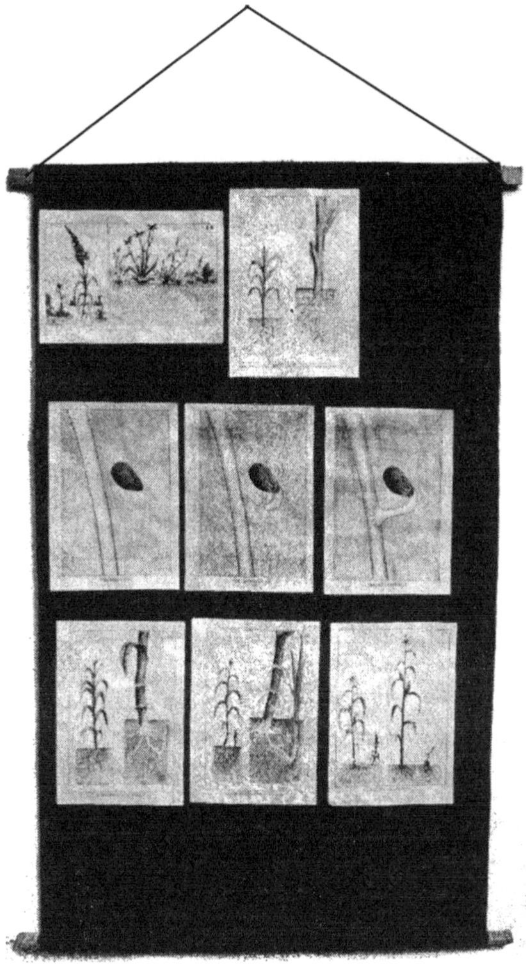

Fig. 35. Pictures from the *Striga* control programme fixed on a cloth board (dimensions 1 m x 1.5 m)

USING PICTURE SERIES

- ☞ A picture series is designed for group extension. The target group is every interested farmer.
- ☞ The extension worker should fix a date and a meeting place in one village and invite the participants personally. The group size should be between 15 and 20 farmers. If there are too many participants it is difficult for the individual to follow the programme. If there are too few the development of a discussion will be hampered. The participants should belong to different age groups.
- ☞ Discussions are an important part of the programme. The active co-operation of the participants is re-

quired. Through added questions the extension worker should achieve an active participation of the farmers and discussions in the group. In that way farmers will be encouraged to exchange their experiences and together express their perception on how to control parasitic weeds.

☞ The extension worker shall not appear to be a teacher who only gives a lecture, but he has to act as a moderator by using suitable questions.

☞ The extension worker should give the farmers enough time to answer. He should not immediately give the answers as are mentioned in the manual. If the group does not know the "right" answer the moderator (extension worker) should repeat the question or re-formulate the question.

☞ Through a clear explanation of the biology of the parasite and the control options, supported by questions of the moderator, the farmers should be able to individually identify possible solutions.

☞ This process is supported by the use of the pictures. First the pictures are shown to the participants, e.g., by going round and asking what they see on the picture. Questions will be asked and the content and message of the picture discussed. Then the picture will be fixed to the flannel board. It is important that the pictures will be fixed in coherent sequences. A sequence is e.g., the underground development of the parasite. After completing a sequence the pictures are removed from the board and the extension worker continues with the next sequence if no remarks arise or further discussion is needed.

☞ During the performance, the extension worker has to ask several times some check questions to verify if every participant has understood the mentioned topics. If something is not clear, the extension worker, in his function as moderator, has to clarify the obscurities through discussion with the group.

☞ Together with the extension officer the farmers should be able to select the most appropriate combination of control methods that fits best into their individual situation.

☞ A field visit should be organised to where farmers already apply one or more of the control methods. As the farmers hear about and see the results obtained by other farmers they will be more convinced of experimenting with new control options.

☞ It seems useful that the programme together with farm visits will be carried out several times during the vegetation period. So the development of the crops and the parasite can be pursued.

☞ Therefore every extension worker should look for farmers in his area who use some of the methods and ask them whether they agree to a presentation of their fields to farmer groups.

☞ Also visits to research stations support the programme.

EVALUATION

✶ The experiences gained through group animation at more than 50 villages includes the following advantages over other media such as video or slide presentations:

- The materials are easy to transport.

- No power source or further technical equipment is needed.
- It is relatively cheap to produce and replicate.

Experiences made during the animation revealed:

* Through visualisation participation and concentration is enhanced.
* The material is suitable for a better understanding of the biology of *Striga*.
* The animation stimulates an immediate urge for action.
* The material can be adjusted to the special requirements of a particular village animation considering the state of knowledge and experience of participants.
* The newly developed optional pictures offer avenues to highlight special issues, e.g., the need for group action or communal labour in the area of *Striga* control.

Nevertheless, shortcomings were also realised, such as:

* Some pictures are still not easy to understand even though pre-testing and further refinement took place. This can be attributed to the fact that they do not reflect common cultural practices, e.g., pictures of sole crops, post harvest weeding.
* The animator needs to be well versed in the programme through practice and coaching. Skills for moderating group discussions are often lacking.
* The inability of some extension staff to communicate fluently in the local language is a major constraint to the use of the material.
* Unsustained participation through the session which is often attributed to the length of the session (1.5 - 3 hours), and partly also to be ascribed to poor moderation quality or ability.
* Effective participation is expected with a crowd of 15-20. However, each time the animation took place a big crowd gathered (up to 100 people) making it difficult for extension staff to animate the large crowd and to ensure adequate participation of all participants.
* In those instances where too many people gathered, a good view onto the pictures could not be ensured for everybody, the size of the picture material being too small.
* In most cases the participation of women was very limited and could only be enhanced by repeatedly addressing them directly.
* Adverse weather conditions, wind or rain, sometimes made the use of the material difficult if not impossible.

COMMENTS

- Combining the picture series in a flip chart block is often proposed. This may reduce some of the technical disadvantages (e.g., no need for a flannel board, pictures always in "correct" order, less prone to adverse weather conditions). On the other hand, by using a flip chart block a major advantage, the "sequencing" of pictures, by fixing them on a flannel board, is then lost.

LITERATURE

Albrecht H, H Bergmann, G Diederich, E Grosser, U Hoffmann, P Keller, G Payr & R Sülzer 1989. Agricultural Extension, Volume 1, Basic concepts and methods. Rural Development Series, Sonderpublikation der GTZ, No. 212.

Albrecht H, H Bergmann, G Diederich, E Grosser, U Hoffmann, P Keller, G Payr & R Sülzer 1990. Agricultural Extension, Volume 2, Examples and background mate-

rial. Rural Development Series, Sonderpublikation der GTZ, No. 238.
Feil P, K Hummler & S Kachelriess 1997. Development of visual extension material for a *Striga* control programme in Northern Ghana through action research. European Journal of Agricultural Education and Extension. European Journal of Agricultural Education and Extension 4(1): 1-16.
Fischer PA 1999. Action research in extension material and message development: The *Striga* problem of northern Ghana revisited. Kommunikation und Beratung (27). Margraf Verlag, Weikersheim, Germany.
Fischer PA, O Abdel-Rahman, S Kachelriess, J Kroschel & V Hoffmann. Promoting *Striga* control measures at on-farm level in Northern Ghana. PASCON Conference, Mali (in press).
Gabathuler E 1979. Le système de vulgarisation CFSME/AE. Bulletin agricole du Rwanda, 12. 10. 1979, pp. 188-199.
GRAAP 1985. Pour une pédagogie de l'autopromotion communautaire, Nouvelle édition pour les animateurs villageois (Animator's book). GRAAP, Bobo-Dioulasso.
Hoffmann V 1991. Bildgestützte Kommunikation in Schwarz-Afrika. Tropical Agriculture 7; Margraf Verlag, Weikersheim, Germany.
Hummler K 1996. *Striga* Beratung. Entwicklung eines visuel unterstützten Beratungsprogrammes zur Kontrolle von *Striga hermonthica* in der Nordregion Ghanas. Kommunikation und Beratung (5). Margraf Verlag, Weikersheim, Germany.
Kachelriess S, J Kroschel, K Hummler, V Hoffmann, B Osterburg & J Sauerborn 1996. Linking farmers and researchers - Developing appropriate extension material to implement a farmer oriented *Striga* control Strategy in Northern Ghana. In: MT Moreno, JI Cubero, D Berner, D Joel, LJ Musselman & C Parker (eds.), Advances in Parasitic Plant Research. Proceedings 6th Parasitic Weed Symposium, Cordoba, Spain, 863-870.
Kroschel J, J Sauerborn, S Kachelriess, V Hoffmann & H Mercer-Quarshie 1997. Possibilities and constraints in implementing *Striga* control methods in African agriculture. In: F Heidhues, A Fadani (eds.), Food security and innovations, successes and lessons learned, International Symposium, Verlag Peter Lang, Frankfurt am Main, Germany, 121-135.

ADDRESSES

Extension Services, MoFA, P.O.B. 14, Tamale N/R, Ghana
Direction Provinciale de l'Agriculture, Fès, CT 3202 Douiet, Maroc.
Groupe de lutte contre l'*Orobanche* sur des légumineuses alimentaires (GLUCOLA), c/o INRA, Centre Régional de la Recherche Agronomique Saïs-Moyen Atlas, Service recherche - Développement, B.P. 578, Meknès, Maroc
SARI Nyankpala, P.O.B. 52, Tamale N/R, Ghana
ACDEP-AIS, P.O.B. 1188, Tamale N/R, Ghana

Authors: S Kachelriess, K Hummler & P Fischer

7.2.2 Printed material

INTRODUCTION

Printed text material (bulletins, brochures, leaflets...) may mainly be targeted for extension staff since many farmers may be illiterate in regions where parasitic weeds are a problem. "*The information value of texts is virtually nil and it is unrealistic in most societies to hope that the son or daughter will read written information aloud*" (Albrecht *et al.* 1989). Such material could serve as basic information for the extension staff to prepare for extension sessions and to refresh the specific knowledge required.

OBJECTIVES

- To sensitise users/readers of the material,
- To provide general or specific information to extension staff,
- To provide guide lines for extension sessions.

MATERIALS

✓ Depending on the layout and quantity of material to be produced the technical requirements will vary. Simple leaflets can be produced using a typewriter and multiplication is done with a photocopier. Illustrations can be hand drawn or cut from other sources and inserted by gluing them into the master copy. Where computers and a scanner are available editing of text and graphics can be done with most standard software. This would facilitate editing and updating the document for new editions and to adapt the material to be used under different conditions, e.g., other regions.

PROCEDURES

☞ The first step would be to identify the need for information of intended users of the material. The type and quantity of information to be included in the object will then determine the content and extent of it. There could be simple leaflets, somewhat more extensive brochures or even small manuals.

☞ Leaflets will be useful for sensitising users about parasitic weeds (e.g., the biology) or to give special information on a limited topic (e.g., one control method). They contain addresses of organisations or persons to contact to get more information. Such leaflets may consist of a single A4-page, tri-folded printed (photocopied) on both sides of the paper (Fig. 36).

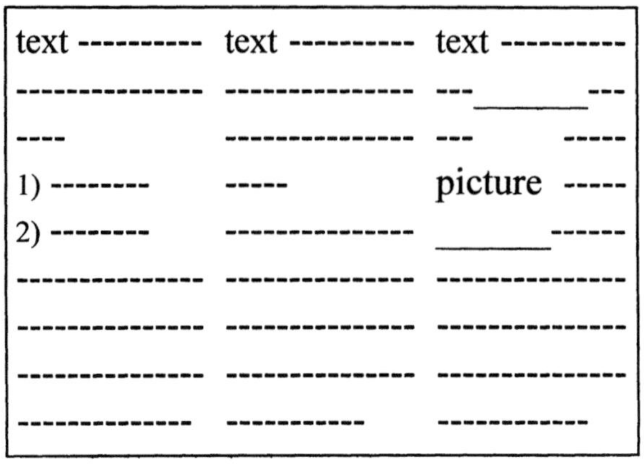

Fig. 36. Possible layout of a leaflet

☞ A limited number of graphics and diagrams could be included. Black and white photographs must be suitable for reproduction like photocopying. Leaflets can then be produced in large quantities at relatively low costs. Where appropriate the leaflet can be produced in local language.

☞ Brochures of several pages contain extended and more detailed information. They can give an overview of the problem with parasitic weeds, give details on their biology and elaborate more on possible options on how to control them. There is more room for illustrations. Background information (e.g., how to calibrate and use a knapsack sprayer) or research results can figure in such brochures.

☞ Technically, the number of pages should be a multiple of 4 to facilitate later binding.

☞ Leaflets and brochures can be part of a series which covers in its entirety all or most aspects related to the ecology, the biology, the control options, on-farm experiments and extension.

☞ Manuals can serve as guide lines for extension sessions containing instructions, e.g., on how to use extension material or how to set up field experiments. Additional information will be useful for the extension staff to prepare activities in the field. Picture books help to know and identify different species of parasitic weeds. Documentation of training material of workshops attended assists the extension staff to recall information.

☞ The copies of printed material could be distributed

- at training sessions related to the topic, and
- through the documentation and information unit of an extension service.

The first option could ensure that the material appears immediately relevant to the trainee and the motivation of reading it is higher.

EVALUATION

✶ The quantity of reprints ordered might be a good indicator for the quality of the material. A survey amongst the intended users could monitor and evaluate the impact and use of printed materials.

COMMENTS

- Leaflets are not very durable and might easily be lost. Although, occasionally very old (sometimes out dated) leaflets can be found in offices or with staff. Brochures may last longer depending on the quality of the envelope and the binding. They could be taken to and used in the field. The quantity produced is often insufficient and reprints are rarely made. Manuals, especially with colour photos are often an object of prestige and stay in offices and with higher level staff and field workers might have less or no access to them.

LITERATURE

Albrecht H, H Bergmann, G Diederich, E Grosser, U Hoffmann, P Keller, G Payr & R Sülzer 1989. Agricultural Extension, Volume 1, Basic concepts and methods. Rural Development Series, Sonderpublikation der GTZ, No. 212.

Albrecht H, H Bergmann, G Diederich, E Grosser, U Hoffmann, P Keller, G Payr & R Sülzer 1990. Agricultural Extension, Volume 2, Examples and background material. Rural Development Series, Sonderpublikation der GTZ, No. 238.

Ramaiah KV, C Parker, MJ Vasudeva Rao & LJ Musselman 1983. *Striga* identification and control handbook. ICRISAT Information Bulletin No. 15, ICRISAT, Patancheru, India, 52 pp.

Linke KH, J Sauerborn & MC Saxena 1989. *Orobanche* field guide. University of Hohenheim, Germany/ICARDA, Aleppo, Syria. (also available in Arabic).

ADDRESSES

ACDEP-Agricultural Information Service, P.O.B. 1188, Tamale N/R, Ghana

Direction Provinciale de l'Agriculture, Fès, CT Douiet, Maroc.

Extension Services, MoFA, P.O.B. 14, Tamale N/R, Ghana

Groupe de lutte contre l'*Orobanche* sur des légumineuses alimentaires (GLUCOLA), c/o INRA, Centre Régional de la Recherche Agronomique Saïs-Moyen Atlas, Service Recherche - Développement, B.P. 578, Meknès, Maroc

INRA, Centre Régional de la Recherche Agronomique Saïs-Moyen Atlas, Service Recherche - Développement, B.P. 578, Meknès, Maroc

SARI Nyankpala, P.O.B. 52, Tamale N/R, Ghana

Author: S Kachelriess

7.3 On-farm experimentation

INTRODUCTION

Recommendations for parasitic weed control are mainly based on on-station research results. Some recommendations have existed for a long time and yet very few of them are practised by farmers. This calls for the verification and validation of recommendations by the farmers themselves. Further, on-farm research may help to adapt research findings to the conditions farmers are facing and increase their understanding of proposed control options and increase their capacity in acting as researchers themselves. A fundamental problem with (on-farm) research regarding parasitic weed control is that positive (and negative) effects may only be seen after several years of applied control measures.

OBJECTIVES

- Farmers see and/or experiment with new farming options and evaluate how they may fit into their farming practice. They become "researchers" themselves.
- Extension staff gain experience with the introduction of control options to and adaptation by farmers. This will help to improve extension messages and the ability of the extension staff to support farmers to test alternatives and to adapt and adopt innovations.
- Researchers may learn from farmers and improve their research agenda to fit better to farmers needs.

These goals may be reached by different ways. Various degrees of farmer - extension staff - researcher interaction may occur within the process and within different approaches. The nature of the information to be sought determines the intensity of farmer - extension staff - researcher collaboration. The purpose of "on-farm" experiments must be clear to all parties involved.

Demonstration plots allow for demonstrating e.g., the effectiveness of control methods, like glyphosate spraying to control broomrape in broad bean, tolerant/resistant varieties or the effect of soil solarisation. Such plots could be maintained by a farmer but the management responsibility is more with the extension staff or the researcher. The uniform layout and quantity of such trials may also allow for the collection of agronomic data. Nevertheless, farmers assessment of the sample technologies will be the most important outcome of the experiment.

Joint experiments would allow and require more collaboration between the researcher/extension staff and the farmer. The ideal, though hardly ever achieved, is farmer participation right from the planning stage up to the assessment of the experiment. This would ensure that:

- Farmers views are incorporated in the planning of the experiment.
- Farmers, conversant with the concept and objectives of the experiment require less support by researchers during implementation, and if experienced with the tested options, have a sound basis for their own independent assessment of the experiment (Werner 1993).

MATERIALS

✓ There is no "standard" list of materials which could be used for the different kinds of experiments. Materials needed (e.g., pegs, note books...) as well as required inputs (e.g., seeds, fertiliser...) vary for different types and purposes of experiments. Apart from physical objects, organisational and logistic aspects are important to consider.

PROCEDURES

☞ It is important to note that the procedure is more of an evolving process of mutual learning which will take more than one season to achieve the objectives stated above. At the beginning the experiments might have more of an exploratory character while at the end adapted control option are verified and disseminated on a larger scale. It is not a linear process but contains feed-back loops which should allow adjustment whenever necessary (Fig. 37). Only some general remarks can be made here and further information can be found in the literature cited at the end of this chapter.

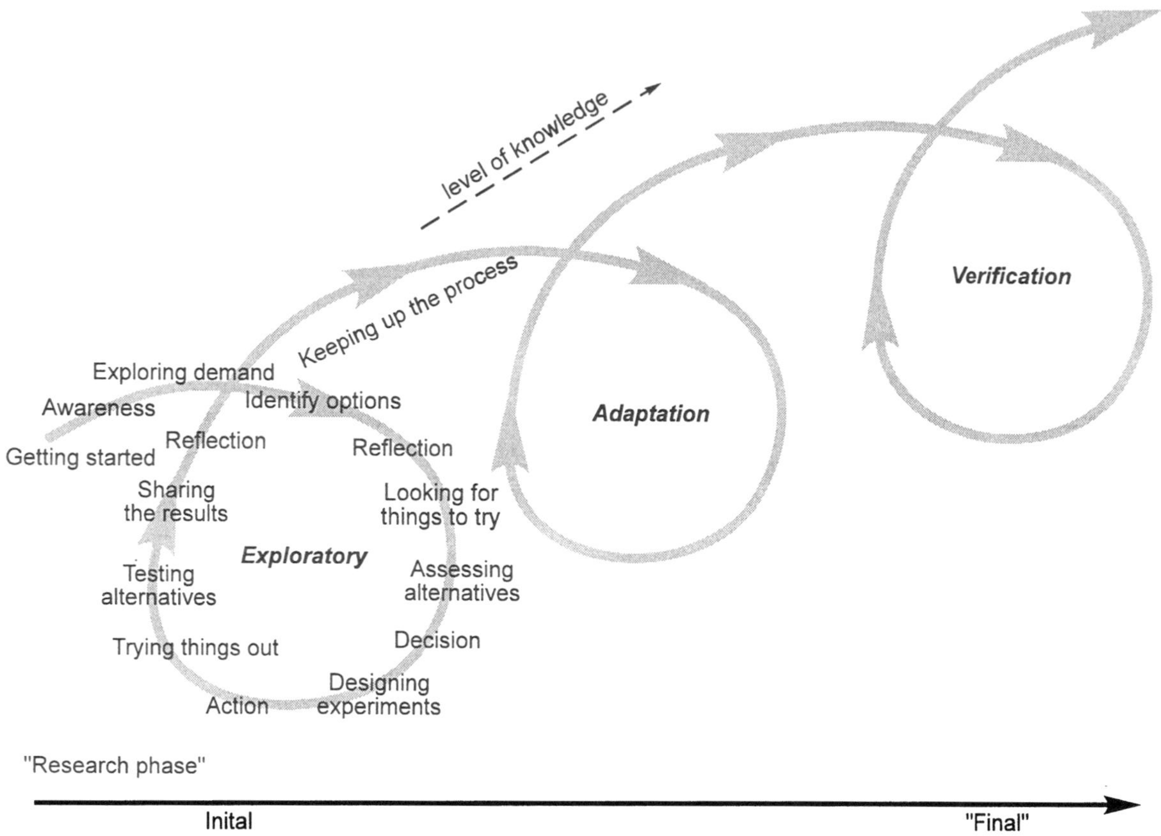

Fig. 37. The sequence of different steps and phases in participatory on-farm research

<u>1) Getting started (*Creating awareness / Exploring the demand*)</u>

☞ (Pre-)selection of site(s): Farmers who “feel a problem/opportunity” related to parasitic weeds should be contacted. It should be discussed whether there is sufficient scope to sustain the initial process. Information about “hot spots” may also be available through surveys and studies conducted earlier or through discussion with key informants like extension staff.

☞ Accessibility during all seasons of the year and logistical and organisational aspects should be considered, but also the bias of such selection should be discussed (e.g., “road bias”, meaning that you may contact only farmers close to the road, which may be better off and their opportunities may differ from their fellows far in the country side).

- ☞ Building relationships for co-operation: e.g., verbal mutual contract about the proposed process, the role of the participants, potential outcomes, and mechanisms for resolving dispute.

2) Looking for things to try (*Reflection / Identification of alternatives*)

From a variety of options existing through scientific research and farmers knowledge those should be selected which seem suitable. After this procedure a selection of options to be tested should exist. The steps could be:

- ☞ Identifying priorities / opportunities:

  Try to determine the criteria for ranking of different priorities (increasing yield, yield security...), look for untapped or under utilised resources, availability and access to resources, constraints using these resources.

- ☞ Identifying local community and review scientific knowledge / information.
- ☞ A joint screening of options could be part of or follow an extension activity. Reflection could be stimulated through farmer - expert (researcher, extension staff) workshop or farmers could be invited to a research station. Identification of selection criteria and (pre-) selection of options to be experimented with could then be done in the form of a "screening workshop" with interested farmers.
- ☞ Decisions could be prepared by reviewing existing experimental practice:
  - What do farmers try out?
  - How do they do it?
  - Why do they do it in this way?

During this phase RRA/PRA (Rural Rapid Appraisal/Participatory Rural Appraisal) tools, (semi-structured interviews with individuals and groups, ranking, time trends...) could be used to tap indigenous knowledge.

3) Designing experiments (*Reflection / Decision*)

The decision about and the design of the experiment comprises several interlinked steps. These require the interaction of all parties involved for implementing and the evaluation of the trial.

- ☞ Defining trial objectives.
- ☞ Determining the required environmental conditions.
- ☞ Choice of treatments, treatment levels and arrangement of treatments.
- ☞ Determining replications within and across farms.
- ☞ Choice of appropriate statistical design(s).
- ☞ Determining the plot sizes.
- ☞ Defining information to be gathered.
- ☞ Determining the required inputs.

The following questions may help in the process of planning and designing experiments:

- ☞ Who will actually do the experiment? (researcher/extension staff or farmer managed trial)
- ☞ What do farmers expect from researchers/extension staff? (advice ...)
- ☞ What do researchers expect from the farmers? (observations, maintenance ...)
- ☞ What treatments (levels) do we want to apply? (preferably one-factorial with few or only one treatment level)

☞ Do we need controls?

☞ How many times should the experiment be repeated and under which conditions?

☞ How do we lay out the trials?

☞ How large do we want the experiment to be?

☞ What inputs do we need and how can we obtain them?

☞ According to what criteria do we select locations? (access, degree of infestation, previous crops...)

Designing evaluation protocols:

☞ What will be our criteria for evaluating what happens in the experiment?

☞ What/when do we need to observe / measure / record / discuss and how do we organise this?

Table 12 contains criteria which will help in designing experiments. It is assumed that at the beginning trials have more of an exploratory character and that finally adapted innovations can be verified (and adopted) by more and more farmers.

Table 12. Changes of significance or extent of selected key criteria during the process of on-farm research.

Criterion	**Research phase**		
	Initial		Final
	Research objectives		
	Exploration	Adaptation	Verification
Level of knowledge/confidence with technology	+	++	+++
Farmer - researcher collaboration	+++	++	+
Possible significance of socio-economic criteria	+++	++	++
Possible no. of treatments	+++	++	+
Management responsibility of farmers	+	++	+++
Number of farmers involved	+	++	+++

Source: Werner (1993) + = low, +++ = high

<u>4) Trying things out (*Decision / Action*)</u>

This is the actual experimental phase in which observations and experiences are made, recorded and evaluated.

☞ Implementing experiments:

Especially innovations should be implemented stepwise and not one "package" given at the start and then the experimenter (the farmer) is left alone. This could only be done if the level of knowledge/confidence with the technology is already quite high. Experiences made during the set-up and implementation of the trial are the first evaluation of the experiment.

☞ Measurement / observation could be organised by:

- regular group meetings, and
- periodical collection and joint discussion of data.

☞ Evaluation of the experiment is continuous by the farmer; the extension staff and the researcher may have less chance to visit and evaluate the trial. During the growing period (at critical times of the experiment) and at or after harvest they should be present.

- Field days and exchange visits could be relevant to farmers to compare experiences and observations.
- Preference ranking and classification (and other RRA/PRA tools) could be useful to jointly evaluate observations.
- Formal statistical analysis is described in Werner (1993).

Conclusions should be drawn together and documented, they will serve as basis for next years experiments.

☞ Strengthening supportive linkages:

Apart from those people directly involved in the experiment contacts with other departments, NGOs and others could be established and a local network created. If possible superiors of staff involved should be invited to the field.

5) Sharing the results (*Evaluation, Reflection, Decision*)

☞ As during the experiment sharing the experiences and results with other "experimenters" broadens the basis to decide on how to further adapt or to adopt a technology.

- Visit to secondary sites.
- Visit of neighbours
- Explaining and discussing experiences.

☞ Experienced farmers can be assisted to help disseminating a proven technology by:

- Farmer to farmer training
- informal individual peer training
- informal group training
  - →farmer as trainer for visiting groups
  - →farmer as key presenter at workshops in other communities
- Formal group training
  - →farmer as trainer in farmer training courses

☞ Results obtained and experiences made should be documented. This documentation serves as basis for e.g.,

- Farmer's manuals and audiovisuals (dramas, shows, role plays, "modern" (electronic) media...).
- Adaptation of extension materials (leaflets, picture series...).

☞ Workshops/presentations can be organised

- For other potential members.
- For senior officials.

6) Keeping up the process

☞ As stated earlier, the process is continuous and needs to be kept going, especially in the case of parasitic weeds on which an impact can only be expected over the long term. Therefore motivation and commitment of farmers, researchers and extension staff should be kept at a high level. Experiences gained after each phase help to:

- Consolidate
  - Farmers groups.
  - Inter-village co-operation, and
  - institutional support.
- Develop locally manageable systems for monitoring:
  - The experiments.
  - The diffusion process.
  - The impact on farming, the agro-ecosystem, and livelihood.

☞ After the initial phases and building up the research capacities of all parties involved, organisational advice and development becomes increasingly important to strengthen the self-organisational capacities.

☞ Sharing is not only limited to research results but by documenting the experimentation and the process support can also be given to others.

☞ Last but least the monitoring of the impact in terms of technology adoption by other farmers is the final mile stone.

☞ Positive assessment and verbal acceptance of options by farmers during or after the trial may not be the ultimate indicator for the adoption of a potential innovation. Spontaneous adoption of innovations is the first indicator for the dissemination of an innovation. Werner (1993) suggests to investigate spontaneous adoption by a survey in the season after the experiment has been carried out. The surveyor should explore
  - How many of the framers exposed to the experiment adopted which of the trial options.
  - What are the reasons for adoption or non-adoption of trial options.
  - What kind of modifications were made to the original experimental treatments.
  - Whether adoption or rejection depends on specific characteristics of individual or groups of farmers.

☞ A high rate of spontaneous adoption will facilitate the promotion of such technology by the extension service. While the reasons for rejection (non-adoption) should be taken as a feedback for looking for things to try.

COMMENTS

- The process described above will not start on its own and sustain its continuous motion without being supported from all parties involved. It requires a shift in attitude from researchers (that scientific research and knowledge is not always superior to those of farmers), from extension staff (from teaching and telling to assistance and facilitating) and from farmers (from expecting to get free solutions as recipients to active commitment).
- All this may require training e.g., in communication skills, RRA/PRA methods, statistical analysis etc. to improve the capacities of the different actors. This needs time and money to be specially dedicated for such an approach.
- Institutional (e.g., hierarchy) and organisational and logistic issues (e.g., access to and cost of transport) may be bottlenecks and hamper a participatory development and testing of innovations. Also it should not be expected that a standard package or message which is often demanded (e.g., in the T&V extension system) is the final result of the process but that a basket of choices is given to the farmer.

LITERATURE

Eponou T 1993. Partners in agricultural technology: Linking research and technology transfer to serve farmers. ISNAR Research Report No. 1. The Hague: International Service for National Agricultural Research.

Gnon T 1995. Experiments with agricultural innovations: Community level farmer organizations and extension. In: Agricultural Extension in Africa, Proceedings of an international workshop, Yaoundé, Cameroon, January 1994. Technical Centre for Agricultural and Rural Co-operation (CTA), Wageningen, 85-94.

Gubbels P 1995. The role of peasant farmer organisations in transforming agricultural research and extension practices in West Africa. In: Agricultural Extension in Africa, Proceedings of an international workshop, Yaoundé, Cameroon, January 1994.

Technical Centre for Agricultural and Rural Co-operation (CTA), Wageningen, 95-120.
Hagmann J, E Chuma & G Nehanda 1995. Development and institutionalising of participatory innovation development and extension in the agricultural extension service in Zimbabwe. GTZ Kurzinfo der OE425, Nr. 21, GTZ, Eschborn, 37-42.
Hamilton NA 1996. Learning to learn with farmers - An adult learning extension project. PhD Thesis Agricultural University Wageningen.
Khon Kaen University 1987. Proceedings of the 1985 International Conference on Rapid Rural Appraisal. Rural Systems Research and Farming Systems Research Projects, Khon Kaen University, Thailand.
Kachelriess S 1997. Training module: Introducing Participatory Technology Development. PTD Circular August 1997, Nr. 7: 3-4. Published by ETC Netherlands, P.O.B. 64, NL-3830 AB Leusden.
Reijntjes C, B Haverkort & A Waters-Bayer 1992. Farming for the future. Macmillan Press, London and Basingstoke.
Schönhuth M & U Kievelitz 1994. Participatory learning approaches, Rapid Rural Appraisal Participatory Appraisal, an introductory guide. Schriftenreihe der GTZ, No. 248. TZ-Verlagsgesellschaft, Rossdorf.
Steiner KG 1987. On-farm experimentation handbook for rural development projects. Guidelines for the development of ecological and socio-economic sound extension messages for farmers. TZ-Verlagsgesellschaft, Rossdorf.
Werner J 1993. Participatory Development of Agricultural Innovations. Procedures and methods of on.farm Research. Schriftenreihe der GTZ No. 234, TZ-Verlags-Gesellschaft, Roßdorf, Germany.

Authors: S Kachelriess & P Fischer

8 Glossary

Note: Arrows [→] are used to refer to related terms or information.

Active ingredient = a.i.; the potent portion of a chemical applied to soil or plant on which the estimation of the chemical effect is based.

Additive gene action Allelic interaction in which dominance is absent; the → heterozygote is intermediate in →phenotype between → homozygotes for the alternative alleles; opposite of → non-additive gene action.

After-ripening Physiological or biochemical processes (probably metabolism of a germination inhibiting substance in the aleuron layer of the seed) of several months (→ dormancy) during which the seed will not germinate.

Agar gel assay Laboratory assay to evaluate host genotypes for low stimulation of *Striga* seed germination.

Agro-ecology Science of the energetic, material and informative interrelation between the living organisms including man and the natural, abiotic and technical elements of the agricultural landscape.

Amplification fragment length polymorphism (AFLP) Variation between individuals in DNA fragments amplified in a polymerase chain reaction; → DNA marker.

Appressorium Thickening of the tip of the germ tube upon its attachment to the host root (the contact to the host root vessels is not yet established).

Alectrol Germination stimulant isolated from cowpea. → strigol.

Allele One of a pair of alternative contrasting forms of a gene, in which one may be dominant and the other recessive. The pair occupies the same site or locus in a pair of homologous chromosomes.

Allogamy Fertilisation which results from cross-fertilisation; opposite of → autogamy.

Aleurone layer The outside layer of endosperm in a grain below the seed coat. It is rich in oil and protein but lacks starch.

Anthesis The period when floral parts open and pollen is dispersed from the anthers. Usually this is the period during which fertilisation occurs.

Antibiosis Unfavourable influence (inhibition of growth or reproduction) of a plant towards a pest (e.g., a parasitic weed).

Autogamy Fertilisation which results from self-pollination; opposite of → allogamy.

Autotrophic A plant that procures its food independently. It is the ability of a plant to form carbohydrates and proteins from carbon dioxide and inorganic compounds; opposite of → heterotrophic.

Biological control The control of crop pests by living organisms harmless to the crops but which destroy or reduce the number of harmful pests

Biotype	Growth form of a particular species varying in aggressiveness to a particular host plant. Synonyms: pathogenic biotype, pathotype, race, physiological race, strain. Usually the same phenotype (morphotype) as other individuals of the species.
Block	In experiments a block is a group of items under treatment or observation. In field experiments, it consists of a group of contiguous plots assigned to a piece of land. In experimental designs for statistical analyses, there are usually two or more blocks, often called replications. Blocking helps remove heterogeneity. The plots or items in a block are supposed to be more homogeneous than these in other blocks.
Border effect	The better or worse performance of yield components at the edges or borders of a plot, especially in field experiments.
Bract	A modified leaf in whose axil a flower arises.
Bud stage	Term used for the early developmental stage of *Orobanche* just prior to the → tubercle stage
Capsule	Fruit of *Striga, Orobanche, Cuscuta* and *Alectra* which develops from a compound pistil. According to the specific type of opening these capsules are loculised.
Catch crop	A crop that gets heavily attacked by parasitic weeds (any host crop). The infested crop has to be destroyed before the parasites set seed.
Check	= control, a row or plot of a well known variety or treatment included in a nursery or test plot for comparison with new entries or treatments.
Checkerboard layout	Experimental design in field experiments where each test variety is surrounded by a susceptible check cultivar.
Chemotropism	Growth of the germ-tube of *Striga, Orobanche* and *Alectra* towards the host root in response to exudates from that root (positive c.); the growth is directed by the non-homogenous distribution of chemical substances.
Circumnutation	→ nutation.
Coefficient of variation	= C.V., a quantity that is used to indicate how precise and reliable an experiment is. It is computed by dividing the sample standard deviation by sample grand mean and multiplying by 100. The C.V. depends also on the type of an experiment, e.g., insecticide or a herbicide trials will have a higher C.V. than yield trials.
Co-evolution	Joint phylogenetic development of two species with parasitic or symbiotic relationship.
Competition	Common demand of several individuals for a limited requisite. „Infra-specific c.“ = between taxa below a species, „intra-specific c.“ = between the individuals of a species, „inter-specific c.“ = between species.

Conditioning Also „preconditioning", „moisture treatment", „pre-treatment" or „sensitization". Period of 5 to 10 days during which a physiological process occurs that enables the seed of some parasitic plants to respond to a stimulant.

Correlation coefficient A measure of the degree to which variables vary together or a measure of the intensity of association , e.g., genotypic correlation. Correlation can be simple when there is only one independent variable and multiple when there is more than one independent variable.

Crop rotation → Rotation. In the context of parasitic weeds this term means the change of host and non-hosts in a rotation.

Crown roots Secondary roots of parasites.

Cultural practices Methods for the control of parasitic weeds that include crop rotation with non-hosts, sowing date, weeding, tillage, fertilisation, harvesting etc.

Curative Healing effect of plant protection methods against pests on already infected plants.

Diallel Mating design used in genetic studies; each parent line is crossed with all other parent lines, the resulting progenies are evaluated together or without the parent lines for the trait of interest.

DNA Deoxyribonucleic acid, the molecule that encodes genetic information.

DNA marker An identifiable physical location on a chromosome of individual genotypes (e.g., RFLP, AFLP) whose inheritance can be monitored. The closer the markers are to an actual gene of interest, the lower the probability that they will be separated during replication processes, and hence the greater the probability that they will be inherited together, and that the marker can be used in → marker-assisted selection.

Dominant gene action Alleles that fully manifest their → phenotype when present in the heterozygous state. The alleles whose phenotypic expressions are masked by dominant alleles are termed recessive alleles.

Dormancy Physiological rest or inactive growth in some matured seeds (and other plant organs) that delays germination or growth for some time even though there is physiological maturity. Primary dormancy = induced by endogenous factors; secondary dormancy = induced by environmental factors. Parasite seeds have a dormant period (after-ripening period) of 0 to 9 (or even more) months. This period is determined largely by internal factors or in *Cuscuta* by the seed coat; certain chemicals (oxidising agents) and physical changes (heating) reduce dormancy.

Double pot technique Technique to collect host plant root exudates.

Ecology The science that deals with the relation of organisms (plants) to their environment and to each other.

Ecological strain	Distinct individuals within a species with genetically and locally different morphological characters.
Embryo	Unlike most other seed the embryogenesis in *Striga, Orobanche* and *Cuscuta* occurs only during germination. In the seed the embryo is still undifferentiated.
Endophyte	(1) Bacteria or fungi living within the plant tissue; (2) with parasitic plant it is that part of a haustorium which is surrounded by host tissue.
Endosperm	The nutritive tissue in the seed which serves for the nutrition of the embryo and the seedling.
Epiphyte	Non-parasitic plant which lives on a host plant, e.g., Orchidaceae.
Epistatic gene action	Non-reciprocal interaction of non-allelic genes: situation where one gene masks the expression of another gene; → non-additive gene action.
Eplee bag	3 × 3 cm nylon bags with a mesh opening of 116 microns, in which *Striga* seeds may be buried at various depths in pots or field plots. The pores in the cloth allow ample passage of gases, moisture, and stimulants into the bags. The percentage viable or germinated *Striga* seeds is determined after defined periods of time.
Ethylene	Gas (C_2H_4) with growth regulation functions in plants. With some parasite seeds (*Striga*) it acts as a germination stimulant.
Facultative parasite	A hemi-parasite (e.g., *Rhinanthus, Odontites*) that can complete a full life cycle, either as a normal plant sustained only by its own photosynthesis or supported additionally by parasitism. In the latter case the facultative parasite shows more vigorous growth.
False host	→ Trap crop.
Foliar herbicide	Herbicide applied on the leaves.
Frequency	In ecological studies the frequency refers to the relative number of observations of a certain plant or insect species.
Fumigation	The destruction of any soil-born organisms (all stages of bacteria, insects, seeds etc.) by exposure to poison gas or smoke (→ soil fumigation).
Gall	Excessive or monstrous growth on or in a plant often caused by an insect and in some cases by nematodes.
Gene	The unit of inheritance passed on to progenies from parents. It is responsible for expression of heritable characters. The site it occupies on a chromosome is called a locus. A character may be controlled by one or several genes.
General combining ability (GCA)	Genetic measure of a parent line from in crossing experiments (→ diallel): mean performance of the parent line in all its crosses, expressed as deviation from the mean of all crosses.
Genetic engineering	Transfer of genes, i.e., resistance genes, from whatever organisms in which they occur into chosen crops (or other plants).

Genotype Genetic constitution of a plant (or organism); it may be expressed or unexpressed, depending on environmental effects on the plant in a given location. It also refers to a group of plants having a common or specific make-up; → phenotype.

Genus A taxonomic group comprising one or more species that are of common origin and are different from other taxonomic groups. Similar genera are grouped in a family, which is the next level above the species.

Germination Initial growth of the embryo from a seed or spore in which the radicle and shoot or hypha protrude from the seed or spore. In the case of parasite seeds any emergence of the germ tube is called germination.

Germination distance Maximal distance between sorghum rootlets and germinated *Striga* seeds in the → agar gel assay.

Germination inhibitors Hormones or chemicals that suppress germination.

Germination stimulants → stimulants.

Germ tube Hyaline root-like structure (formed by the embryo) that protrudes through the seed testa at the mycropylar end upon germination of *Striga, Orobanche* and *Alectra* seed. Its length is up to 3 mm, its width 0.12 -0.25 mm. It is also called „radicle", „germ tube like organ" or „procaulôme". In strict botanical sense it is not a root, as the calyptra is lacking.

Gibberellic acid Growth regulating substance (plant hormone); germination stimulant for *Orobanche*.

Glyphosate N-(phosphonomethyl)glycine, systemic herbicide.

GR 24 Synthetic → germination stimulant.

Hand-pulling Handweeding of emerged parasite shoots.

Haustorium Specialized contact organ of parasitic plants to obtain water and nutrients form their host plant; plural: haustoria. This term is usually applied to describe the complete intra- and extramatricular part of the contact organ, which morphologically represents a tuber of the hypocotyl (occasionally rather large in *Orobanche).* According to their contact zone they are called root-, leaf- or stem-haustoria.

Haustorium inducing factor Compound inducing the formation of haustoria, e.g., 2,6-dimethoxyparabenzoquinone for *Striga.*

Hemi-parasite Parasitic plant that has photosynthetic ability („green parasites" like *Striga, Rhinanthus, Viscum* etc.); synonym: semi-parasite. → holoparasite.

Herbicide-resistance By classical breeding or genetic engineering a certain degree of herbicide resistance of a crop plant can be attained that allows the application of heretofore phyto-toxic rates of a herbicide to a certain crop.

Heritability in the broad sense	Degree of genetic determination, i.e. the relative importance of heredity in determining phenotypic values.
Heterotrophic	Dependence of an organism (man, animal, chlorophyll-less plants, many fungi and bacteria) on other organisms or plants for at least part of it's nutrition.
Heterozygote	A diploid individual that has different alleles at the corresponding loci of homologous and therefore does not breed true.
Holo-parasite	Parasitic plant lacking any photosynthetic capacity (*Orobanche, Lathraea, Cytinus, Cistanche* etc.) and which hence depends totally on a host, i.e. it is an → obligate parasite. (Nevertheless, *Orobanche* and *Cuscuta* have still some few chlorophyll containing plastids). → hemi-parasite.
Homeostasis	The tendency of a system to maintain a dynamic equilibrium and in case of disturbance to restore the equilibrium by its own regulatory mechanisms. **Physiological homeostasis:** A type of phenotypic reaction buffering the organism against fluctual external conditions by internal self-regulatory mechanisms. **Genetic homeostasis:** Tendenc of a population to equilibrate its gene pool and to maintain a genetic composition leading to an optimal balance /adaptation to the environment in which the population lives.
Homozygote	A diploid individual that has identical alleles at the corresponding loci of homologous chromosomes and therefore does breed true.
Host (plant)	A host (of parasitic plants) is an organism that supports the development of the parasite by contributing water and nutrients.
Host preference	Within the host range of a species, some hosts are attacked more heavily or more frequently than others.
Host range	The plant species that can be attacked by a parasite species.
Host root exudates	Various chemical substances which are released from the host root into its rhizosphere. Some of them stimulate germination of parasite seed.
Host specificity	When a host range is limited to a single host species, or a very narrow range of specified hosts.
Hydroponic	A method of raising a crop by supplying all required nutrients through a nutrient solution to the root system.
Hyper-parasite	A parasite parasitized by another parasite, e.g., *Orobanche* by the insect *Phytomyza* (synonym: secondary parasite). The comparative term would be „superhyperparasite", e.g., insects parasitizing *Phytomyza*, e.g., *Tetrastichus phytomyzae* (Eulophidae).
Hypersensitivity	Resistance mechanism: locally limited cell death in response to a parasite attack; resulting in disruption of the contact between host and parasite, and therefore parasite death.
Immunity	Total resistance, i.e. no infestation despite favourable conditions for parasitism. → resistance; → susceptibility, → tolerance.

Incubation A process of keeping plants, disease pathogens or seed under a certain temperature to hasten growth or to ascertain a specific environment.

Induced dormancy Inactivity or rest period in response to climatic conditions, e.g., → wet dormancy in *Striga.*

Inflorescence The flowering shoot on which flowers are arranged on the floral axis.

Infra-specific variation Range of bio-types/genotypes between taxa below the species level. → competition.

Inoculum Spores, bacteria, eggs of insects or fragments of mycelium of pathogens which can infect plants.

In situ In natural or original position, not removing a plant from it's original position for study or for any other operation.

Integrated control Combination of methods aiming at reducing chemical plant protection to the necessary level by using primarily biological, biotechnical, cultural and breeding methods.

Inter-cropping Growing two or more crops simultaneously in a field in different but proximate stands.

Inter-specific variation Range of types between species. → competition; → specific resistance.

Intra-specific variation Range of types between the individuals of a species. → competition; → biotype; → specific resistance.

***In vitro* (cultivation)** Artificial cultivation (within glass) of a plant, usually on specific nutrient media and under controlled environment.

In vivo Process operated or occurring in/at the body of a living organism.

Labelling studies Radioactive markers that are used in studies on translocation of herbicides in plants, e.g., accumulation of labeled glyphosate applied to the host of *Orobanche* in the tissue of the parasite.

Larva Stage in the live cycle of most insects. Sometimes also known as „caterpillar“, „maggot“, „grub“ or „worm“; Plural: larvae.

Life-cycle With parasitic plants: Successive developmental stages from seed to mature plant.

Linear regression → regression.

Location Site or place where a plant is found or an experiment is carried out.

Marker-assisted selection Identifying and selecting minor genes of interest by linkage with markers which could be scored more easily.

Mean = average; the total of observations, measurements, scores etc., being divided by the number of such observations.

Methyl bromide → soil fumigant.

Mixed cropping	Growing two or more crops simultaneously on the same land with no distinct row arrangements. →inter-cropping.
Model	Simplified presentation of the function of a thing or a fact. „Conceptual m.“ = verbal description of a fact; „mathematical m.“ = description of a development by mathematical structures; „deterministic m.“ = the initial conditions determine the development in time; „stochastic m.“ = contains elements reflecting the variability of certain developments; „simulation m“. = elaborate model for the simulation of complex developments (includes normally deterministic and stochastic elements).
Monogenic resistance	Resistance due to one single resistance gene; opposite of → polygenic resistance.
Morphotype	Different appearance (habit) but same genetic background of individuals of a parasite species due to exogenous factors (e.g., host plant); synonym: phenotype.
Mortality	Reduction of individuals in a population due to death.
NaOCl	Sodium hypochlorite, cleaning and bleaching agent; used for sterilisation of the seed surface against fungi and bacteria, and for the decoloration of the seed testa of *Orobanche*.
Nodules	Swellings on roots of most *Fabaceae*; produced by bacteria (e.g., *Rhizobium* spp.) associated with the roots of these plants. These bacteria are able to fix atmospheric nitrogen and to accumulate it in the nodules. Different from the term „nodule“ applied to early stages of attachment of parasitic plants.
Non-additive gene action	Sum of → dominant and → epistatic gene action.
Nutation	Movement of the seedling of *Cuscuta* in order to reach a suitable host stem. This coiling, which is counterclockwise, is also called → circumnutation.
Obligate parasite	Parasite that needs a host plant to survive, e.g., *Striga, Orobanche, Cuscuta, Alectra* (→ facultative parasite).
Osmotic pressure	Pressure exerted in living bodies as a result of unequal concentration of salts separated by a cell wall or membrane.
Outcrossing	A cross of unrelated plants or of different genotypes (within one species), usually under natural conditions.
Parasitic plant	Specialized plant types that take a proportion or all of the required water and nutrients from other plants.
Parasitism	Coexistence of two different organisms, which supports one of them (parasite) at the expense of the other (host).
Pasteur pipette technique	Technique to collect host plant root exudates.
Pathogen	The causal agent of a disease
Pathotype	→ biotype.

Phenotype The observable properties of an organism, produced by the genotype in conjunction with the environment. → genotype.

Phloem Transport system for assimilates (especially sucrose) from source to sink.

Photosynthesis Process during which green plants (chlorophyll) manufacture carbohydrates from water and carbon dioxide using energy absorbed from sunlight. The by-products are water and oxygen.

Phytoalexins Low molecular weight secondary metabolites which serve the plant in self-defence. They are produced in response to the attack by a pathogen.

Polygenic resistance Resistance due to several resistance genes; opposite of → monogenic resistance.

Population All individuals of a species in a geographic and ecologically defined area.

Population dynamics Shifts in time and space of density and distribution of a population in response to the environment and endogenous factors.

Population structure Indices for the characterisation of a population, e.g., density, dispersal, age structure, fertility, biotypes, frequency of → homozygotes and → heterozygotes.

Post-emergence herbicide Herbicide applied after the emergence of the crop. → pre-emergence.

Preconditioning → conditioning (induction of physiological changes in the parasite seeds, so that they are able to respond to a germination stimulant).

Pre-emergence herbicide Herbicide applied before the emergence of the crop. → post-emergence.

Primary haustorium Haustorium originating from the first contact of the germtube with a host root. → secondary haustorium.

Primary hosts These are hosts on which *Cuscuta* spp. can establish and develop from the seedling stage. → secondary hosts.

QTL Quantitative trait locus or loci: region(s) of a chromosome contributing to the phenotypic expression of a certain polygenic trait, i.e., resistance.

Race → bio-type.

Radicle The young root of a seedling.→ germ-tube.

Recessive gene In diploid organisms, a gene that is phenotypically manifest in the homozygous state but is masked in the presence of its dominant allele.

Regression In biometrics it is the dependence of a variable called „dependent“ on another variable called „independent variable“. The regression coefficient is the numerical measure of the rate of change of the independent variable on the dependent variable. The straight or curved regression line indicates the nature of a regression on a graph with the dependent on the Y axis and the independent on the X axis.

Resistance — Crop genotypes or varieties that show less attack, which usually is measured in terms of number of parasites attached or emerged. Total resistance would be → immunity. The converse is → susceptibility.

Restriction fragment length polymorphism (RFLP) — Variation between individuals in DNA fragment sizes cut by specific restriction enzymes; polymorphic sequences that result in RFLPs are used as → DNA markers.

Root chamber — A host plant is grown in small flat ($20 \times 10 \times 3$ cm^3) boxes filled with soil, and inoculated with parasite seeds. The technique allows a observation of the underground development of parasites, therefore it is suitable to study germination, attachment and shoot development.

Root exudates — → host root exudates.

Saprophyte — Non-pathogenic plant that obtains nourishment from the products of organic breakdown and decay.

Screening — Evaluation of host cultivars for specific traits, e.g., resistance in the field or individual resistance mechanisms in the laboratory.

Secondary dormancy — → wet dormancy.

Secondary haustoria — The first contact organ of the parasite with it's host develops into the → primary haustorium; all further contact organs are called secondary haustoria.

Secondary hosts — These are hosts on which a parasite (e.g., *Cuscuta* spp.) is apparently unable to establish from seedlings but on which it is able to make attachements and haustoria formation after it is well established on a primary host. → wet dormancy.

Seed bank — The total amount of seed in a soil; it can be described horizontally or vertically in the profile of the soil.

Seed coat — = testa; the outer protective covering of the seed. Covers the endosperm of the seed except at the mycropylar end.

Seed viability — Capability of living, i. e. a spore or seed is able to germinate and grow, if favourable conditions occur. Different (higher or equal) from the actual germination rate. → Tetrazolium test.

Semi-parasite — → hemi-parasite.

Shoot — The part of the plant that is usually above the ground. → spike.

Soil fumigation — Chemical method to disinfect the soil. → methyl bromide. → fumigation.

Soil map — A map showing the distribution of soil types or other soil mapping units.

Soil sampling Collection of representative soil samples for further evaluation (e.g., for the number of parasite seed).

Soil texture Describes the particles of varying sizes that constitute the soil. The sizes of the particles vary from gravel, coarse sand, fine sand and silt down to clay. These in varying proportion form the mechanical aspect of the soil.

Solarization Physical method to disinfect the soil with the aid of sun's energy. The moist ground is covered with a transparent polyethylene foil. Radiation heats the soil, and at the same time the plastic foil preserves the soil moisture and diminishes heat loss by evaporation. The method has proven effective against various soil-borne pests.

Sorgoleone Germination stimulant for *Striga*, isolated in 1989 from sorghum. → strigol.

Specific resistance Resistance to parasitic weeds can be species-specific, i.e., resistance to *S. asiatica* does not necessarily hold against *S. hermonthica* and *vice versa*, or biotype/race-specific, i.e., resistance present at one location may be overcome by a different parasite biotype at another location. → inter-specific variation and → intra-specific variation.

Spike Upper part of the stem carrying the inflorescence.

Split-application Chemical or fertiliser application where the whole rate per unit area is divided into two or more doses and applied at two or more given intervals.

Split-plot design Design used for factorial experiments, involving at least two factors not equally important and divided into main and sub-plots.

Spore reproductive body of fungi and other lower plants, containing one or more cells modified to survive adverse environments.

Sterilisation →surface sterilisation.

Stimulant In the context of parasitic plants: Germination inducing compound (root exudates, cytokinins, gibberellins, strigol-analogues like GR 24, ethylene gas).

Stomata Pores in the epidermis of leaves and stems. They regulate gas exchange and → transpiration. In parasitic plants stomata mostly have a rudimentary structure.

***Striga* syndrome rating** Simple and quick assessment of the host plant's damage due to parasite attack, particularly useful in maize breeding for *Striga* resistance. The 1 to 9 scale reflects damages in the form of leaf chlorosis or firing, poor ear or panicle development, and stunting.

Strigol Germination stimulant for *Striga*, isolated in 1966 from cotton roots. The chemical substance dl-strigol was first synthesized in 1974. Concentrations of only 10^{-12} to 10^{-6} of this compound induce *Striga* seed germination. Different synthetic analogues were identified so far, but commercial production is still too expensive. → sorgoleone; → alectrol.

Suicidal germination Obligate parasites need a host plant to survive. If germination of such parasites` seed occurs, the seedling will die if no suitable host is around. Suicidal germination can be induced by the injection of → ethylene gas into the soil or by growing → trap crops.

Surface sterilisation Seed treatment with e.g., NaOCl in order to eliminate fungal and bacterial growth in laboratory tests.

Survey Collection of data on the occurrence of certain characters, e.g., infestation level with seeds of parasitic weeds in the soil or parasitic weeds in a certain region.

Susceptibility Reaction of varieties that are parasitized and substantially suffer from the parasite. → tolerance; → resistance; → immunity.

Symbiosis Association of dissimilar organisms to their mutual advantage, e.g., the association of nitrogen fixing bacteria with leguminous plants.

Synthetic varieties Open-pollinated varieties composed of several specifically selected parent lines, e.g., lines with differing resistance mechanisms.

Testa → seed coat.

Tetrazolium test A colour test for evaluation of → seed viability. In viable cells of the seed endosperm the colourless Tetrazolium chloride (TTC) gets converted to the red formazan.

Thinning The act of reducing the number of plants, especially at seedling stage to reduce the plant number per stand or per unit area of land.

Threshold level Intensity of attack at which control methods (due to epidemiological or technical reasons) are required to avoid the attainment of the economic threshold.

Tolerance Reaction of varieties that are parasitized to the same extent but suffer less damage than standard varieties. The converse is „sensitivity". → resistance; → immunity.

Toxin Poisonous substance of biological origin.

Transpiration In plants it is the physically and physiologically controlled emission of water vapour through → stomata. It supports the transport of water and nutrients in the vessel.

Trap crop Plant that stimulates with root exudates the germination of seeds of parasitic plants, but do not support their growth (= false host).

Tubercle The appressorium is already larger than the parasite seed, the contact to the host vessels is established. The justification of the term „tubercle" from a morphological point of view is doubtful, as the term „haustorium" covers this developmental stage. From the practical side this terminology however is useful in the case of the chemical control of *Orobanche.*

Variance Extent of dispersion of data around a mean value. The symbol is s^2.

Viability Capability to live; capability of a spore or seed to germinate and grow.

Vigor score Simple and quick assessment of the host plant's or the parasite's growth/vitality on a scale from 1 to 9.

Virulence Aggressiveness of the parasite. → biotype.

Wet dormancy Inactivity of *Striga* seed (no germination occurs in spite of favourable conditions) as a response to extended exposure to moisture stress. → dormancy.

Xylem Main transport system for water, minerals and some organic compounds.

Yield potential The maximum yield that can be produced under favourable conditions.

LITERATURE

Abifarin AO 1984. A dictionary of commonly used terms in crop improvement with particular reference to rice. IITA, Ibadan, Nigeria, 178pp.

Anonym 1999. A glossary of genetics. In the World Wide Web at: http://linkage.rockefeller.edu/wli/glossary/genetics.html.

Aust HJ, H Buchenauer, F Klingauf, P Niemann, HM Pöhling & F Schönbeck 1991. Glossar phytomedizinischer Begriffe. Schriftenreihe der Deutschen Phytomedizinischen Gesellschaft. Band 3; Verlag Ulmer, Stuttgart, 123 pp.

King CR & WD Stansfield 1990. A dictionary of genetics. Fourth Edition. Oxford University Press, New York, 336pp.

Parker C & CR Riches 1993. Parasitic Weeds of the World - Biology and Control. CAB International, Wallingford, UK, 332 pp.

Rieger R, A Michaelis & M Green 1991. Glossary of genetics. 5th Edition. Springer Verlag, Berlin-Heidelberg, Germany, 553pp.

Strasburger E 1979. Lehrbuch der Botanik. Verlag G. Fischer, Stuttgart, 31. Auflage, 1078 pp.

9 List of crop scientific names

Alfalfa = lucerne	*Medicago sativa* L.
Bambara groundnut	*Voandzeia subterranea* (L.) Thou. ex DC.
Berseem	*Trifolium alexandrinum L.*
Carrot	*Daucus carota* L.
Clover	*Trifolium spp.*
Chickpea	*Cicer arietinum* L.
Chickling	*Lathyrus cicera* L.
Coffee	*Coffea arabica* L.
Common vetch	*Vicia sativa* L.
Cotton	*Gossypium* spp.
Cowpea	*Vigna unguiculata* (L.) Walp.
Cranberry	*Vaccinium macrocarpon* Ait and Schult.
Eggplant	*Solanum melongena* L.
Faba bean	*Vicia faba* L.
Flax, linseed	*Linum usitatissium* L.
Finger millet	*Eleusine coracan* (L.) Gaertn.
Fonio	*Digitaria exilis* (Kipp.) Stapf
Groundnut	*Arachis hypogae* L.
Hemp	*Cannabis sativa* L.
Indian mustard	*Brassica juncea* (L.) Czern et Coss.
Lentil	*Lens culinaris* Medik.
Litchi	*Litchi chinensis* Sonn.
Maize	*Zea mays* L.
Narbon vetch	*Vicia narbonensis* L.
Noog	*Guizotia abyssinica* (L. f.) Cass.
Ochrus pea	*Lathyrus ochrus* L.(DC.)
Pea	*Pisum sativum* L.
Peach	*Prunus persica* Batsch
Pearl millet	*Pennisetum americanum* (L.) Leeke
Potato	*Solanum tuberosum* L.
Rice	*Oryza sativa* L.
Safflower	*Carthamus tinctorius* L.
Sorghum	*Sorghum bicolor* (L.) Moench
Sugarbeet	*Beta vulgaris* L.
Sugar cane	*Saccharum officinale* L.
Sunflower	*Helianthus annuus* L.
Soybean	*Glycine max* (L.) Merr.
Tobacco	*Nicotiana tabacum* L.
Topinambur	*Helianthus tuberosus* L.
Tomato	*Lycopersicon esculentum* Mill.
Trefoil	*Lotus corniculatus* L.
Turnip rape (wild bio-type)	*Brassica campestris* L.
Vine	*Vitis vinifera* L.
Winter vetch	*Vicia dasycarpa* auct.

10 Literature

Aalders AJG & R Pieters 1985. *In vitro* testing with 2,3,5-triphenyl tetrazolium chloride (TTC) of *Orobanche crenata* seed metabolism. FABIS Newsletter 13: 35-37.

Abayo GO, JK Ransom, J Gressel & GD Odhiambo 1996. *Striga hermonthica* control with acetolactate synthase inhibiting herbicides applied to maize seed with target-site resistance. In: MT Moreno, JI Cubero, D Joel, LJ Musselman & C Parker (eds.), Advances in Parasitic Plant Research. 6th International Parasitic Weed Symposium, Cordoba, Spain, 761-768.

Abbas HK & CD Boyette 1992. Phytotoxicity of fumonisin B_1 on weed and crop species. Weed Technology 6: 548-552.

Abbas HK, CD Boyette, RE Hoagland & RF Vesonder 1991. Bioherbicidal potential of *Fusarium moniliforme* and its phytotoxin, fumonisin. Weed Science 39: 673-677.

Abbasher AA & J Sauerborn 1992. *Fusarium nygamai*, a potential bioherbicide for *Striga hermonthica* control in sorghum. Biological Control 2: 291-296.

Abbasher AA, J Kroschel & J Sauerborn 1995. Microorganisms of *Striga hermonthica* in Northern Ghana with potential as biocontrol agents. Biocontrol Science and Technology 5: 157-161.

Abbasher AA, J Sauerborn 1995. Pathogens attacking *Striga hermonthica* and their potential as biological control agents. In: ES Delfosse & RR Scott (eds.), Proc. 8th Int. Symp. On Biological Control of Weeds, Lincoln University, Canterbury, New Zealand, 527-533.

Abdeev Y & BM Sherbinin 1978. [Investigating tomato resistance to broomrape and the inheritance of this trait]. Weed Abstracts 27: 3074.

Abifarin AO 1984. A dictionary of commonly used terms in crop improvement with particular reference to rice. IITA, Ibadan, Nigeria, 178pp.

Adams ME 1988. Agricultural extension in developing countries. Longman Scientific & Technical, ELBS edition, Singapore.

Adrianaivo AP, S Kachelriess, J Kroschel & W Zehrer 1998. Biologie et gestion du *Striga* à Madagascar. Projet „Ecologie et Gestion des Plantes Parasites“ / Direction de la Protection des Végétaux, Nanisana, Madagascar, 61pp.

Agbobli CA & B Hugenin 1987. Evaluation agronomique du probleme du parasitisme du *Striga* sur maïs dans le Sud Togo. In: C Weber & W Forstreuter (eds.), Parasitic Flowering Plants. Proceedings 4th International Symposium on Parasitic Flowering Plants, Marburg, Germany, 11-25.

Agbobli CA 1991. Effect of nitrogen rates on *Striga asiatica* emergence in maize culture in Togo. In: JK Ransom, LJ Musselman, AD Worsham & C Parker (eds.), Proceedings of the 5th International Symposium of Parasitic Weeds. CIMMYT, Nairobi, Kenya, 28-30.

Aggarwal VD, N Muleba, I Drabo, J Souma & M Mbewe 1984. Inheritance of *Striga gesnerioides* resistance in cowpea. In: C Parker, LJ Musselman, RM Polhill & AK Wilson (eds.), Proceedings 3rd International Symposium on Parasitic Weeds, ICARDA/International Parasitic Seed Plant Research Groupe, 7-9 May 1984, Aleppo, Syria. ICARDA, P.O.Box 5466, Aleppo, Syria,, 143-147.

Akobundu IO 1991. Integrated weed management for *Striga* control in cropping systems in Africa. In: SK Kim (ed.), Combating *Striga* in Africa. Proceedings, International Workshop organized by IITA, ICRISAT and ICRC, 1988 IITA, Ibadan, Nigeria, 122-125.

Albrecht H, H Bergmann, G Diederich, E Grosser, U Hoffmann, P Keller, G Payr & R Sülzer 1989. Agricultural Extension, Volume 1, Basic concepts and methods. Rural Development Series, Sonderpublikation der GTZ, No. 212.

Albrecht H, H Bergmann, G Diederich, E Grosser, U Hoffmann, P Keller, G Payr & R Sülzer 1990. Agricultural Extension, Volume 2, Examples and background material. Rural Development Series, Sonderpublikation der GTZ, No. 238.

Aleksiev A 1967. Trench ploughing as a control measure against broomrape on tobacco. B`lgarski Tyutyun (Bulgaria), 12(3): 13-16.

Al-Menoufi OA & FM Ashton 1991. Studies on the parasitism of *Cuscuta* spp. Series 8: susceptibility and resistance of some *Lycopersicon* species to *Cuscuta campestris* infection. In: JK Ransom, LJ Musselman, AD Worsham & C Parker (eds.), Proceedings of the 5th International Symposium of Parasitic Weeds. CIMMYT, Nairobi, Kenya, 293-297.

Altomare C, A Lorgieco, A Bottalico, G Mulè, A Moretti & A Evidente (1995). Production of type A trichothecenes and enniatin B by *Fusarium sambucinum* Fuckel *sensu lato.* Mycopathologia 129: 177-181.

Anderson DM & ML Cox 1997. *Smicronyx* species (Coleoptera: Curculionidae), economically important seed predators of witchweeds (*Striga* spp.) (Scrophulariceae) in sub-saharan Africa. Bulletin of Entomological Research 87, 3-17.

Anderson DM 1974. Some species of *Smicronyx* (Coleoptera: Curculionidae) associated with *Cuscuta* species (Convolvulaceae) in Pakistan. In: Proceedings of the Entomological Society Washington 76, 359-374.

Andrews FW 1945. The parasitism of *Striga hermonthica* Benth. on *Sorghum* spp. under irrigation. Applied Biology 32: 193-200.

Anonym. Guidelines for the safe and effective use of pesticides. G.I.F.A.P. Brussels, Belgium.

Anonymous 1981. Agricultural compendium for rural development in the tropics and subtropics. Elsevier Scientific Publishing Company, Amsterdam, 739 pp.

Anonym 1999. A glossary of genetics. In the World Wide Web at: http://linkage.rockefeller.edu/wli/glossary/genetics.html.

Antonova TS 1994. Biochemical aspects of the development of new virulent forms in the Moldavian population (race C) of *Orobanche cumana* Wallr. against the background of resistant sunflower cultivars. In: AH Pieterse, JAC Verkleij & SJ ter Borg (eds.), Biology and Management of *Orobanche*. Proceedings of the 3rd International Workshop on *Orobanche* and related *Striga* research. Amsterdam, The Netherlands, Royal Tropical Institute, 290-292.

Ashworth LJ 1976. Quantitative detection of seed of branched broomrape in California tomato soils. Plant Disease Reporter 60: 380-383.

Attawi F 1977. Morphologisch-anatomische Untersuchungen an den Haustorien einiger *Orobanche*-Arten. Berichte der Deutschen Botanischen Gesellschaft 90: 173-182.

Aust HJ, H Buchenauer, F Klingauf, P Niemann, HM Pöhling & F Schönbeck 1991. Glossar phytomedizinischer Begriffe. Schriftenreihe der Deutschen Phytomedizinischen Gesellschaft. Band 3; Verlag Ulmer, Stuttgart, 123 pp.

Babiker AGT & AM Hamdoun 1983. Factors affecting the activity of ethephon in stimulating seed germination of *Striga hermonthica* (Del.) Benth. Weed Research 23: 125-131.

Babiker AGT, AM Hamdoun, A Rudwan, NG Mansi & HH Faki 1987. Influence of soil moisture on activity and persistence of the strigol analogue GR24. Weed Research 27: 173-178.

Babiker AGT, NE Ibrahim & WG Edward 1988. Persistence of GR7 and *Striga* germination stimulant(s) from *Euphorbia aegyptiaca* Boiss. in soils and in solutions. Weed Research 28: 1-6.

Barralis G, R Chadoeuf & JP Gouet 1986. Essai de détermination de la taille de l` échantillon pour l` étude du potentiel semencier d` un sol. Weed Research 26: 291-297.

Bashir MO & LJ Musselman 1984. Some natural enemies of *Striga hermonthica* in the Sudan. Tropical Pest Management 30, 211.

Bebawi FF & AF Farah 1981. Effects of parasitic and non-parasitic weeds on sorghum. Experimental Agriculture 17: 415-418.

Bebawi FF & GA Elhag 1983. Nutritive value of the parasitic weed *Striga hermonthica.* Tropical Agriculture 60: 44.

Bebawi FF, RE Eplee, CE Harris & RS Norris 1984. Longevity of witchweed (*Striga asiatica*) seed. Weed Science 32: 494-497.

Bebawi FF & RE Eplee 1986. Efficacy of ethylene gas as a germination stimulant for *Striga hermonthica* seed. Weed Science 34: 694-698.

Bebawi FF 1987. Cultural practices in witchweed management. In: LJ Musselman (ed.), Parasitic Weeds in Agriculture, Vol. I *Striga*, CRC Press, Florida, USA, 159-171.

Bebawi FF 1987. Cultural practices in witchweed management. In: LJ Musselmann (ed.), Parasitic Weeds in Agriculture, Volume I *Striga.* CRC Press, Boca Raton, Florida, 159-171.

Bedi JS & N Donchev 1991. Results on mycoherbicides control of sunflower broomrape *(Orobanche cumana* Wall.) under glasshouse and field conditions. In: JK Ransom, LJ Musselman, AD Worsham & C Parker (eds.), Proc. 5th Int. Symp. on Parasitic Weeds, CIMMYT, Nairobi, Kenya, 76-82.

Bengaly MP, T Defoer & WA Stoop 1996. On-farm studies of *Striga* in South Mali: Contributions of a farmer participatory approach. In: MT Moreno, JI Cubero, D Berner, D Joel, LJ Musselman & C Parker (eds.), Advances in Parasitic Plant Research. Proceedings VI Parasitic Weed Symposium. Cordoba, Spain, 855-861.

Ben-Hod G, D Losner, DM Joel & AM Mayer 1991. *In vitro* culture of *Orobanche aegyptiaca*. Annales of Botany 68: 413-416.

Ben-Hod G, D Losner, DM Joel & AM Mayer 1991. Infection of tomato roots using callii of broomrape. Physiologia Plantarum 83: 353-356.

Benz W, W Koch & A Moosmann 1984. Ein Extraktionsverfahren zur Bestimmung des Unkrautsamenpotentials in Böden. Zeitschrift für Pflanzenkrankheiten und Pflanzenschutz, Sonderheft 10: 109-114.

Berner DK, MD Winslow, AE Awad, KF Cardwell & DR Mohan Raj 1993. *Striga* Research Methods - A manual (Draft Edition). IITA, Ibadan, Nigeria. 71 pp.

Berner DK, AE Awad & EI Aigbokhan 1994. Potential of imazaquin seed treatment for controls of *Striga gesnerioides* and *Alectra vogelii* in cowpea (*Vigna unguiculata*). Plant Disease 78: 18-23.

Berner DK, KF Cardwell, BO Faturoti, FO Ikie & OA Williams 1994. Relative roles of wind, crop seeds, and cattle in the dispersal of *Striga* species. Plant Disease 78: 402-406.

Berner DK, JG Kling & BB Singh 1995. *Striga* research and control, a perspective from Africa. Plant Disease 79: 652-660.

Berner DK, MO Alabi, U Di-Umba & FO Ikie 1996. Proposed integrated control program for *Striga hermonthica* in Africa. In: MT Moreno, JI Cubero, D Berner, D Joel, LJ Musselman & C Parker (eds.), Advances in Parasitic Plant Research; 6th International Parasitic Weed Symposium, Cordoba, Spain, 817-825.

Bischof F 1984. Untersuchung zur Bestimmung der Keimfähigkeit von Samen einiger *Orobanche*-Arten. Zeitschrift für Pflanzenkrankheiten und Pflanzenschutz, Sonderheft 10: 105-107.

Bondarenko NV 1986. [Biologische Bekämpfung von *Orobanche*]. Biologiceskaja Zascita Rasrenij, Moskva, 226-230.

Booth C 1971. The genus *Fusarium*. CMI, Kew, Surrey, England, 237pp.

Borchardt D 1998. Population genetic structure of *Setosphaeria turcica*, the causal agent of Northern corn leaf blight, investigated with molecular markers. Ph. D. Thesis, University of Hohenheim, Stuttgart, Germany.

Botha PJ 1948. The parasitism of *Alectra vogelii*, with special reference to the germination of it's seed. South African Journal of Botany 14: 63-80.

Bottalico A, A Lerario & A Visconti 1983. Production of mycotoxins (zearalenone, trichothecenes and moniliformin) by *Fusarium* species in Italy. Microbiologie-Aliments-Nutrition 1: 133-142.

Bottalico A, A Logrieco & A Visconti 1989. *Fusarium* species and their mycotoxin in infected cereals in the field and in stored grains. In: J Chelkowsky (ed.), *Fusarium*, mycotoxins, taxonomy and pathogenicity, Elsevier, Amsterdam, 85-119.

Bottalico A, A Logrieco, A Ritieni, A Moretti, G Randazzo & P Corda 1995. Beauvericin and fumonisin B_1 in preharvest *Fusarium moniliforme* maize ear rot in Sardinia. Food Additives and Contaminants 12: 599-607.

Braun M, W Koch & M Stiefvater 1987. Solarisation for soil sanitation. - Possibilities and limitations. Gesunde Pflanzen 7: 301-309.

Brown R & M Edwards 1944. The germination of the seed of *Striga lutea*. I. Host influence and the progress of germination. Annals of Botany 8: 131-148.

Bruckner G, F Flindt & T Kunsch 1979. Biologische Feldmethoden. Quelle & Meyer, Heidelberg, Germany.

Bulat H 1956. Instructions for making the tetrazolium test. Proceedings of the International Seed Testing Association 21: 52-55.

Caemmerer S von & GD Farquhar 1981. Some relations between the biochemistry of photosynthesis and the gas exchange of leaves. Planta 153: 376-387.

Capasso R, A Evidente, A Cutignano, M Vurro MC Zonno & A Bottalico 1996. Fusaric and 9,10-dehydrofusaric acids and their methyl esters from *Fusarium nygamai*. Phytochemistry, 41(4): 1035-1039.

Capdepon M, A Fer & P Ozenda 1985. Sur un système inédit rejet d'un parasite: exemple de la *Cuscuta* sur cotonnier (*C. lupuliformis* Krock. sur *Gossypium hirsutum* L.). Comptes Rendues de l'Académie des Sciences III. Sciences de la vie 300(6): 227-237.

Carsky RJ, L Singh & Ndikawa 1994. Suppression of *Striga hermonthica* on sorghum using cowpea intercrop. Experimental Agriculture 30: 349-358.

Carson AG 1986. Research and development strategies for the control of *Striga hermonthica* in the Gambia. In: Proceedings of the FAO/OAU All-African Government Consultation on *Striga* control, 20-24 October 1986, Maroua, Cameroon, 100-117.

Carson AG 1988 Detailed survey of the parasitic weed *Striga hermonthica* in The Gambia. Tropical Pest Management 34(2): 162-164.

Castiglione S, G Wang, G Damiani, C Bandi, S Bisoffi & F Sala 1993. RAPD fingerprints for identification and for taxonomic studies of elite poplar (*Populus* spp.) clones. Theoretical and Applied Genetics 87: 54-59.

Chang M, DH Netzly, LG Butler & DG Lynn 1986. Chemical regulation of distance: Characterisation of the first natural host germination stimulant for *Striga asiatica.* Journal of the American Chemistry Society 108: 7858-7860.

Chapman JA & JM Kinghorn 1955. Window flight traps for insects. Canadian Entomologist 87: 46-47.

Charudattan R 1991. The mycoherbicide approach with plant pathogens. In: OD Te-Beest (ed.), Microbial control of weeds. Chapman and Hall, New York, USA, London; UK, 24-57.

Chater AO & DA Webb 1972. *Orobanche*. In: TG Tutin, VH Heywood, NA Burgess, SM Walters & DA Webb (eds.), Flora Europaea, III. University Press, Cambridge, England, 286-393.

Chu CC 1978. The N_6 Medium and its applications to anther culture of cereal crops. In: Proc. of Symp. on Plant Tissue Culture, Beijing, Science Press.

Ciotola M, AK Watson & SG Hallett 1995. Discovery of an isolate of *Fusarium oxysporum* with potential to control *Striga hermonthica* in Africa. Weed Research 35, 303-309.

Cottenie A 1980. Soil and plant testing as a basis for fertilizer recommendations. FAO soils bulletin 38/2 Rome.

Coutts HH 1984. The application of pesticides. - Assessment techniques for spray pattern analysis. GTZ, Eschborn, Germany.

Cubero JI & A Martinez 1980. The genetics of the resistance of faba beans to *Orobanche crenata.* FABIS 2: 19-20.

Cubero JI 1986. Breeding for resistance to *Orobanche* and *Striga*: a review. In: SJ ter Borg (ed.), Proceedings, Workshop on Biology and Control of *Orobanche.* LH/VPO Wageningen, The Netherlands, 127-139.

Cubero JI 1991. Breeding for resistance to *Orobanche* species: a review. In: K Wegmann & LJ Musselman (eds.), Progress in *Orobanche* research. Eberhard-Karls-Universität, Tübingen, Germany, 257-277.

Cubero JI, AH Pieterse, SA Khalil & J Sauerborn 1994. Screening techniques and sources of resistance to parasitic angiosperms. Euphytica 73: 51-58.

Cubero, JI, MT Moreno, D Rubiales & J Sillero 1999. Resistance to *Orobanche*: The State of the Art. Congresos y Jornadas 51/99. Junta de Andalucia, Consejeria de Agricultura y Pesca, Spain, 199 pp.

Davidson A 1963. Insects attacking *Striga* in Kenya. Nature 197, 923.

Dawoud DA & J Sauerborn 1994. Impact of drought stress and temperature on *Striga hermonthica* and *Alectra vogelii* at early growth stages. Experimental Agriculture 30: 249-257.

Dawoud DA 1995. Autecological studies on phytoparasites of the genera *Alectra, Orobanche* and *Striga* with special reference to their early developmental stages. PLITS 13(1), 112 pp.

Dawson JH & JR Holstun 1971. Estimating losses from weeds in crops. In: L Chiarappa (ed.), Crop loss assessment methods: FAO Manual on the evaluation and prevention of losses by pests, diseases and weeds. FAO and C.A.B., Farnham Royal, UK.

Dawson JH, FM Ashton, WV Welker, JR Frank & GA Buchanan 1984. Dodder and its control. USDA, Farmer Bulletin No. 2276.

Dawson JH 1987. *Cuscuta* and it's control. In: HChr Weber & W Forstreuter (eds.), Proceedings,. 4th Int. Symposium on Parasitic

Flowering Plants, Marburg, Germany, 137-149.

Debrah SK 1994. Socio-economic constraints to the adoption of weed control techniques: the case of *Striga* control in the West African Semi-Arid Tropics. International Journal of Pest Management 40(2), 153-158.

Dembélé B, A Raynal-Roques, G Sallé & C Tuquet 1994. Plantes Parasites des Cultures et des Essences Forestières au Sahel. Institut du Sahel, Bamako, Mali, CTA, Wageningen, The Netherlands, 43 pp.

Dessaint F 1991. La répartition spatiale du stock semencier: comparaison de techniques statistiques. Weed Research 31: 41-48.

Dhingra OD & JB Sinclair 1986. Basic plant pathology methods. CRC Press, Boca Baton, USA, 355pp.

Dietrich WF, J Miller, R Steen, MA Merchant, D Damron Boles, Z Husain, R Dredge, MJ Daly, KA Ingalls, TJ OConnor, CA Evans, MM DeAngelis, DM Levinson, L Kruglyak, N Goodman, NG Copeland, NA Jenkins, TL Hawkins, L Stein, DC Page & ES Lander 1996. A comprehensive genetic map of the mouse genome. Nature 380: 149.

Doggett H 1965. *Striga hermonthica* on sorghum in East Africa. Journal of Agricultural Science 65: 183-194.

Doggett H 1988. Sorghum. 2nd edition, John Wiley, New York. 392-394.

Dörr I, A Staack & R Kollmann 1994. Resistance of *Helianthus* to *Orobanche* – histological and cytological studies. In: AH Pieterse, JAC Verkleij & SJ ter Borg (eds.), Biology and Management of *Orobanche*. Proceedings of the 3rd International Workshop on *Orobanche* and related *Striga* research. Amsterdam, The Netherlands, Royal Tropical Institute, 276-289.

Doyle FJ & JL Doyle 1990. Isolation of plant DNA from fresh tissue. Focus 12: 13-15.

Duke SO 1986. Microbially produced phytotoxins as herbicides - a perspective. In A Putnam & CS Tang (eds.), The Science of Allelopathy. John Wiley & Sons, Inc., 287-304.

Efron Y 1993. Screening maize for tolerance to *Striga hermonthica*. Plant Breeding 110: 192-200.

Egley GH & JE Dale 1970. Ethylene, 2-chloroethylphosphonic acid and witchweed germination. Weed Science 18: 586.

Egley GH, Eplee RE & Norris RS 1990. Discovery and development of ethylene as a witchweed seed germination stimulant. In: PF Sand, RE Eplee & RG Westbrooks (eds.), Witchweed research and control in the United States. Weed Science Society of America, Champaign, IL, USA. 56-66.

Ejeta G, LG Butler & AG Babiker 1992. New approaches to the control of *Striga*. *Striga* Research at Purdue University. Research Bulletin RB-991. Agricultural Experiment Station. Purdue University. West Lafayette, IN.

Ejeta G & LG Butler 1993. Host plant resistance to *Striga*. In: International Crop Science I. 1993. Crop Science Society of America, 677 pp. Segoe Rd, Madison, WI53711, USA. 561-569.

Ejeta G & LG Butler 1996. Biotechnological approaches for understanding mechanisms of resistance to *Striga*. In: MT Moreno, JI Cubero, D Berner, D Joel, LJ Musselman & C Parker (eds.), Advances in Parasitic Plant Research; 6th International Parasitic Weed Symposium, Cordoba, Spain, 567-572.

Ellis MB 1971. Dematiaceous Hyphomycetes. CMI, Kew, Surrey, England, 608pp.

Ellis MB 1976. More Dematiaceous Hyphomycetes. CMI, Kew, Surrey, England, 507pp.

Eplee RE 1975. Ethylene: a witchweed seed germination stimulant. Weed Science 23: 433.

Eplee GH 1976. Determination of *Striga* seeds in soil. USDA witchweed laboratory, P.O. Box 269, Whiteville, North Carolina 28472.

Eplee RE & RS Norris 1987. Chemical control of *Striga*. In: LJ Musselman (ed.), Parasitic Weeds in Agriculture, Vol. I, *Striga*. CRC Press, Boca Raton, Florida, USA, 173-182.

Eplee RE & RS Norris 1987. Field research techniques,. In: LJ Musselman (ed.), Parasitic Weeds in Agriculture, Vol. 1: *Striga*. CRC Press, Boca Raton, Florida, USA, 271-280.

Eplee RE & RG Norris 1990. Soil sampling collection equipment and equipment to separate seeds from soil. In: PF Sands, RE Eplee & RG Westbrooks (eds.), Witchweed Research and Control in the United States. Weed Science Society of America, Champaign, IL, 136-140.

Eplee R & M Langston 1991. Integrated control methodology for *Striga* in the USA. In:

JR Ransom, LJ Musselman, AD Worsham & C Parker (eds.), Proceedings of the 5th International Symposium of Parasitic Weeds; CIMMYT, Nairobi, Kenya, 397-399.

Eplee R & R Norris 1996. Prevention: a management strategy for parasitic weeds. In: MT Moreno, JI Cubero, D Berner, D Joel, LJ Musselman & C Parker (eds.), Advances in Parasitic Plant Research; 6th International Parasitic Weed Symposium, Cordoba, Spain, 757-759.

Eponou T 1993. Partners in agricultural technology: Linking research and technology transfer to serve farmers. ISNAR Research Report No. 1. The Hague: International Service for National Agricultural Research.

Esilaba AO, F Reda, T Mulatu, JK Ransom, G. Woldewahid, A. Tesfaye, I Fitiwy & G Abate 1997. Participatory rural appraisal on *Striga* in the northern Ethiopian highlands. African Crop Science Conference Proceedings, Vol. 3.

EWRC 1973. Proceedings Symposium on Parsaitic Weeds, Malta, 1973. European Weed Research Council, Wageningen, 295 pp.

Farina MPW, PEL Thomas & P Channon 1985. Nitrogen, phosphorus and potassium effects on the incidence of *Striga asiatica* (L.) Kuntze in maize. Weed Research 25: 443-447.

Farquhar HH 1937. Witchweed - new light on the means by which it is spread. Rhodesian Journal of Agriculture 34: 563.

Feil P, K Hummler & S Kachelriess 1997. Development of visual extension material for a *Striga* control programme in Northern Ghana through action research. European Journal of Agricultural Education and Extension. European Journal of Agricultural Education and Extension 4(1): 1-16.

Fischer E, HH Geiger & S Deimling 1996. *In vitro*-Kultur von *Striga hermonthica*. In: International Association For Plant Tissue Culture, 5. Tagung der Deutschen Sektion, Universität Hohenheim, 10.-12. Oktober 1996.

Fischer PA 1999. Action research in extension material and message development: The *Striga* problem of northern Ghana revisited. Kommunikation und Beratung (27). Margraf Verlag, Weikersheim, Germany.

Fischer PA, O Abdel-Rahman, S Kachelriess, J Kroschel & V Hoffmann. Promoting *Striga* control measures at on-farm level in Northern Ghana. PASCON Conference, Mali (in press).

Francisco-Ortega J, HJ Newbury & BV Ford-Lloyd 1993. Numerical analysis of RAPD data highlight the origin of cultivated tagasate (*Chamaecytisus proliferus* ssp. *Palmensis*) in the Canary Islands. Theoretical and Applied Genetics 87: 264-270.

Freitag J, DE Hess, HG Welz & HH Geiger 1996. Pearl millet and sorghum specific races of *Striga hermonthica* in Niger. In: MT Moreno, JI Cubero, D Berner, D Joel, LJ Musselman & C Parker (eds.), Advances in Parasitic Plant Research. Proceedings of the 6th International Parasitic Weed Symposium, April 16-18, 1996, Cordoba, Spain, 472-478.

Frost HM 1995. *Striga hermonthica* surveys in Western Kenya. Brighton Crop Protection Conference Weeds, Brighton, UK, Vol. 1, 145-150.

Fulton TM, J Chunwongse & SD Tanksley 1995. Microprep protocol for extraction of DNA from tomato and other herbaceous plants. Plant Molecular Biology Reporter 13: 207-209.

Gabathuler E 1979. Le système de vulgarisation CFSME/AE. Bulletin agricole du Rwanda, 12. 10. 1979, pp. 188-199.

Gaertner EE 1950. Studies of seed germination, seed identification, and host relationships in dodders, *Cuscuta* spp. Cornell University, Agricultural Experiment Station, Ithaca, New York, Memoir 294, 56 pp.

Garcia-Torres L, M Jurado-Esposito, M Castejon-Munoz & F Lopez-Granados 1996. Herbicide treated crop seeds for control of *Orobanche* spp. In: MT Moreno, JI Cubero, D Berner, D Joel, LJ Musselman & C Parker (eds.), Advances in Parasitic Plant Research; 6th International Parasitic Weed Symposium, Cordoba, Spain, 700-705.

Gehri A, S Kachelriess, P Fischer 1998. Assessing women's needs in *Striga* control. In: J Kroschel, H Mercer-Quarshie, J Sauerborn (eds.), Advances in Parasitic Weed Control at on-farm level. Vol. I. Joint Action to Control *Striga* in Africa. Margraf Verlag, Weikersheim, Germany, 45-56.

Geiger U, J Kroschel & J Sauerborn 1996. *Striga asiatica*, a problem in the Middle West of Madagascar. In: MT Moreno, JI Cubero, D Berner, DM Joel, LJ Musselman

& C Parker (eds.), Advances in Parasitic Plant Research. Proceedings VI Parasitic Weed Symposium. Cordoba, Spain, 479-486.

Geipert S, J Kroschel & J Sauerborn 1996. Distribution and economic importance of *Orobanche crenata* in faba bean production of Morocco. In: W Bertenbreiter & M Sadiki (1996): Rehabilitation of Faba Bean. Proceedings 1st Maghrebian Seminar on Faba Bean Research, 24.-27. Mai 1995, Rabat/Marokko, 113-120.

Geipert S 1997. Potentiale und Grenzen der Bekämpfung von *Orobanche crenata* Forsk. im Ackerbohnenanbau (*Vicia faba* L.) Marokkos. PLITS 15(5).

Gelderblom WCA, K Jaskiewicz, WFO Marasas, PG Thiel, RM Horak, R Vleggaar & NPJ Kriek 1988. Fumonisins - Novel mycotoxins with cancer-promoting activity produced by *Fusarium moniliforme*. Applied and Environmental Microbiology 54: 1806-1811.

Gerlach W & H Nirenberg 1982. The genus *Fusarium* - a pictoral atlas. Biologische Bundesanstalt für Land- und Forstwirtschaft, Berlin-Dahlem, Kommisionsverlag Paul Parey, Berlin, Germany.

Gil J, LM Martin & JI Cubero 1987. Genetics of resistance in *Vicia sativa* L. to *Orobanche crenata* Forsk. Plant Breeding 99: 134-143.

Gil JT 1986. Variacion intraespecifica en *Vicia sativa*. Ph. D. Thesis, University of Cordoba, Spain.

Gnon T 1995. Experiments with agricultural innovations: Community level farmer organizations and extension. In: Agricultural Extension in Africa, Proceedings of an international workshop, Yaoundé, Cameroon, January 1994. Technical Centre for Agricultural and Rural Co-operation (CTA), Wageningen, 85-94.

Gomez KA & AA Gomez 1984. Statistical procedures for agricultural research. Second Edition. John Wiley & Sons, New York, USA, 680 pp.

GRAAP 1985. Pour une pédagogie de l'autopromotion communautaire, Nouvelle édition pour les animateurs villageois (Animator's book). GRAAP, Bobo-Dioulasso.

Greathead DJ & JED Milner 1971. A survey of *Striga* spp. (Scrophulariacaea) and their natural enemies in East Africa with a discussion on the possibilities of biological control. Tropical Agriculture (Trinidad) 48, 111-124.

Greathead DJ 1984. The natural enemies of *Striga* spp. and their utlization as biological control agents. In: ES Ayensu, H Doggett, RD Keynes, J Marton-Lefèvre, LJ Musselman, C Parker & A Pickering (eds.), *Striga* Biology and Control. IRL Press Ltd Oxford, UK, 133-160.

Gressel J, Y Kleifeld & DM Joel 1994. Genetic engineering can help control parasitic weeds. In: AH Pieterse, JAC Verkleij & SJ ter Borg (eds.), Biology and Management of *Orobanche*, Proceedings of the 3rd International Workshop on *Orobanche* and Related *Striga* Research, Royal Tropical Institute, Amsterdam, The Netherlands. 406-418.

Gressel J 1995. The potential role of herbicide resistant crops in world agriculture. In: SO Duke (ed.), Herbicide Resistant crops. Lewis Publishers, Boca-Raton, USA, 231-250.

Gressel J, L Segel & JK Ransom 1996. Managing the delay of evolution of herbicide resistance in parasitic weeds. International Journal of Pest Management 42(2): 113-129.

Gubbels P 1995. The role of peasant farmer organisations in transforming agricultural research and extension practices in West Africa. In: Agricultural Extension in Africa, Proceedings of an international workshop, Yaoundé, Cameroon, January 1994. Technical Centre for Agricultural and Rural Co-operation (CTA), Wageningen, 95-120.

Gurney AL, MC Press & JK Ransom 1995. The parasitic angiosperm *Striga hermonthica* can reduce photosynthesis of its sorghum and maize host in the field. Journal of Experimental Botany 46: 1817-1823.

Gwinner J 1997. Curriculumentwicklung - Aufgabe von Pflanzenschutzprojekten? GTZ-423 Info Nr. 1/97. GTZ, Eschborn, Germany.

Hagmann J, E Chuma & G Nehanda 1995. Development and institutionalising of participatory innovation development and extension in the agricultural extension service in Zimbabwe. GTZ Kurzinfo der OE425, Nr. 21, GTZ, Eschborn, 37-42.

Hamilton NA 1996. Learning to learn with farmers - An adult learning extension project. PhD Thesis Agricultural University Wageningen.

Hammad S, H El-Arosi & O Al-Menoufi 1967. *Phytomyza orobanchia* Kalt (Diptera: Agromyzidae) feeding on *Orobanche crenata* Forsk. in Egypt. Bulletin de la Société Entomologique d'Egypte 51: 141-144.

Hamrick JL 1989. Isozymes and the analysis of genetic structure in plant populations. In: DE Soltis & PS Soltis (eds), Isozymes in Plant Biology. Chapman and Hall Ltd., 87-105.

Hanlin RT 1990. Illustrated genera of Ascomycetes. APS Press, St. Paul, Minn., USA, 236 pp.

Harloff HJ & K Wegmann 1993. Evidence for a mannitol cycle in *Orobanche ramosa* and *O. crenata*. Journal of Plant Physiology 141: 513-520.

Haussmann BIG, DE Hess, BVS Reddy, HG Welz & HH Geiger 1996a. Quantitative-genetic parameters for resistance to *Striga hermonthica* in sorghum. In: MT Moreno, JI Cubero, D Berner, D Joel, LJ Musselman & C Parker (eds.), Advances in Parasitic Plant Research. Proceedings of the 6th International Parasitic Weed Symposium, April 16-18, 1996, Cordoba, Spain, 681-688.

Haussmann BIG, DE Hess, BVS Reddy, HG Welz & HH Geiger 1996b. Analysis of resistance to *Striga hermonthica* in diallel crosses of sorghum. Proceedings of the 4th General Workshop of the Pan-African *Striga* Control Network (PASCON), October 28 - November 1, 1996, Bamako, Mali (in press).

Haussmann BIG, DE Hess, ML Koyama, L Grivet, HFW Rattunde & HH Geiger 2000. Breeding for *Striga* Resistance in Cereals. Proceedings, Workshop held at IITA, Ibadan, Nigeria. Margraf Verlag, Weikersheim, Germany, 376 pp.

Hepper FN 1984. Species of *Striga* in East Africa. In: C Parker et al (eds.) Proceedings of the 3rd International Symposium on Parasitic Weeds, ICARDA/ISPRG, Aleppo, Syria, 262-265.

Hernandez L, MT Moreno & JI Cubero 1984. Studies on the resistance of *Vicia faba* to *Orobanche crenata* Forskall. In: Parker C, LJ Musselman, RM Polhill & AK Wilson (eds.), Proceedings 3rd International Symposium on Parasitic Weeds, Aleppo, Syria, 210-220.

Hess DE & G Ejeta 1987. Effects of cultural treatments on infestation of *Striga hermonthica* (Del.) Benth. (Scrophularia-ceae) on sorghum in Niger. In: HChr Weber & W Forstreuter (eds.), Parasitic Flowering Plants. Proceedings of the 4th ISPFP, Marburg, Germany, 367-375.

Hess DE & G Ejeta 1992. Inheritance of resistance to *Striga* in sorghum genotype SRN 39. Plant Breeding 109: 233-241.

Hess DE, G Ejeta & LG Butler 1992. Selecting sorghum genotypes expressing a quantitative biosynthetic trait that confers resistance to *Striga*. Phytochemistry 31: 493-497.

Hess DE & BIG Haussmann 1999. Status quo of *Striga* control: Prevention, mechanical and biological control methods and host plant resistance. In: J Kroschel, H Mercer-Quarshie, J Sauerborn (eds.), Advances in Parasitic Weed Control at On-farm Level. Vol. I. Joint Action to Control *Striga* in Africa. Margraf Verlag, Weikersheim, Germany, 75-87.

Hezewijk van MJ, AH Pieterse, MC Saxena & SJ ter Borg 1987. Relationship between sowing date and *Orobanche* development on faba been (*Vicia faba* L.) and lentil (*Lens culinaris* Medik.) in Syria. In: HChr Weber & W Forstreuter (eds.), Proceedings of the 4th International Symposium on Parasitic Flowering Plants, Marburg, Germany, 377-390.

Hoagland RE 1990. Microbes and microbial products as herbicides, an overview. In: RE Hoagland (ed.), Microbes and microbial products as herbicides. American Chemical Society Washington DC, 2-52.

Hoffmann V 1991. Bildgestützte Kommunikation in Schwarz-Afrika. Tropical Agriculture 7; Margraf Verlag, Weikersheim, Germany.

Hummler K 1996. *Striga* Beratung. Entwicklung eines visuel unterstützten Beratungsprogrammes zur Kontrolle von *Striga hermonthica* in der Nordregion Ghanas. Kommunikation und Beratung (5). Margraf Verlag, Weikersheim, Germany.

Hunter J & JH Visser 1986. The nitrate reductase activity (NRA) of cultivated Scrophulariacean root parasites. South African Journal of Botany 52(3): 246-248.

Hussain HSN & EI Aigbokhan 1991. Growth stages of *Striga gesnerioides* on cowpea roots. In: JR Ransom, LJ Musselman, AD Worsham & C Parker (eds.), Proceedings of the 5th International Symposium of

Parasitic Weeds, CIMMYT, Nairobi, Kenya, 279-286.

Ihl B & I Miersch 1996. Susceptibility and resistance of *Lycopersicon* to infection by *Cuscuta*. In: MT Moreno, JI Cubero, D Berner, D Joel, LJ Musselman & C Parker (eds.), Advances in Parasitic Plant Research; 6th International Parasitic Weed Symposium, Cordoba, Spain, 599-607.

Jacobsohn R, A Greenberger, J Katan, M Levi & H Alon 1980. Control of Egyptian broomrape (*Orobanche aegyptiaca*) and other weeds by means of solar heating of the soil by polyethylene mulching. Weed Science 28: 312-316.

Jacobsohn R, D Ben-Ghedalia & K Marton 1987. Effect of the animal's digestive system on the infectivity of *Orobanche* seeds. Weed Research 27: 87-90.

Jacobsohn R & R Marcus 1988: Quantitative determination of broomrape (*Orobanche* spp.) seeds in vetch seeds. Weed Research 28: 159-162.

Jacobsohn R, Y Kleifeld, FP Agrawal, PJ & K Marton 1991. Soil fumigation with Telone II for broomrape (*Orobanche* spp.) control. In: K Wegmann & LJ Musselman (eds.), Progress in *Orobanche* research. Eberhard-Karls-Universität, Tübingen, Germany, 185-190.

Jain R & CL Foy 1987. Influence of various nutrients and growth regulators on germination and parasitism of *Orobanche aegyptiaca*. In: HChr Weber & W Forstreuter (eds.), Parasitic Flowering Plants. Marburg, Germany, 427-436.

Janetschek H 1982. Ökologische Feldmethoden. Ulmer Verlag Stuttgart, Germany.

Joel DM, Y Kleifeld, Y Losner-Goshen, Y Herzlinger & J Gressel 1995. Transgenic crops against parasites. Nature 374: 220-221.

Joel DM, V Portnoy & N Katzir 1996. Identification of single tiny seeds of *Orobanche* using RAPD analysis. Plant Molecular Biology Reporter 14: 6-11.

Joel DM, V Portnoy, G Tzuri, R Greenberg & N Katzir 1996. Molecular markers for the identification of *Orobanche* species. In: MT Moreno, JI Cubero, D Berner, D Joel, LJ Musselman & C Parker (eds.), Advances in Parasitic Plant Research; 6th International Parasitic Weed Symposium, Cordoba, Spain, 151-160.

Johnson AW, G Rosebery & D Parker 1976. A novel approach to *Striga* and *Orobanche* control using synthetic stimulants. Weed Research 16: 223-227.

Jones M 1991. Seed mass and seed production in British *Orobanche* species. In: Proceedings of the 5th International Symposium of Parasitic Weeds; JR Ransom, LJ Musselman, AD Worsham & C Parker (eds.), CIMMYT, Nairobi, Kenya, 447-453.

Jost A 1997. Integrierter Getreideanbau in Nord-Ghana unter besonderer Berücksichtigung der *Striga*-Problematik. PLITS 15(4).

Jost A, J Kroschel & J Sauerborn 1996. Studies on *Smicronyx* Spp. and *Junonia orithya* and their potential for biological control of *Striga hermonthica* in northern Ghana. In: MT Moreno, JI Cubero, D Berner, D Joel, LJ Musselman & C Parker (eds.), Advances in Parasitic Plant Research. Proceedings 6th Parasitic Weed Symposium. Cordoba, Spain, 888-897.

Kachelriess S 1987. Estimation of *Orobanche* seeds in soil. FABIS Newsletter 19: 17-18.

Kachelriess S 1988. Methode zur Bestimmung von *Orobanche*-Samen im Boden. Zeitschrift für Pflanzenkrankheiten und Pflanzenschutz, Sonderheft 11: 55-60.

Kachelriess S, J Kroschel, K Hummler, V Hoffmann, B Osterburg & J Sauerborn 1996. Linking farmers and researchers - Developing appropriate extension material to implement a farmer oriented *Striga* control Strategy in Northern Ghana. In: MT Moreno, JI Cubero, D Berner, D Joel, LJ Musselman & C Parker (eds.), Advances in Parasitic Plant Research. Proceedings 6th Parasitic Weed Symposium, Cordoba, Spain, 863-870.

Kachelriess S 1997. Training module: Introducing Participatory Technology Development. PTD Circular August 1997, Nr. 7: 3-4. Published by ETC Netherlands, P.O.B. 64, NL-3830 AB Leusden.

Kapralov SI 1974. [*Phytomyza* against broomrape]. Zernovoe Chozjajstvo, Moskva 13(7), 43-44.

Kapralov SI 1975. [*Phytomyza* Einsatz im Gebiet Rostov]. Zascita Rastenij 20(6), 22-23.

Kasasian L 1973. Miscellaneous observations on the biology of *Orobanche crenata* and *O. aegyptiaca*. In: Symposium on Parasitic

Weeds. European Weed Research Council/ Royal University of Malta, 68-75.

Katzir N, V Portnoy, G Tzuri, M Castejon-Muñoz & DM Joel 1996. Use of random amplified polymorphic DNA (RAPD) markers for the study of the parasitic weed *Orobanche.* Theoretical and Applied Genetics 93: 367-372.

Khan MQ & DV Murthy 1955. *Smicronyx albovariegatus* Faust (Curculionidae: Coleoptera) on *Striga* spp. Indian Journal of Entomology 17, 362.

Khon Kaen University 1987. Proceedings of the 1985 International Conference on Rapid Rural Appraisal. Rural Systems Research and Farming Systems Research Projects, Khon Kaen University, Thailand.

Kim SK 1991. Breeding for tolerance and general resistance in maize: a novel approach to combating *Striga* in Africa. In: Improving *Striga* management in Africa, Proceedings, 2nd General Workshop of the Pan-African *Striga* Control Network (PASCON), 23-29 June 1991, Nairobi, Kenya. 168-176.

Kim SK 1994. Genetics of maize tolerance of *Striga hermonthica.* Crop Science 34: 900-907.

Kim SK, AY Akintunde & P Walker 1994. Responses of maize, sorghum and millet host plants to infestation by *Striga hermonthica.* Crop Protection 13(8): 582-589.

King CR & WD Stansfield 1990. A dictionary of genetics. Fourth Edition. Oxford University Press, New York, 336pp.

Kirk AA 1993. A fungal pathogen with potential for biocontrol of *Striga hermonthica (Scrophulariaceae).* Entomophaga 38: 459-460.

Kleifeld Y, Y Goldwasser, G Herzlinger, S Golan, T Blumenfeld & H Buxbaum 1995. Selective control of *Orobanche* in tomatoes with rimsulfruron (DPXE-9636). In: AH Pieterse, JAC Verkleij & SJ ter Borg (eds.), Biology and Management of *Orobanche.* Royal Tropical Institute, Amsterdam, The Netherlands, 561-571.

Klein O, J Kroschel & J Sauerborn 1999. Éfficacité de lachers supplementaires de *Phytomyza orobanchia* Kalt. [Diptera: Agromyzidae] pour la lutte biologique contre l'*Orobanche* au Maroc. In: J Kroschel, M Abderabihi & H Betz (eds.), Advances in Parasitic Weed Control at On-farm Level. Vol. II. Joint Action to Control *Orobanche* in the WANA Region. Margraf Verlag, Weikersheim, Germany, 161-171.

Klyueva MP & GV Pamikchi 1982. [Technologie des *Phytomyza* Einsatzes]. Zascita Rastenij 27(1): 33-34.

Knapp M 1990. [Study on transpiration of *Orobanche crenata* Forsk. and on the influence of parasitism on transpiration and photosynthesis of different host plants]. Diploma thesis (in German), University of Hohenheim, Germany.

Konate A 1986. *Striga* in Mali. In: Proceedings of the FAO/OAU All-African Government Consultation on *Striga* control, 20-24 October 1986, Maroua, Cameroon, 58-61.

Koyama ML 1998. Genetic variability of *Striga hermonthica* and stability of resistance in sorghum. Ph.D. thesis, Wye College, University of London, UK, 30 pp.

Koyama ML 2000. Molecular markers for the study of pathogen variability: implications for breeding resistance to *Striga hermonthica.* In: Haussmann BIG, DE Hess, ML Koyama, L Grivet, HFW Rattunde & HH Geiger (eds.), Breeding for *Striga* resistance in cereals. Proceedings of a workshop held at IITA, Ibadan, Nigeria. Margraf Verlag, Weikersheim, Germany, 227-246.

Kranz B 1997. Die Entwicklung der parasitischen Blütenpflanze *Striga hermonthica* (Del.) Benth. unter dem Einfluß von organischer Düngung. Supra-regional GTZ-project "Ecology and Management of Parasitic Weeds", Internal Report, University of Hohenheim, Germany, 110pp.

Kranz B, D Fugger, J Kroschel & J Sauerborn 1998. The influence of organic manure on *Striga hermonthica* (Del.) Benth. infestation in northern Ghana. Advances in GeoEcology 31: 615-619.

Kroschel J 1989. [The effect of nitrogen on phanerogamic root parasites and their host plants with special refernce to *Striga hermonthica* (Del.) Benth. (Scrophulariaceae)]. PLITS 7(3).

Kroschel J 1995. The influence of different nitrogen sources on the development of *Striga hermonthica* and studies into the ecological behaviour of parasitic flowering plants. International Workshop on Parasitic Plants – Biology and Resistance. 30 May – 2 June 1995. Long Ashton Research Station, Bristol, UK, Book of Abstracts, 11.

Kroschel J, AA Abbasher & J Sauerborn 1995. Herbivores of *Striga hermonthica* in northern Ghana and approaches to their use as biocontrol agents. Biocontrol Science and Technology 5, 163-164.

Kroschel J 1996. Bioherbicidal effect of fumonisin B1 on parasitic weeds of the genus *Striga*. Internal Report Supra-regional GTZ-project "Ecology and Management of Parasitic Weeds", University of Hohenheim, Germany.

Kroschel J & Sauerborn J 1996. Farming systems and the problems of applying *Striga* control methods - a comparison of case studies from northern Ghana, Tanzania and Malawi. In: MT Moreno, JI Cubero, D Berner, D Joel, LJ Musselman & C Parker (eds.), Advances in Parasitic Plant Research. Proceedings 6th Parasitic Weed Symposium. Cordoba, Spain, 843-853.

Kroschel J, A Hundt, AA Abbasher & J Sauerborn 1996. Pathogenicity of fungi collected in Northern Ghana to *Striga* hermonthica. Weed Research, 36, 515-520.

Kroschel J, B Mössner & J Sauerborn 1996. Estimating maize yield losses caused by *Striga asiatica* in Malawi. In: MT Moreno, JI Cubero, D Berner, D Joel, LJ Musselman & C Parker (eds.), Advances in Parasitic Plant Research. Proceedings VI Parasitic Weed Symposium. Cordoba, Spain, 335-346.

Kroschel J, J Sauerborn, S Kachelriess, V Hoffmann & H Mercer-Quarshie 1997. Possibilities and constraints in implementing *Striga* control methods in African agriculture. In: F Heidhues, A Fadani (eds.), Food security and innovations, successes and lessons learned, International Symposium, Verlag Peter Lang, Frankfurt am Main, Germany, 121-135.

Kroschel J 1998. *Striga* – How will it affect African agriculture in the future? An ecological perspective. In: K Martin, J Müther & A Auffarth (eds.), Agroecology, Plant Protection and the Human Environment: Views and Concepts. PLITS 16(2): 137-158.

Kroschel J 1999. Analysis of the *Striga* problem, the first step towards future joint action. In: J Kroschel, H Mercer-Quarshie & J Sauerborn (eds.), Advances in Parasitic Weed Control at On-farm Level. Vol. I. Joint Action to Control *Striga* in Africa. Margraf Verlag, Weikersheim, Germany, 3-25.

Kroschel J, H Mercer-Quarshie & J Sauerborn 1999. Advances in Parasitic Weed Control at On-farm Level. Vol. I. Joint Action to Control *Striga* in Africa. Margraf Verlag, Weikersheim, Germany, 324 pp.

Kroschel J, M Abderabihi & H Betz 1999. Advances in Parasitic Weed Control at On-farm Level. Vol. II. Joint Action to Control *Orobanche* in the WANA Region. Margraf Verlag, Weikersheim, Germany, 347 pp.

Kroschel J & O Klein 1999. Biological control of *Orobanche* spp. with *Phytomyza orobanchia* Kalt., a review. In: J Kroschel, M Abderabihi & H Betz (eds.), Advances in Parasitic Weed Control at On-farm Level. Vol. II. Joint Action to Control *Orobanche* in the WANA Region. Margraf Verlag, Weikersheim, Germany, 135-159.

Kroschel J, A Jost & J Sauerborn 1999. Insects for *Striga* control – Possibilities and constraints. In: J Kroschel, H Mercer-Quarshie & J Sauerborn (eds.), Advances in Parasitic Weed Control at On-farm Level. Vol. I. Joint Action to Control *Striga* in Africa. Margraf Verlag, Weikersheim, Germany, 117-132.

Kukula ST & H Masri 1984. Integrated control practices and chemical control of *Orobanche crenata* in faba bean. In: C Parker, LJ Musselman, RM Polhill & AK Wilson (eds.), Proceedings of the 3rd International Symposium on Parasitic Weeds, ICARDA, Aleppo, Syria, 1984. 256-261.

Kukula S & H Masri 1985. A greenhouse technique for screening faba beans (*Vicia faba* L.) for resistance to *Orobanche* spp. FABIS Newsletter 12: 20-23.

Kuijt J 1969. The Biology of Parasitic Flowering Plants. University of California Press, Berkeley, 346 pp.

Kumar U & Rangaswamy NS 1977. Regulation of seed germination and polarity in seed development in *O. aegyptiaca* by growth substances. Biologia Plantarum (Praha) 19: 353-359.

Kust CA 1963. Dormancy and viability of witchweed seeds as affected by temperature and relative humidity during storage. Weed Science 11, 247-250.

Lane JA, JA Bailey & PJ Terry 1991a. An *in vitro* growth system for studying the parasitism of cowpea (*Vigna unguiculata*) by

Striga gesnerioides. Weed Research 31: 211-217.

Lane JA, MJ Kershaw, THM Moore, DV Child, PS Terry & JA Bailey 1991b. An *in vitro* system to investigate the parasitism of cereals by *Striga* species and resistance of cowpea to *Alectra vogelii*. In: JK Ransom, LJ Musselman, AD Worsham & C Parker (eds.), Proceedings of the 5th International Symposium of Parasitic Weeds. CIMMYT, Nairobi, Kenya, 237-240.

Lane JA, DV Child, THM Moore & JA Bailey 1995. Wild relatives of cereals as new sources of resistance to *Striga*. In: Parasitic Plants: Biology and Resistance. International Workshop, 30 May - 2 June 1995, IACR-Long Ashton Research Station, Bristol, UK, 13.

Langston MA & TJ English 1990. Depletion of witchweed (*Striga asiatica*) in the soil. In: PF Sand, RE Eplee & RG Westbrooks (eds.), Witchweed Research and Control in the United States of America. Weed Science Society of America, Champaign, 114-125.

Lanjouw J & Stafleu FA 1964. Index Herbariorum. Utrecht.

Last FT 1960. Effect of cultural treatments on the incidence of *Striga hermonthica* (Del.) Benth. and yields of sorghum in the Sudan: Field Experiments 1957/58. Annals of Applied Biology 48: 207-229.

Last FT 1960. Incidence of *Striga hermonthica* (Del.) Benth. on two varieties of irrigated sorghum differently manured, spaced and thinned. Tropical Agriculture, Trinidad 37: 309-319.

Last FT 1961. Direct and residual effects of *Striga* control treatments on sorghum yields. Tropical Agriculture, Trinidad 38: 49-55.

Lekic MB 1970c. The role of the *dipteran Phytomyza orobanchia* Kalt. (Agromyzidae) in reducing parasitic phanerogam populations of the *Orobanche* genus in Vojvodina. Contemporary Agriculture 18(7/8), 59-68.

Lekic MB 1974. [Investigations of the dipteran *Phytomyza orobanchia* Kalt. as a controler of parasitic phanerogams of the genus *Orobanche*]. Savremena Poljoprivreda, Novi Sad 22(1-2), 93-99.

Lewontin RC & JL Hubby 1966. Genetics 54: 595.

Linke KH & W Vogt 1987. A method and its application for observing germination and early development of *Striga* (Scrophulariaceae) and *Orobanche* (Orobanchaceae). In: HChr Weber & W Forstreuter (eds.), Proceedings, 4th Int. Symposium on Parasitic Flowering Plants, Marburg, Germany, 501-509.

Linke KH 1987. Untersuchungen über Keimung und Jugendentwicklung von *Striga* und *Orobanche*. PLITS 5(3), 95 pp.

Linke KH 1989. *Orobanche* studies. In: Food legume improvement program - Annual report for 1988. ICARDA, Aleppo, Syria. 236-239.

Linke KH 1989. *Orobanche* studies. In:. Food legume improvement program - Annual report for 1988. ICARDA, Aleppo, Syria, 214-242.

Linke KH, J Sauerborn & MC Saxena 1989. *Orobanche* field guide. University of Hohenheim, Germany/ICARDA, Aleppo, Syria. (also available in Arabic).

Linke KH 1990. *Orobanche* studies. In: Food legume improvement program - Annual report for 1989. ICARDA, Aleppo, Syria, 214-242.

Linke KH 1990. *Orobanche* studies. In: Food legume improvement program - Annual report for 1989. ICARDA, Aleppo, Syria, 243-266.

Linke KH, C Vorländer, MC Saxena 1990. Occurence and impact of *Phytomyza orobanchia* (Diptera: Agromyzidae) on *Orobanche crenata* (Orobanchaceae) in Syria. Entomophaga 35(4), 633-639.

Linke KH, H Schnell & MC Saxena 1991. Factors affecting the seed bank of *Orobanche crenata* in fields under lentil based cropping systems in Northern Syria. In: JK Ransom, LJ Musselman, AD Worsham & C Parker (eds.), Proceedings of the 5th International Symposium of Parasitic Weeds. CIMMYT, Nairobi, Kenya, 321-327.

Linke KH, KB Singh & MC Saxena 1991. Screening technique for resistance to *Orobanche crenata* Forsk. in chickpea. Chickpea Newsletter 24: 32-34.

Linke KH, MC Saxena, J Sauerborn & H Masri 1991. Effect of soil solarization on the yield of food legumes and on pest control. In: J DeVay, JJ Stapleton & CE Elmore (eds.), Proceedings of the 1st International Conference on Soil Solarization,

Amman, Jordan. FAO Plant Production and Protection Paper 109: 139-154.

Linke KH 1991. *Orobanche* studies. In: Food legume improvement program - Annual report for 1990. ICARDA, Aleppo, Syria, 216-242.

Linke KH 1991. *Orobanche* studies. In: Food legume improvement program - Annual report for 1990. ICARDA, Aleppo, Syria, 222-242.

Linke KH & MC Saxena 1991. Studies on the longevity of *Orobanche crenata* seeds in a lentil-wheat-barley rotation. Legume Program; research and training plans 1991/1992 season. ICARDA; Aleppo, Syria, 29-30.

Linke KH & MC Saxena 1991. Towards an integrated control of *Orobanche* spp. in some legume crops. In: K Wegmann & LJ Musselman (eds.), Progress in *Orobanche* research. Eberhard-Karls-Universität, Tübingen, Germany, 248-256

Linke KH 1992. Biology and control of *Orobanche* in legume crops. PLITS 10(2), 33-36.

Linke KH, A Abdel-Moneim & MC Saxena 1992. Inter- and intra-specific variation in susceptibility of forage legumes to the parasitic weed *Orobanche crenata* Forsk. Field Crops Research 32: 277-285.

Linke KH, A Abdel-Moneim & MC Saxena 1992. Inter- and intra-specific variation in susceptibility of forage legumes to the parasitic weed *Orobanche crenata* Forsk. Field Crops Research 32: 277-285.

Linke KH 1999. Status quo of *Orobanche* management: Preventive, cultural, and physical control. In: J Kroschel, M Abderabihi, H Betz (eds.), Advances in Parasitic Weed Control at On-farm Level. Vol. II. Joint Action to Control *Orobanche* in the WANA Region. Margraf Verlag, Weikersheim, Germany, 107-133.

Liu Z, A Fer & FM Lecoq 1991. L'imazaquine: un herbicide prometteur pour la lutte curative contre la cuscute (*Cuscuta* spp.) dans les cultures de soja (*Glycine max*). Weed Research 31: 33-40.

Logrieco A, A Moretti, A Ritieni, J Chelkowsky, C Altomare, A Bottalico & G Randazzo 1993. Natural occurrence of Beauvericin in preharvest *Fusarium subglutinans* infected corn ears in Poland. Journal of Agricultural Food Chemistry 41: 2149-2152.

Lopez-Granados F & L Garcia-Torres 1991. Estimation of *Orobanche crenata* seed production. In: K Wegmann & LJ Musselman (eds.), Progress in *Orobanche* research. Eberhard-Karls-Universität, Tübingen, Ger-many, 92-98.

Losner D, G Ben-Hod, AM Mayer & DM Joel 1996. Aseptic broomrape infection of tomato root culture. Israel Journal of Plant Sciences 44: 89-94.

Luft JH 1961. Improvement in epoxy resin embedding methods. Journal of Biophysical and Biochemical Cytology 9: 409-414.

Mack RN 1991. The commercial seed trade: an early disperser of weeds in the United States. Economic Botany 45: 257-273.

Mangnus E & B Zwanenburg 1991. Design and synthesis of germination stimulants for seeds of *Striga* and *Orobanche* spp. In: K Wegmann & LJ Musselman (eds.), Progress in *Orobanche* Research. Eberhard-Karls-Universität, Tübingen, Germany, 157-166.

Marx JL 1985. Plant gene transfer becomes a fertile field. Science 230: 1148-1150.

Mathews LJ 1989. Improved package for *Striga* control for the small-scale farmer. In: TO Robinson & HR Broad (eds.), *Striga*-Improved Management in Africa. Proceedings FAO/OAU All-Africa Government Consultation on *Striga* control, Maroua, 1986. FAO Plant Production and Protection Paper 96, 183-186.

Matthews GA & EW Thornhill 1994. Pesticide application equipment for use in agriculture, Vol. 1, Manually carried equipment. FAO Agricultural Services Bulletin 112/1. FAO, Rome.

Mead R & RN Curnow 1990. Statistical Methods in Agriculture and Experimental Biology. Chapman and Hall, London, UK. 335 pp.

Melchinger AE 1990. Use of molecular markers in breeding for oligogenic disease resistance. Plant Breeding 104: 1-19.

Michel BE & MR Kaufmann 1973. The potential use of polyethylene glycol 6000. Plant Physiology 51: 914-916.

Millan T, S Cobos & AM Torres 1996. The use of molecular markers in *Orobanche*. In: MT Moreno, JI Cubero, D Berner, D Joel, LJ Musselman & C Parker (eds.), Advances in Parasitic Plant Research; 6th International Parasitic Weed Symposium, Cordoba, Spain, 161-166.

Mitchell .D 1985. Goats in land and pasture. In: J.W. Copland (ed.), Goat production and research in the Tropics. Canberra, Australia, ACIAR 115-116.

Mohamed AH, G Ejeta, LG Butler & TL Housley 1998. Moisture content and dormancy in *Striga asiatica* seeds. Weed Research 38: 257-265.

Moore RP 1985. Handbook on tetrazolium testing. International Seed Testing Association, Zurich, Switzerland, 99 pp.

Moore RP 1986. ISTA Handbuch für die Tetrazoliumuntersuchung. Internationale Vereinigung für Saatgutprüfung, Zürich, Schweiz, 118 pp.

Moreno MT, JI Cubero, D Berner, D Joel, LJ Musselman & C Parker 1996. Advances in Parasitic Plant Research. Proceedings 6th International Parasitic Weed Symposium, Cordoba, Spain, 929 pp.

Mühlenberg M 1993. Freilandökologie. UTB Quelle & Meyer Heidelberg, Germany.

Müller S, C Hauck & H Schildknecht 1992. Germination stimulants produced by *Vigna unguiculata* Walp cv. Saunders Upright. Journal of Plant Growth Regulation 11: 77-84.

Murashige T & F Skoog 1962. A revised medium for plant growth and bioassay with tobacco tissue culture. Physiologia Plantarum 15: 473-497.

Murthy DV & AS Rao 1949. *Precis orithya* feeding on *Striga* spp. Current Science 9 (18), 342.

Murthy DV 1960. Insects attacking *Striga*. Nature 187, 959.

Musselman LJ, AB Worsham & RE Eplee 1979. Proceedings 2nd International Symposium on Parsitic Weeds, Raleigh, 1979. North Carolina State University, Raleigh, 296 pp.

Musselman LJ, C Parker & N Dixon 1982. Notes on autogamy and flower structure in agronomically important species of *Striga* (Scrophulariaceae) and *Orobanche* (Orobanchaceae). Beiträge zur Biologie der Pflanzen 56: 329-343.

Musselman LJ, PC Matteson & S Fortune 1983. Potential pollen vectors of *Striga hermonthica* (Scrophulariaceae) in West Africa. Annals of Botany 51, 859 - 862.

Musselman LJ 1986. Taxonomy of *Orobanche*. In: SJ ter Borg (ed.), Proceedings of a workshop on biology and control of *Orobanche*, LH/VPO, Wageningen, 2-10.

Musselman LJ 1987. Parasitic Weeds in Agriculture. Volume I. *Striga*. CRC Press, Boca Raton, 317 pp.

Musselman LJ 1987. Taxonomy of witchweeds. In: LJ Musselman (ed.), Parasitic weeds in Agriculture. CRC Press, Boca Raton, Florida, 3-12.

Nagar N & GC Sanwal 1984. Biochemical aspects of parasitism in *Cuscuta reflexa*: Inhibition of cell wall degrading enzymes of *Cuscuta* by non-susceptible plants. In: C Parker et al (eds.), Proceedings of the 3rd International Symposium on Parasitic Weeds, ICARDA/ISPRG, Aleppo, Syria, 175-183.

Nassib, AM, AA Ibrahim & SA Khalil 1982. Breeding for resistance to *Orobanche*. In: G Hawtin & C Webb (eds.), Faba Bean Improvement. Martinus Nijhoff, The Hague, 199-206.

Ndungu D, GD Odhiambo & JK Ransom 1994. Methodology for quantifying *Striga* seed numbers in soils of East Africa. In: E Adipala, MA Dekunda, JS Tenywa, MW Ogenga-Latigo, & JO Mugah (eds.), Africa Crop Science Conference Proceedings Vol. 1: 220-221.

Neeley D, J Matata & F Kisyombe 1991. Statistical techniques for the analysis of on-farm crop trials - a manual. CIMMYT, Harare, Zimbabwe and Nairobi, Kenya. 197 pp.

Netzly DH, JL Riopel, G Ejeta & LG Butler 1988. Germination stimulants of witchweed (*Striga asiatica*) from hydrophobic root exudate of sorghum (*Sorghum bicolor*). Weed Science 36: 441-446.

Nickrent DL & RJ Duff 1996. Molecular studies of parasitic plants using ribosomal RNA. In: MT Moreno, JI Cubero, D Berner, D Joel, LJ Musselman & C Parker (eds.), Advances in Parasitic Plant Research; 6th International Parasitic Weed Symposium, Cordoba, Spain, 27-52.

Nir E, B Rubin & SW Zharasov 1996. On the biology and selective control of field dodder (*Cuscuta campestris*). In: MT Moreno, JI Cubero, D Berner, D Joel, LJ Musselman & C Parker (eds.), Advances in Parasitic Plant Research; 6th International Parasitic Weed Symposium, Cordoba, Spain, 809-816.

Nirenberg H 1981. A simplified method for identifying *Fusarium* spp. occurring on wheat. Canadian Journal of Botany 59: 1599-1609.

Nishida T & T Torii 1970. A handbook of field methods for research on rice stemborers and their natural enemies. Davis Company, Philadelphia, UK.

Norambuena H, J Kroschel, O Klein 1998. Introduction of *Phytomyza orobanchia* for biocontrol of *Orobanche* spp. in Chile. In: J Kroschel, H Betz, M Abderahibi (eds.), Advances in Parasitic Weed Control at Onfarm level. Joint Action to control *Orobanche* in the WANA region. Margraf Verlag, Weikersheim, Germany, 197-204.

Ogborn JEA 1984. Research priorities in agronomy. In: ES Ayensu, H Doggett, RD Keynes, J. Marton-Lefevre, LJ Musselman, C Parker & A Pickering (eds.), *Striga*, Biology and Control. IRL Press Ltd Oxford, UK, 195-212.

Okonkwo SNC 1966a. Studies on *Striga senegalensis* III. *In vitro* culture of seedlings. Establishment of cultures. American Journal of Botany 53(7): 679-687.

Okonkwo SNC 1966b. Studies on *Striga senegalensis* IV. *In vitro* culture of seedlings. Effect of various sugars and Glutamine. American Journal of Botany 53(7): 687-693.

Okonkwo SNC 1970. Studies on *Striga senegalensis* V. Origin and development of buds from roots of seedlings reared *in vitro*. Phytomorphology 20: 144-150.

Okonkwo SNC 1987. Research techniques - laboratory. In: LJ Musselman (ed.), Parasitic Weeds in Agriculture Vol. I - *Striga* CRC Press. Boca Raton, Florida, 257-267.

Okonkwo SNC & DN Okafor 1988. Stimulation of germination of seeds of root parasitic weeds by strigol analogues GR7 and GR24. In: SNC Okonkwo, IO Akobundu & AO Ayeni (eds.), Proceedings of the 3rd Biennial Conference of the West African Weed Science Society, Ibadan, 31-41.

Okonkwo SNC 1996. Aseptic culture of parasitic weeds - an overview. In: MT Moreno, JI Cubero, D Berner, D Joel, LJ Musselman & C Parker (eds.), Advances in Parasitic Plant Research; 6th International Parasitic Weed Symposium, Cordoba, Spain, 283-292.

Olivier A, N Benhamou & GD Leroux 1991. Cell surface interactions between sorghum root and the parasitic weed *Striga hermonthica*: cytochemical aspects of cellulose distribution in resistant and susceptible host tissues. Canadian Journal of Botany 69: 1679-1690.

Olivier A, JC Glaszmann, C Lanaud, G Salle & GD Leroux 1996. An insight into the population structure and genetic diversity of *Striga hermonthica* in West Africa. In: MT Moreno, JI Cubero, D Berner, D Joel, LJ Musselman & C Parker (eds.), Advances in Parasitic Plant Research; 6th International Parasitic Weed Symposium, Cordoba, Spain, 113-121.

Olroyd H 1958. Collecting, preserving and studying insects. Hutchinson & Co. Ltd London, UK.

Orita M, Y Suzuki, T Sekita & K Hayashi 1989. Rapid and sensitive detection of point mutations and DNA polymorphisms using the polymerase chain reaction. Genomics 5: 874-879.

Oswald A, H Frost, JK Ransom, J Kroschel, K Shepherd & J Sauerborn 1996. In: MT Moreno, JI Cubero, D Berner, D Joel, LJ Musselman & C Parker (eds.), Advances in Parasitic Plant Research; 6th International Parasitic Weed Symposium, Cordoba, Spain, 795-800.

Parker C, AM Hitchcock & KV Ramaiah 1977. The germination of *Striga* species by crop root exudates; techniques for selecting resistant crop cultivars. In: Proceedings, Asian-Pacific Weed Science Society 6th Conference, Jakarta, 67-74.

Parker C & N Dixon 1983. The use of polyethylene bags in the culture and study of *Striga* spp. and other organisms on crop roots. Annals of Applied Biology 103: 485-488.

Parker C 1984. The influence of *Striga* species on sorghum under varying nitrogen fertilization. In: Parker C, LJ Musselman, RM Polhill & AK Wilson 1984. Proceedings 3rd International Symposium on Parasitic Weeds, ICARDA/International Parasitic Seed Plant Research Groupe, 7-9 May 1984, Aleppo, Syria. ICARDA, P.O.Box 5466, Aleppo, Syria, 90-98.

Parker C, LJ Musselman, RM Polhill & AK Wilson 1984. Proceedings 3rd International Symposium on Parasitic Weeds, ICARDA/ International Parasitic Seed Plant Research Groupe, 7-9 May 1984, Aleppo, Syria. ICARDA, P.O.Box 5466, Aleppo, Syria, 265 pp.

Parker C & CR Riches 1993. Parasitic Weeds of the World - Biology and Control. CAB International, Wallingford, UK, 332 pp.

Parkinson V, Y Efron, L Bello & K Dashiell 1987. Trap crops as a cultural measure in *Striga* control in Africa. FAO Plant Protection Bulletin 35: 51-54.

Paterson AH, SD Tanksley & ME Sorrells 1991. DNA markers in plant improvement. Advanced Agronomy 46: 39-90.

Patterson DT 1987. Environmental factors affecting witchweed growth and development. - In: LJ Musselman (ed.), Parasitic weeds in Agriculture, Vol. 1, *Striga*. CRC Press, Boca Raton, USA, 27-41.

Pavlista AD 1981. Why hasn't witchweed spread in the United States? - Weeds Today 12(1): 16-19.

Petzoldt K, Y Nemli & J Sneyd 1994. Integrated control of *Orobanche cumana* in sunflower. In: AH Pieterse, JAC Verkleij & SJ ter Borg (eds.), Biology and Management of *Orobanche*. Proceedings of the 3rd International Workshop on *Orobanche* and related *Striga* research. Amsterdam, The Netherlands, Royal Tropical Institute, 442-449.

Pieterse AH 1979. The broomrapes (Orobanchaceae) - a review. Abstracts on Tropical Agriculture 5(3): 9-35.

Pieterse AH 1991. The effect of nitrogenous fertilisers on the germination of seeds of *Striga hermonthica* and *Orobanche crenata*. In: K Wegmann & LJ Musselman (eds.), Progress in *Orobanche* Research, Proceedings of the International Workshop on *Orobanche* Research, Eberhard-Karls Universität, Tübingen, Germany, 115-124.

Pieterse AH, L Garcia-Torres, OA Al-Menoufi, KH Linke & SJ ter Borg 1992. Integrated control of the parasitic angiosperm *Orobanche* (broomrape). 2nd International Food Legume Research Conference, Cairo, Egypt.

Pieterse AH, JA Verkleij & S-J ter Borg 1994. Biology and management of *Orobanche*. Proceedings of the 3rd International Workshop on *Orobanche* and related *Striga* research, Amsterdam, The Netherlands, November 8-12, 1993. Amsterdam. Royal.Tropical Institute.

Pieterse AH 1996. The effect of nitrogen on *Orobanche* and *Striga* - State of the art. In: MT Moreno, JI Cubero, D Berner, D Joel, LJ Musselman & C Parker (eds.), Advances in Parasitic Plant Research; 6th International Parasitic Weed Symposium, Cordoba, Spain, 273-282.

Pinkerton F & G Strobel 1976. Serinol as an activator of toxin production in attenuated cultures of *Helminthosporium sacchari*. Proc. Natl. Acad. Sci.USA, 73, 4007-4011.

Pogorletsky BK & EE Geshele 1976. Sunflower immunity to broomrape, downy mildew and rust. Proceedings of the 7th International Sunflower Conference, 238-243.

Polniaszek TI, C Parker & CR Riches 1991. Variation in the virulence of *Alectra vogelii* populations on cowpea. Tropical Pest Management 37: 152-154.

Portères R 1952. Linear cultural sequences in primitive systems of agriculture in Africa and their significance. Sols Africains 2: 15.

Press MC, JM Tuohy & GR Stewart 1987. Gas exchange characteristics of the sorghum-*Striga* host-parasite association. Plant Physiology 84: 814-819.

Press MC, JJ Nour, FF Bebawi & GR Stewart 1989. Antitranspirant-induced heat stress in the parasitic plant *Striga hermonthica* - A novel method of control. Journal of Experimental Botany 40: 585-598.

Pülschen L 1990. Natural site factors and agronomic aspects in the grain-plough complex of the Shewa Province, Ethiopia. PLITS 8(3), 84 pp.

Raju CA 1996. Variability in tobacco germplasm towards *Orobanche* infection. In: MT Moreno, JI Cubero, D Berner, D Joel, LJ Musselman & C Parker (eds.), Advances in Parasitic Plant Research; 6th International Parasitic Weed Symposium, Cordoba, Spain, 609-613.

Ramaiah KV, C Parker, MJ Vasudeva Rao & LJ Musselman 1983. *Striga* identification and control handbook. ICRISAT Information Bulletin No. 15, ICRISAT, Patancheru, India, 52 pp.

Ramaiah KV 1984. Patterns of *Striga* resistance in sorghum and millets with special emphasis on Africa. In: Proceedings of the International Workshop on the Biology and Control of *Striga*, held at Dakar, Senegal, 14 - 17 November 1983. ICSU Press, France, 71-92.

Ramaiah KV 1985. Hand pulling of *Striga hermonthica* on pearl millet. Tropical Pest Management 31(4): 326-327.

Ramaiah KV 1987. Breeding cereal grains for resistance to witchweed. In: LJ Musselmann (ed.), Parasitic Weeds in Agriculture, Volume I *Striga*. CRC Press, Boca Raton, Florida, 227-242.
Ramaiah KV 1987. Breeding cereal grains for resistance to witchweed. In: LJ Musselmann (ed.), Parasitic Weeds in Agriculture, Volume I *Striga*. CRC Press, Boca Raton, Florida, 227-242.
Ramaiah KV, VL Chidley & LR House 1990. Inheritance of *Striga* seed germination stimulant in sorghum. Euphytica 45: 33-38.

Rangaswamy NS 1963. Studies on culturing seeds of *Orobanche aegyptiaca* Pers. In Maheshwari, P., Rangaswamy, NS (eds.), Plant Tissue and Organ Culture. A Symposium. 345-354.

Ransom JK, RE Eplee, MA Langston & RS Norris 1990. Methodology for establishing witchweed (*Striga asiatica*) in research plots. Weed Technology, Volume 4, 581-584.

Ransom JK, LJ Musselman, AD Worsham & C Parker 1991. Proceedings 5th International Symposium on Parasitic Weeds, Nairobi, 1991. CIMMYT, Nairobi, 550 pp.

Ransom JK & GD Odhiambo 1994. Long-term effects of fertility and hand-weeding on *Striga* in maize. In: AH Pieterse, JAC Verkleij & SJ ter Borg (eds.), Biology and Management of *Orobanche*. Proceedings of the 3rd International Workshop on *Orobanche* and related *Striga* research. Amsterdam, The Netherlands, Royal Tropical Institute, 513-519.

Ransom JK, GD Odhiambo, RE Eplee & AO Diallo 1996. Estimates from field studies of the phytotoxic effects of *Striga spp.* on maize. In: MT Moreno, JI Cubero, D Berner, D Joel, LJ Musselman & C Parker (eds.), Advances in Parasitic Plant Research. Proceedings 6th Parasitic Weed Symposium. Cordoba, Spain, 327-333.

Rao KN & RSN Rao 1991.Studies on tolerance of black gram and green gram varieties to dodder (*Cuscuta chinensis* Lam.). In: JK Ransom, LJ Musselman, AD Worsham & C Parker (eds.), Proceedings of the 5th International Symposium of Parasitic Weeds. CIMMYT, Nairobi, Kenya, 170-175.

Reichmann S, J Kroschel & J Sauerborn 1995. Distribution and infestation of *Striga* species in Shinyanga region of Tanzania and evaluation of control methods. Brighton Crop Protection Conference Weeds, Brighton, UK, Vol. 1, 151-156.

Reijntjes C, B Haverkort & A Waters-Bayer 1992. Farming for the future. Macmillan Press, London and Basingstoke.

Reuter, BC 1986. The habitat, reproductive ecology and host relations of *Orobanche fasciculata* Nutt. (Orobanchaceae) in Wisconsin. Bulletin of the Torrey Botanical Club 113(2): 110-117.

Riches CR 1989. The biology and control of *Alectra vogelii* Benth., (Scrophulariaceae) in Botswana. Ph.D. Thesis, University of Reading, UK, 208 pp.

Riches CR, KA Hamilton & C Parker 1992. Parasitism of *Alectra* species by grain legumes. Annals of Applied Biology 121, 361-370.

Rieger R, A Michaelis & M Green 1991. Glossary of genetics. 5th Edition. Springer Verlag, Berlin-Heidelberg, Germany, 553pp.

Robinson EL & CC Dowler 1966. Investigations of catch and trap crops to eradicate witchweed (*Striga asiatica*). Weeds 14: 275-276.

Rohrmoser K 1985. Handbook for field trials in technical cooperation. TZ-Verlagsgesellschaft, Roßdorf, Germany.

Runge-Metzger A, J Kroschel, J Sauerborn 1997. The economics of *Striga* control in different farming systems in Northern Ghana. In: STO Lagoke, LE van der Straten & SS M'Boob (eds.), 3rd General Workshop of the Pan-African *Striga* Control Network (PASCON) „Integrated Management of *Striga* for the African Farmer", Oct. 18-23, 1993, Harare, Zimbabwe. FAO Accra, Ghana, 134-143.

Saghir AB 1986. Dormancy and germination of *Orobanche* seeds in relation to control methods. In: SJ ter Borg (ed.), Proceedings of a workshop on biology and control of *Orobanche*. LH/VPO, Wageningen, The Netherlands, 25-34.

Saghir AR, M Kurban & B Budayr 1980. Studies on the control of *Orobanche* in Lebanon. Tropical Pest Management 26(1): 51-55.

Sambrook J, EF Fritsch & T Maniatis 1989. Molecular Cloning - a laboratory manual (2nd edition). Cold Spring Harbour Laboratory Press, U.S.

Sand PF 1987. The American witchweed quarantine and eradication program. In: LJ Musselman (ed.), Parasitic weeds in Agriculture. CRC Press, Boca Raton, Florida, 207-223.

Sand PF, RE Eplee & RG Westbrooks 1990. Witchweed Research and Control in the United States of America. Weed Science Society of America, Champaign, 154 pp.

Sankaran T & VP Rao 1966. Insects attacking witchweed (*Striga*) in India. Commonwealth Institute Biological Control, Technical Bulletin 7, 63-73.

Sankaran T, MK Rajendran & Y Ranganath Bhat 1969. The biology of *Eulocastra argentisparsa* Hampson (Lep.: Noctuidae) with special reference to the occurrence of dimorphism. Commonwealth Instiute of Biological Control, Technical Bulletin 11, 31-43.

Sauerborn J, H Masri, MC Saxena & KH Linke 1987. Control of *Orobanche* with Scepter. FABIS Newsletter 19: 14-17.

Sauerborn J, H Masri, MC Saxena & W Erskine 1987. A rapid test to screen lentil under laboratory conditions for susceptibility to *Orobanche*. Lens 14: 15-16.

Sauerborn J, KH Linke, MC Saxena & W Koch 1989. Solarization; a physical control method for weeds and parasitic plants (*Orobanche* spp.) in Mediterranean agriculture. Weed Research, Volume 29: 391-397.

Sauerborn J, MC Saxena & A Mayer 1989. Broomrape control in faba bean (*Vicia faba* L.) with glyphosate and imazaquin. Weed Research 29: 97-102.

Sauerborn J 1989. The influence of temperature on germination and attachment of the parasitic weed *Orobanche* spp. on lentil and sunflower. Angewandte Botanik 63: 543-550.

Sauerborn J, G Wyrwal & KH Linke 1989. Soil sampling for broomrape seeds in naturally infested fields. In: K Wegmann & LJ Musselman (eds.), Progress in *Orobanche* research. Eberhard-Karls-Universität, Tübingen, Germany, 35-43.

Sauerborn J, G Wyrwal & KH Linke 1991. Soil sampling for broomrape seeds in naturally infested fields. In: K Wegmann & LJ Musselman (eds.), Progress in *Orobanche* Research. Eberhard-Karls-Universität, Tübingen, Germany, 35-43.

Sauerborn J 1991. Parasitic flowering plants - Ecology and Management. Margraf Verlag, Weikersheim, Germany, 127 pp.

Sauerborn J 1991. The economic importance of the phytoparasites *Orobanche* and *Striga*. In: JK Ransom, LJ Musselman, AD Worsham & C Parker (eds.), Proc. 5th International Symposium on Parasitic Weeds, Nairobi, Kenya, 137-143.

Sauerborn J, I Dörr, AA Abasher, H Thomas & J Kroschel 1996. Electron microscopic analysis of the penetration process of *Fusarium nygamai*, a hyperparasite of *Striga hermonthica*. Biological Control 7: 53-59.

Sauerborn J, AA Abbasher, J Kroschel, DW Cornes, A Zoschke & KT Hine 1996. *Striga hermonthica* control with *Fusarium nygamai* in maize. In: VC Moran & JH Hoffmann (eds.), Proc 9th Int. Symp. on Biological Control of Weeds, Stellenbosch, South Africa, 461-466.

Sauerborn J, H Sprich & H Mercer-Quarshie 1999. Parasitic weed infestation of maize and sorghum as influenced by crop rotation. In: J Kroschel, H Mercer-Quarshie & J Sauerborn (eds.), Advances in Parasitic Weed Control at On-farm Level. Vol. I. Joint Action to Control *Striga* in Africa. Margraf Verlag, Weikersheim, Germany, 161-169.

Sax K 1923. The association of size differences with seed coat pattern and pigmentation in *Phaseolus vulgaris*. Genetics 8: 552-560.

Schlichting E & H-P Blume 1966. Bodenkundliches Praktikum. Parey Verlag, Berlin, Hamburg, Germany, 205 pp.

Schmitt U 1981. Untersuchungen zur Verbreitung und Bekämpfung von *Orobanche crenata* Forsk. an Ackerbohnen in Marokko. Ph.D. Thesis, University of Bonn, Germany, 194 pp.

Schnell H 1993. Aspekte zur Kontrolle von *Orobanche crenata* Forsk. in Nordsyrien durch Fruchtfolgemaßnahmen unter Berücksichtigung der Dynamik der *Orobanche*-Samenpopulation im Boden. PLITS 11(2).

Schönhuth M & U Kievelitz 1994. Participatory learning approaches, Rapid Rural Appraisal Participatory Appraisal, an introductory guide. Schriftenreihe der GTZ, No. 248. TZ-Verlagsgesellschaft, Rossdorf.

Schwab A 1989. Pestizideinsatz in Entwicklungsländern. Ein Leitfaden für Entwick-

lungshelfer. Tropical Agroecology 3. Margraf, Weikersheim.
Sene D 1995. The role of research and training in agricultural extension. In: Agricultural Extension in Africa, Proceedings of an international workshop, Yaoundé, Cameroon, January 1994. Technical Centre for Agricultural and Rural Co-operation (CTA), Wageningen, 21-28.
Severova E 1991. New taxonomic characters in the genus *Cuscuta* (*Cuscuta*ceae). In: JR Ransom, LJ Musselman, AD Worsham & C Parker (eds.), Proceedings of the 5th International Symposium of Parasitic Weeds; CIMMYT, Nairobi, Kenya, 514-518.
Shawe KG & MJ Ingrouille 1993. Isozyme analysis demonstrates host selection of parasite pathotypes in the association between cowpea and *S. gesnerioides*. Brighton Crop Protection Conference-Weeds 2: 919-924.
Shaxson LJ, CR Riches & JH Seyani 1993. Incorporating farmer knowledge in the design of weed control strategies for smallholders. Brighton Crop Protection Conference Weeds, Brighton, UK, 1149-1154.
Shinde VK & N Kulkarni 1982. Genetics of resistance to *Striga asiatica* in sorghum. In: Proceedings, ICRISAT-AICSIP (ICAR) Working Group Meeting on *Striga* Control. ICRISAT Center 30 Sept. - 1 Oct. 1982. ICRISAT, Patancheru PO, Andhra Pradesh 502 324, India. 134-140.
Sillero JC, D Rubiales & JI Cubero 1996. Risks of *Orobanche* resistance screenings based only on number of emerged shoots per plant. In: MT Moreno, JI Cubero, D Berner, D Joel, LJ Musselman & C Parker (eds.), Advances in Parasitic Plant Research; 6th International Parasitic Weed Symposium, Cordoba, Spain, 651-657.
Singh BB & AM Emechebe 1991. Breeding for resistance to *Striga* and *Alectra* in cowpea. In: JK Ransom, LJ Musselman, AD Worsham & C Parker (eds.), Proceedings of the 5th International Symposium of Parasitic Weeds. CIMMYT, Nairobi, Kenya, 303-305.
Singh L, R Ndikawa & MR Rao 1991. Integrated approach to *Striga* management on sorghum in north Cameroon. In: JR Ransom, LJ Musselman, AD Worsham & C Parker (eds.), Proceedings of the 5th International Symposium of Parasitic Weeds CIMMYT, Nairobi, Kenya, 223-231.
Singh P & RF Moore 1985. Handbook of insect rearing. Vol. I. Elsevier, Amsterdam, Oxford, New York, Tokyo.
Singleton LL, J D Mihail & CM Rush 1993. Methods for research on soil borne phytopathogenic fungi. APS Press, St. Paul, Minn., USA.
Sivanesan A 1987. Graminicolous species of *Bipolaris, Curvularia, Drechslera, Exerohilum*, and their teleomorphs. Mycological Papers 158, 216 pp.
Smith MC, J Holt & M Webb 1993. Population model of the parasitic weed *Striga hermonthica* (Srophulariaceae) to investigate the potential of *Smicronyx umbrinus* (Coleoptera: Curculionidae) for biological control in Mali. Crop Protection 12 (6), 470-476.
Snelder Y, MT Moreno, A Martin & J Gil 1995. Screening for resistance to *Orobanche crenata* Forsk. in *Vicia faba* L.: a comparison between methods. In: Anonymous, 2nd European Conference of Grain Legumes. Copenhagen, 230-231.
Southwood TRE 1989. Ecological methods with particular reference to the study of insect populations. University Press, Cambridge, UK.
Sprich H 1994. Bedeutung der Fruchtfolge zur Ertragssicherung in getreidebetonten Produktionssystemen der Guinea-Savanne unter besonderer Berücksichtigung des parasitischen Unkrauts *Striga hermonthica* (Del.) Benth. PLITS 12(3).
Steiner KG 1987. On-farm experimentation handbook for rural development projects. Guidelines for the development of ecological and socio-economic sound extension messages for farmers. TZ-Verlagsgesellschaft, Rossdorf.
Stewart GR, MC Press, JD Graves, JJ Nour & A Wylde 1991. A physiological characterization of the host-parasite association between *Sorghum bicolor* and *Striga hermonthica* and its implications for *Striga* control. In: SK Kim (ed.), Combating *Striga* in Africa. Proceedings, International Workshop organized by IITA, ICRISAT and ICRC, 1988 IITA, Ibadan, Nigeria, 48-54.
Stierle A, G Strobel, D Stierle & F Sugawara 1992. Analytical methods for phytotoxins. In: Modern methods of plant analysis. New Series, Vol. 13. Plant Toxin Analysis. Springer-Verlag, Berlin, Germany, 1-32.

Strasburger E 1979. Lehrbuch der Botanik. Verlag G. Fischer, Stuttgart, 31. Auflage, 1078 pp.

Strobel G, D Kenfield, G Bunkers, F Sugawara & J Clardy 1991. Phytotoxins as potential herbicides. Experientia 47: 819-826.

Stuber CW 1990. Biochemical and molecular markers in plant breeding. Plant Breeding Review 9: 37-61.

Sukno S, J Rusao, J Melero-Varo & J Fernandez-Martinez 1996. Evaluation of wild *Helianthus* germplasm for resistance to highly virulent races of *Orobanche cernua* Loefl. and transfer to cultivated sunflower. In: MT Moreno, JI Cubero, D Berner, D Joel, LJ Musselman & C Parker (eds.), Advances in Parasitic Plant Research; 6th International Parasitic Weed Symposium, Cordoba, Spain, 625-632.

Tanksley SD, ND Young, AH Paterson & MW Bonierbale 1989. RFLP mapping in plant breeding: new tools for an old science. Biotechnology 7: 257-264.

Tayler LR 1961. Aggregation, variance and the mean. Nature 189: 732-735.

Tennant D 1975. A test of a modified line intersect method of estimating root length. Journal of Ecology 63: 995-1001.

Thalouarn P, MC Arnaud, C Théodet & L Rey 1991. Cytological, biochemical and genetic aspects of carbon fixation in *Striga hermonthica* and *Striga gesnerioides*. In: Ransom JK, LJ Musselman, AD Worsham, & C Parker (eds). Proceedings of the 5th International Symposium on Parasitic Weeds. CIMMYT, Nairobi, Kenya, pp. 51-57.

Theodet C, P Thalouarn, C Figureau & P Ferard 1996. A new approach to *Orobanche* species identification. In: MT Moreno, JI Cubero, D Berner, D Joel, LJ Musselman & C Parker (eds.), Advances in Parasitic Plant Research; 6th International Parasitic Weed Symposium, Cordoba, Spain, 143-150.

Thiel PG, WFO Marasas, EW Sydenham, GS Shephard & WCA Gelderblom 1992. The implications of naturally occurring levels of fumonisins in corn for human and animal health. Mycopathologie 117: 3-9.

Thomas H, J Sauerborn, D Müller-Stöver, A Ziegler, JS Bedi & J Kroschel 1998. The potential of *Fusarium oxysporum* f.sp. *orthoceras* as a biological control agent for *Orobanche cumana* in sunflower. Biological Control 13, 41-48.

Thomas H, A Heller, J Sauerborn & D Müller-Stöver 1999. *Fusarium oxysporum* f.sp. *orthoceras*, a potential mycoherbicide, parasitizes seeds of *Orobanche cumana* (sunflower broomrape): a cytological study. Annals of Botany, 83, 453-458.

Thormann CE, ME Ferreira, LEA Camargo, JG Tivang & TC Osborn 1994. Comparison of RFLP and RAPD markers to estimating genetic relationships within and among cruciferous species. Theoretical Applied Genetics 88: 973-980.

Thuring JWJF 1996. Synthesis and bioactivity of germination stimulants for the seeds of parasitic weeds - a molecular study. PhD-Thesis, University of Nijmegen, The Netherlands.

Torres AM, ND Weeden & A Martin 1992. Linkage between isozymes, RFLP and RAPD markers in *Vicia faba*. Theoretical and Applied Genetics 85: 937-938.

Toure M, A Olivier, BR Ntare & CA St-Pierre 1996. The influence of sowing date and irrigation of cowpea on *Striga gesnerioides* emergence. In: MT Moreno, JI Cubero, D Berner, D Joel, LJ Musselman & C Parker (eds.), Advances in Parasitic Plant Research; 6th International Parasitic Weed Symposium, Cordoba, Spain, 787-794.

Townes H 1972. A light-weight Malaise trap. Entomology News 83: 239-247.

Traoré D 1995. Bionomics of *Smicronyx guineanus* Voss and *S. umbrinus* Hustache (Coleoptera: Curculionidae), potential agents of *Striga hermonthica* (Del.) Benth. PhD thesis, Macdonald Campus, McGill University, Quebec, Canada, (unpublished).

Traoré D, C Vincent & RK Stewart 1995. Life history of *Smicronyx guineanus* and *S. umbrinus* (Col.: Curculionidae) on *Striga hermonthica* (Scrophulariceae). Entomophaga 40 (2), 211-221.

Traoré D, C Vincent & RK Stewart 1996. Association and synchrony of *Smicronyx guineanus* Voss, *S. umbrinus* Hustache (Coleoptera: Curculionidae), and the parasitic weed *Striga hermonthica* (Del.) Benth. (Scrophulariceae). Biological Control 7, 307-315.

Tsivion Y 1979. Morphogenetic sequences in the formation of the haustorium of *Cuscuta campestris*. In: Musselman LJ, AD Worsham & RE Eplee (eds.), Proceedings of the 2nd International Symposium on Para-

sitic Weeds. Raleigh, North Carolina, USA, 174-181.
Tuohy JM & JS Choinski 1991. Water relations and photosynthesis in the mistletoe *Tieghemia bolusii* and its host *Dipolrhynchus condylocarpon*. In: JK Ransom, LJ Musselman, AD Worsham & C Parker (eds.), Proceedings of the 5th International Symposium of Parasitic Weeds. CIMMYT, Nairobi, Kenya, 232-236.
Tuquet C & G Sallé 1991. Biochemical features of chloropasts isolated from hemiparasitic angiosperms. In: Ransom JK, LJ Musselman, AD Worsham & C Parker (eds), Proceedings of the 5th International Symposium on Parasitic Weeds. CIMMYT, Nairobi, Kenya, pp. 262-269.
UNICEF 1993. VIPP Visualisation in participatory programmes. A manual for facilitators and trainers involved in participatory group events. UNICEF Bangladesh. P.O.B. 58, Dhaka 1000, Bangladesh.
USDA 1967. Soil survey laboratory methods and procedures for collecting soil samples. Soil survey investigations rep. no. 1. Soil Cons. Serv. Washington D.C.
Vallance K 1950. Studies on the germination of seeds of *Striga hermonthica*. I. The influence of moisture treatment, stimulant dilution and after-ripening on germination. Annals of Botany 14: 347.
Van Hezewijk MJ, KH Linke, JAC Verkleij & AH Pieterse 1991. The effect of ammonium and urea fertilizer in combination with nitrification inhibitors on *Orobanche crenata* infestation in faba bean and lentil - field experiments. In: JK Ransom, LJ Musselman, AD Worsham & C Parker (eds.), Proceedings of the 5th International Symposium of Parasitic weeds, June 1991, Nairobi, Kenya, 470-483.
Van Hezewijk MJ, KH Linke, F Lopez-Granados, OA Al-Menoufi, L Garcia-Torres, JAC Verkleij & AH Pieterse 1994. Seasonal changes in germination response of buried seeds of *Orobanche crenata*. Weed Research 34(5): 369-376.
Vasudeva Rao MJ 1985. Techniques for screening sorghums for resistance to *Striga*. ICRISAT Information Bulletin No. 20, Andhra Pradesh, India.
Vasudeva Rao MJ & LJ Musselman 1987. Host specificity in *Striga* spp. and physiological "strains". In: LJ Musselmann (ed.), Parasitic weeds in Agriculture, Volume I *Striga*. CRC Press, Boca Raton, Florida, 13-25.
Vasudeva Rao MJ, VL Chidley & LR House 1989. Estimates of grain yield losses caused in sorghum (*Sorghum bicolor* (L.) Moench) by *Striga asiatica* (L.) Kuntze obtained using the regression approach. Agriculture, Ecosystems and Environment 25: 139-149.
Visser JH & LF Wentzel 1980. Quantitative estimation of *Alectra* and *Striga* seed in soil. Weed Research 20: 77-81.
Visser JH, I Dörr & D Kollmann 1990. Compatibility of *Alectra vogelii* with different leguminous host species. Journal of Plant Physiology 135: 737-745.
Vogler RK, G Ejeta & LG Butler 1991. Inheritance of low stimulation of *Striga* seed germination in sorghum cultivar SRN39. Agronomy Abstracts. Denver, CO, 120.
Vogt W, J Sauerborn & M Honisch 1991. *Striga hermonthica*, distribution and infestation in Ghana and Togo on grain crops. In: JK Ransom, LJ Musselman, AD Worsham & C Parker (eds.), Proceedings of the 5th International Symposium of Parasitic weeds, June 1991, Nairobi, Kenya, 372-377.
Vogt W 1993. Entwicklung der Wirt/Parasit-Beziehung *Sorghum bicolor/Striga hermonthica* unter dem Einfluß verschiedener Stickstofformen und Standortfaktoren. PLITS 11(3).
Vos P, R Hogers, M Bleeker, M Reijans, T Van der Lee, M Hornes, A Frijters, J Pot, J Peleman, M Kuiper & M Zabeau 1995. AFLP: a new technique for DNA fingerprinting. Nucleic Acids Research 23: 4407-4414.
Vranceanu AV, N Pirvu, FM Stoenescu & M Pacureanu 1986. Some aspects of the interaction *Helianthus annuus* L./*Orobanche cumana* Wallr. and it's implications on sunflower breeding. In: SJ ter Borg (ed.), Proceedings of a workshop on biology and control of *Orobanche*. LH/VPO, Wageningen, The Netherlands, 181-189.
Vyas BP 1966. Control of *Orobanche* (on tobacco). Indian Farming 16(2): 51.
Walker PT 1983. The assessment of crop losses in cereals. Insect Science Application, 4(1/2): 97-104.
Walter H 1983. Weed sampling in the field and interpretation of sampling data. In:

Weed management in the Philippines, Reports of seminars. PLITS 1(1): 65-80.

Waters-Bayer A 1989. Participatory technology development in ecological-oriented agriculture: some approaches and tools. ODI Agricultural Administration Unit. London, U.K.

Weber HC & W Forstreuter 1987. Parasitic Flowering Plants. Proceedings 4th International Symposium on Parasitic Flowering Plants, Marburg, Germany. Philipps-Universität, Marburg, 848 pp.

Wegmann K & LJ Musselman 1991. Progress in *Orobanche* Research. Eberhards-Karls-Universität, Tübingen, Germany, 360 pp.

Wegmann K, LJ Musselman & DM Joel 1998. Current Problems of *Orobanche* Research. Proceedings of the 4th International *Orobanche* Workshop, September 23-26 1998, Albena, Bulgaria. Institute of Wheat and Sunflower Dobroudja, Bulgaria.

Werner J 1993. Participatory Development of Agricultural Innovations. Procedures and methods of on-farm research. Schriftenreihe der GTZ No. 234, TZ-Verlags-Gesellschaft, Roßdorf, Germany.

Wettstein R 1893. Scrophulariaceae. In: A Engler & K Prantl (eds.), Die Natürlichen Pflanzenfamilien 4(36), 39-107.

Wetzel T 1984. Diagnosemethoden. Springer Verlag, Berlin, Heidelberg, New York, Tokyo.

Wetzel T 1984. Methoden der Schaderreger- und Bestandesüberwachung. In: T Wetzel (ed.), Diagnosemethoden. Springer Verlag, Berlin, Heidelberg, New York, Tokyo, 15-17.

Wilhelm S & LC Benson 1954. Vertical distribution of broomrape seed in tomato field soils. Plant Disease Reporter 38: 553-554.

Williams CN & GH Caswell 1959. An insect attacking *Striga*. Nature 184, 1668.

Williams JGK, AR Kubelic, KJ Livak, JA Rafalski & SV Tingey 1990. DNA polymorphisms amplified by arbitrary primers are useful as genetic markers. Nucleic Acids Research 18: 6531-6535.

Wilson AK & C Parker 1984. Techniques for setting up pot experiments involving the parasitic weed *Striga*. In: C Parker, LJ Musselman, RM Polhill & AK Wilson (eds.) Proceedings of the 3rd International Symposium on Parasitic Weeds, ICARDA/ISPRG, Aleppo, Syria, 133-142.

Wilson-Jones K 1953. Relation of witchweed (*Striga*) to fertility in tropical soils. Nature 172: 128.

Worsham AD & GH Egley 1990. Physiology of witchweed seed dormancy and germination. In: Witchweed Research and Control in the United States, Monograph no. 5, Weed Science Society of America, Champaign, IL, 11-26.

Wyniger R 1974. Insektenzucht. Ulmer Verlag Stuttgart, Germany.

Yaduraju NT & MM Hosmani 1979. *Striga asiatica* control in sorghum. PANS 25(2): 163-167.

Zermane N, G Sallé, J Kroschel & Z Bouznad 1998. Prospects for biological control of the parasitic weed *Orobanche* in Algeria. In: J Kroschel, M Abderabihi & H Betz (eds.), Advances in parasitic weed control at on-farm level. Volume II. Joint action to control *Orobanche* in the WANA region. Margraf Verlag, Weikersheim, Germany, 173-184.

Zonno MC & M Vurro 1999. Effect of fungal toxins on germination of *Striga hermonthica* seeds. Weed Research, 39,15-20.

Zonno MC, M Vurro, R Capasso, A Evidente, A Cutignano, J Sauerborn & H Thomas 1996. Phytotoxic metabolites produced by *Fusarium nygamai* from *Striga hermonthica*. In: VC Moran & JH Hoffmann (eds.), Proceedings of the IX International Symposium on Biological Control of Weeds, 19-26 January 1996, Stellenbosch, South Africa, 223-226.

Zwanenburg B, GL Mhehe, GK `t Lam, FJ Dommerholt & AA Kishimba 1986. The search for new germination stimulants of *Striga* species. In: SJ ter Borg (ed.), Proceedings of a workshop on biology and control of *Orobanche*. LH/VPO, Wageningen, 35-41.

11 Annex

11.1 Key to the main *Striga* spp. (according to Parker & Riches 1993)

A. Calyx ribs usually 5 or 6 (1 per calyx lobe)

Plant about 50 cm high (sometimes up to 1 m), spike of may pink flowers, 1-2 cm across, corolla tube 1.5-2 cm long, bent at or just above halfway (two-thirds up in East Africa), bracts below each flower 2-3 mm wide, fringed with hairs. Parasitic on cereal crops in Africa and Arabia. .... *S. hermonthica* (Del.) Benth.

Plant up to 50 cm high, spike of few to many pink flowers about 1 cm across, corolla tube about 1-1.5 cm long, bent at least two-thirds up from the base, bracts 1-2 mm wide only, mostly without a fringe of hairs. Parasitic on cereals in Africa only .... *S. aspera* (Willd.) Benth

Plant 15-30 cm high, flowers white in relatively dense spike, 2-6 open at a time, 0.5-1cm across, drying bluish. Parasitic on cereal crops in Asia only.. .... *S. densiflora* (Benth.) Benth.

Plant 15-30 cm high, flowers creamy white to yellow, occasionally pale pink, drying black, bracts much longer than calyx, occasionally parasitic on cereal crops in Africa only .... *S. passargei* Engl.

Plant 15-30 cm high, leaves reduced to scales and whole plant often pale green or purplish with little chlorophyll. Flowers usually mauve/purple but sometimes white, in spike with only 2-6 open at a time, 0.5-1 cm across. Parasitic on cowpea, tobacco, sweet potato in Africa only .... *S. gesnerioides* (Willd.) Vatke

B. Calyx ribs usually 10-14 (mainly 2 per calyx lobe)

Plant 15-30 cm high, flowers red, white or yellow, in spike with up to 6 open at a time, 0.5 –1 cm across. Parasitic on cereals in Africa, Asia and USA...*S. asiatica* (L.) Kuntze

C. Calyx ribs usually 15 (3 per calyx lobe)

Plant 15-30 cm high, leaves narrow without conspicuous teeth, flowers white, 2-6 open at a time, 0.5-1 cm across. Parasitic on cereals in Asia only .... *S. angustifolia* (Don) Saldanha

Plant about 50 cm high, annual, leaves up to 1.5 cm wide, coarsely toothed, flowers 1-2 cm across, pale salmon-pink, occasionally white; corolla tube 2 cm long bent just below the corolla lobes, sparsely, coarsely hairy on the calyx ribs. Parasitic on some cereals in Africa only. .... *S. forbesii* Benth.

Shoots up to 1 m high, arising from a perennial rootstock, leaves up to 1 cm wide with some obscure teeth, flowers pink/brick red 2 cm across, corolla tube 2 cm long, bent just below the corolla lobes, calyx covered in short fine hairs. Parasitic on sugarcane in Africa only .... *S. latericea* Vatke

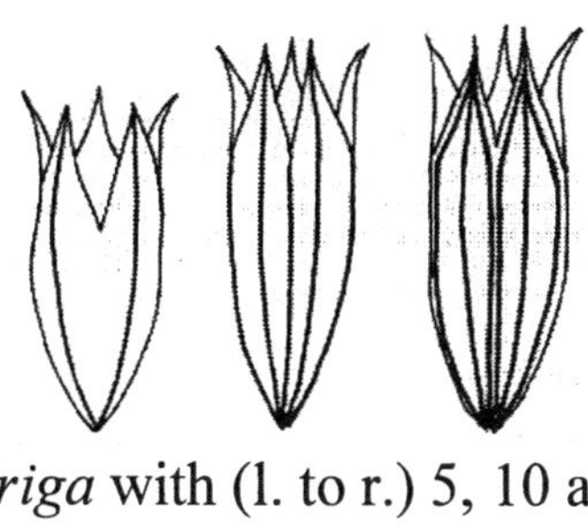

Calyces of *Striga* with (l. to r.) 5, 10 and 15 ribs

11.2 Key to the main *Alectra* spp. (according to Parker & Riches 1993)

A. Plants with well developed leaves

Plants 15-50 cm tall, sharply toothed leaves 14-25 mm long by 8-18 mm wide, calyx 6-8 mm long with hairs on nerves and margins only, anther lobes apiculate. Parasitic on grasses and broadleaf species, particularly Compositae. Recorded from niger seed in Ethiopia .... *A. sessiliflora* (Vahl) Kuntze

Plants 15-30 cm tall, leaves 15-36 mm long by 3-15 mm wide with two to five widely spaced blunt teeth, calyx 7-8.5 mm long, covered with short hairs, anther lobes not apiculate, stamen filaments glabrous. Parasitic on cowpea, groundnut and other cultivated legumes...*A. vogelii* (Benth.)

Plants 15-45 cm tall, leaves 8-20 mm long by 3-15 mm wide with up to two shallow teeth on each margin, calyx 5-7 mm long, with obtuse lobes and covered with short hairs, anther lobes not apiculate, stamen filaments hairy, sometimes densely so. Parasitic on cowpea and groundnut ....*A. picta* (Hiern) Hemsl.

B. Plants without well developed leaves

Plant 10-45 cm tall, leaves scale-like up to 2 mm wide and often appressed to the stem. Parasitic on tobacco and sunflower in Southern Africa only.....*A. orobanchoides* Benth.

11.3 Key to the main *Orobanche* spp. (according to Parker & Riches 1993)

A. Stems usually branched above ground, bracteoles present

Plant branched (except very small individuals), usually 15-20 cm high but up to 30 cm. Flowers mainly uniform pale blue/purple outside but with two white cushions on lower lobes, tube narrowest at base and widening towards open mouth, length usually less than 20 mm. Anthers glabrous or sparsely hairy at the base. Parasitic on a wide range of crops but especially tomato .... *O. ramosa* L.

As above but flowers over 20 mm, anthers glabrous or sparsely hairy. Host range as above .... *O. mutelii* or *O. ramosa* L. sub-species *mutelii*

Plant branched, usually 20-30 cm high, sometimes larger. Flowers as above but longer, always more than 20 mm. Anthers densely hairy. Parasitic on a wide range of crops, especially tomato, also cucurbits .... *O. aegyptiaca* Pers.

B. Stems usually unbranched above ground, bracteoles absent

Plants unbranched above ground though sometimes branching from the base below, 20-30 cm high. Corolla up to 20 mm long, tube uniformly pale, becoming inflated below, with a narrow mouth and small, 1-2 mm lobes with contrasting deep blue or pale coloration. Stigma whitish. Stamens inserted well up the corolla tube, at least 4 mm from the base. Parasitic on Solanaceae (especially tomato, tobacco, eggplant) and Compositae (especially sunflower)..*O. cernua* Loefl.

Plants unbranched, up to 50 cm (occasionally 100 cm). Flowers up to 25 mm long, the tube often streaked purplish at least along the veins, opening out into lobes 2-3 mm long, normally white. Stigma usually purple. Stamens inserted low down in the corolla tube, not more than 3 mm from the base. Parasitic on a wide range of hosts, especially Leguminosae ....*O. minor* Sm.

Plants unbranched, robust, commonly up to 100 cm high. Flowers 20-30 mm long, the tube usually white, opening out to widely divergent lobes 5 mm long, making total width of the flower 1.5-2 cm. Distinctly fragant when fresh. Stigma white, yellow or pinkish. Stamens inserted 2-4 mm from the base. Parasitic on several crops, especially faba bean and lentil .... *O. crenata* Forsk.

11.4 Key to the main *Cuscuta* spp. (according to Parker & Riches 1993)

A. One style, supporting two stigmas (section *Monogyna*)

Style shorter than the elongated stigmas, flowers 6-8 mm long, white with purplish rim. Calyx very short. Capsule conical 5-8 mm long, seeds 3-3.5 mm. Mainly Central to East Asia.. *C. reflexa* Roxb.

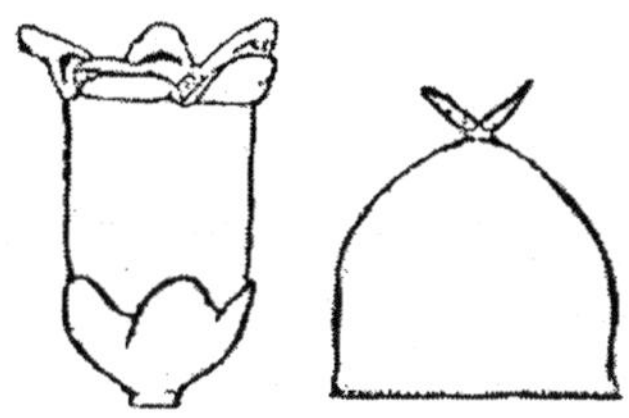

Cuscuta reflexa: (l.) flower, (r.) capsule

Style about as long as stigmas, all extremely short, flowers 3-4 mm, calyx with broad fleshy lobes, almost equalling corolla tube. Capsule elongated, cone-shaped, 6 mm long. Seeds 3-3.5 mm. Mainly in the Middle East............................ *C. monogyna* Vahl.

Cuscuta monogyna: (l.) flower, (r.) capsule

B. Two styles, stigmas linear, without knobs (section Cuscuta)

Perianth mostly 4-parted

Flowers 2-3 mm, pedicelled, in loose heads of 3-8 flowers. Stigmas subsessile. Capsule round, closely enclosed by corolla. Seeds about 1.25 mm. Mainly West and Central Asia.. *C. pedicellata* Ledeb.

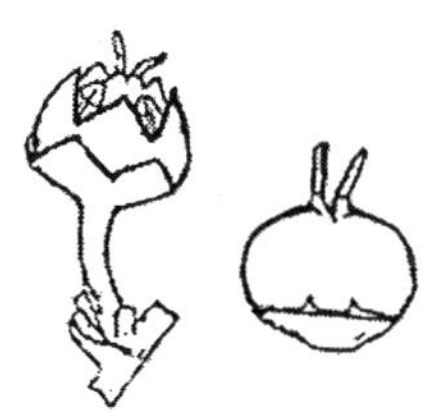

Cuscuta pedicellata: (l.) flower (r.) capsule

Flowers 1.5-2 mm, sessile in very small, dense heads 4-6 mm across; corolla lobes with erect hooded tips. Capsule round. Seeds about 1 mm. Mainly eastern Mediterranean. .... *C. palaestina* Boiss.

Cuscuta palaestina: (l.) flower (r.) capsule

Perianth mostly 5-parted

Calyx lobes fleshy at least at the tip, flowers 1.5-2.5 mm, sessile in heads 5-6 mm across. Capsule round, enveloped in corolla. Seeds about 1 mm. Widespread........ ....*C. planiflora* Ten.

Cuscuta planiflora: (l.) flower, (r.) capsule

Calyx lobes membraneous:

Flowers 3 mm long in heads 10-15 mm across; styles plus stigmas shorter than the ovary. Capsule roughly round. Seeds about 1.2 mm. Only in flax and linseed fields. Widespread. .... *C. epilinum* Weike

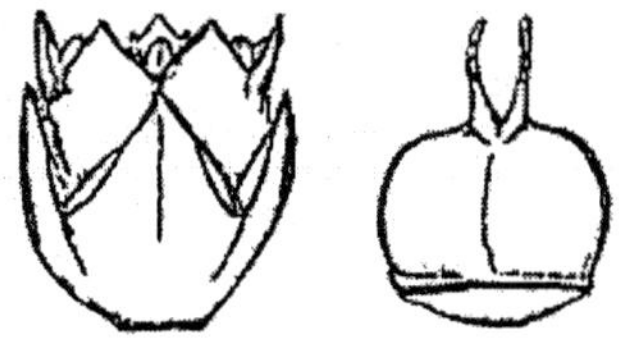

Cuscuta epilinum: (l.) flower, (r.) capsule

Stems slender, reddish. Flowers 3-4 mm in dense heads 7-10 mm across, styles plus stigmas slightly longer than ovary. Seeds about 1 mm. Mainly Europe ....*C. epithymum* (L.) Murr.

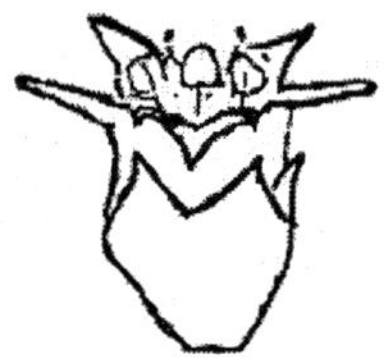

Cuscuta epithymum: flower

C. Two styles, capitate, with knobs (section Grammica)

Flowers granulate, covered with minute protuberances, 2-2.5 mm long on distinct pedicels. Seeds about 1.5 mm. Mainly North and Central America and Caribbean .... *C. indecora* Choisy

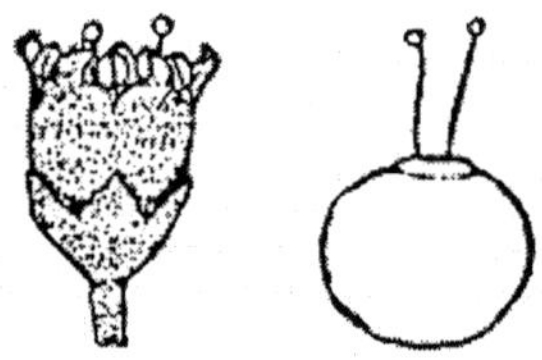

Cuscuta indecora: (l.) flower, (r.) capsule

Flowers not granulate

Capsule enclosed in corolla:

Flowers 2-4 mm long, pedicelled, in a loose head, somewhat glandular, corolla lobes deflexed. Corolla persisting as a cap on the capsule. Seeds about 1.5 mm. Mainly North America ....*C. gronovii* Willd. ex Roem. and Schult.

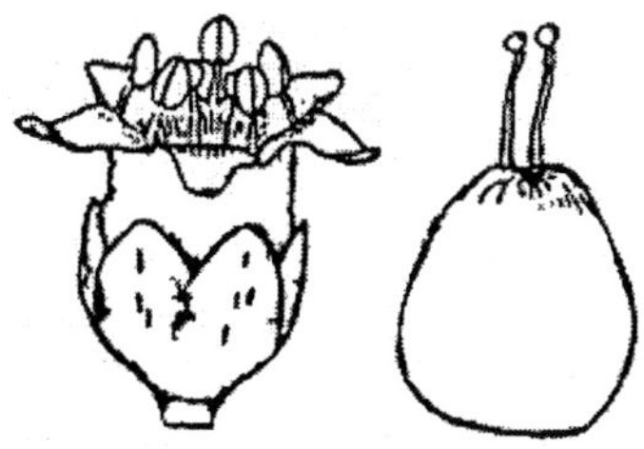

Cuscuta gronovii: (l.) flower, (r.) capsule

Flowers 2-3.5 mm in dense heads. Corolla lobes fleshy at the tip. Capsule 3-4 mm across, enclosed tightly by corolla. circumscissile. Seeds about 1.2 mm. Mainly in East Asia.. *C. chinensis* Lamarck

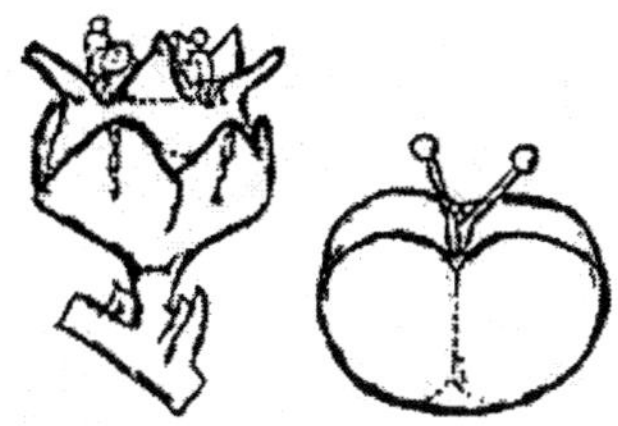

Cuscuta chinensis: (l.) flower, (r.) capsule

Capsule exposed:
Flowers about 2 mm, in compact heads. Corolla lobes obtuse. Seeds about 1.5 mm. Sometimes reddish-glandular on capsule. Distinct crater between styles. Infra-staminal scales bifid. Widespread through Europe and Asia.............................. *C. australis R.Br.*

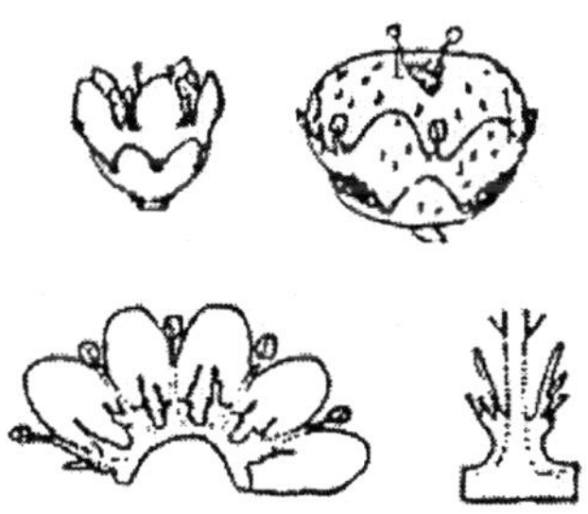

Cuscuta australis: (l.) flower, (r.) capsule,
and (below) dissected flower and enlargement of infra-staminal scale

Flowers 2-3 mm, in compact heads 10-12 mm across. Corolla lobes actue, often flexed upwards. Capsule round, 2-3 mm across, not concealed by corolla. Infra-staminal scales exserted, fimbriate, not bifid. Seeds 1-1.5 mm. Almost worldwide

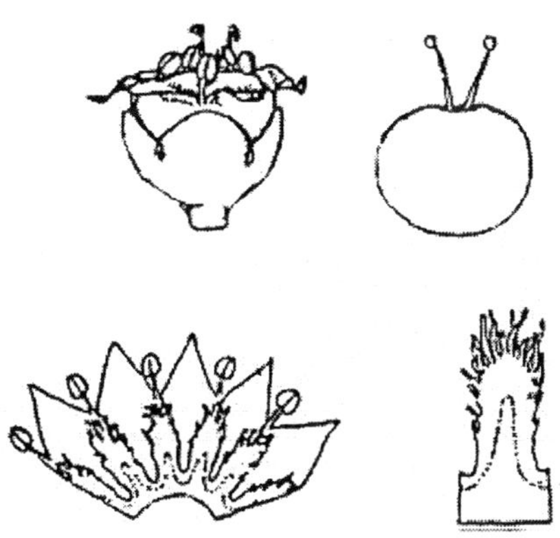

.... *C. campestris* Yuncker

Cuscuta campestris: (l.) flower, (r.) capsule,
and (below) dissected flower and enlargement of infra-staminal scale

11.5 The life-cycles of *Striga, Orobanche* and *Cuscuta*

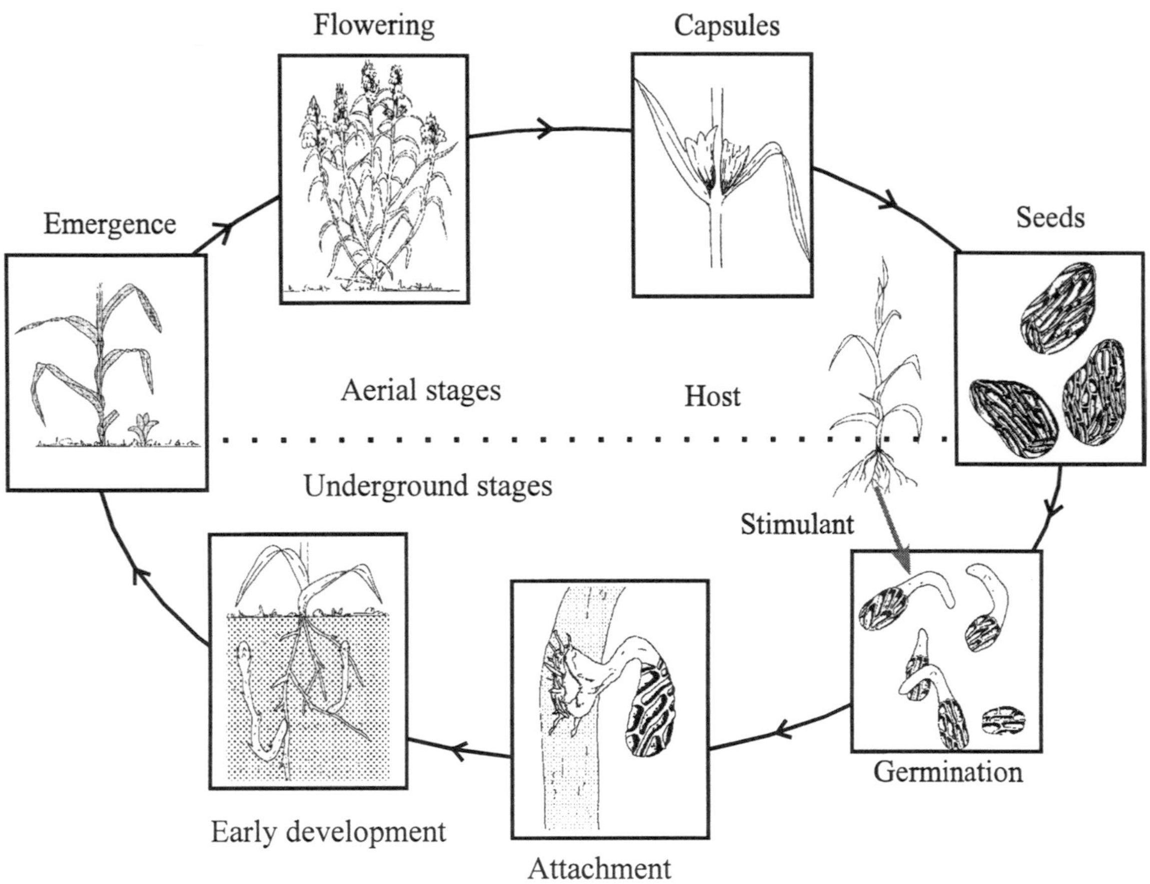

The life cycle of *Striga hermonthica*

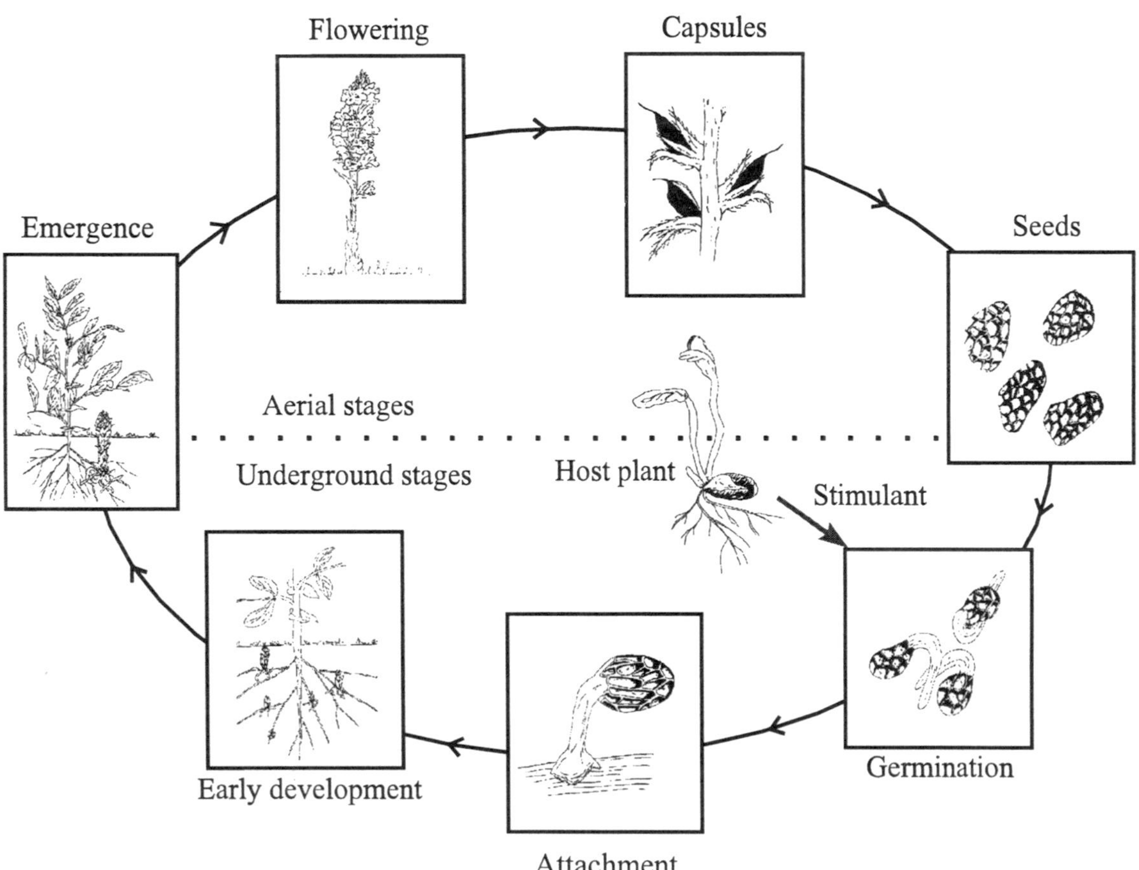

The life cycle of *Orobanche crenata*

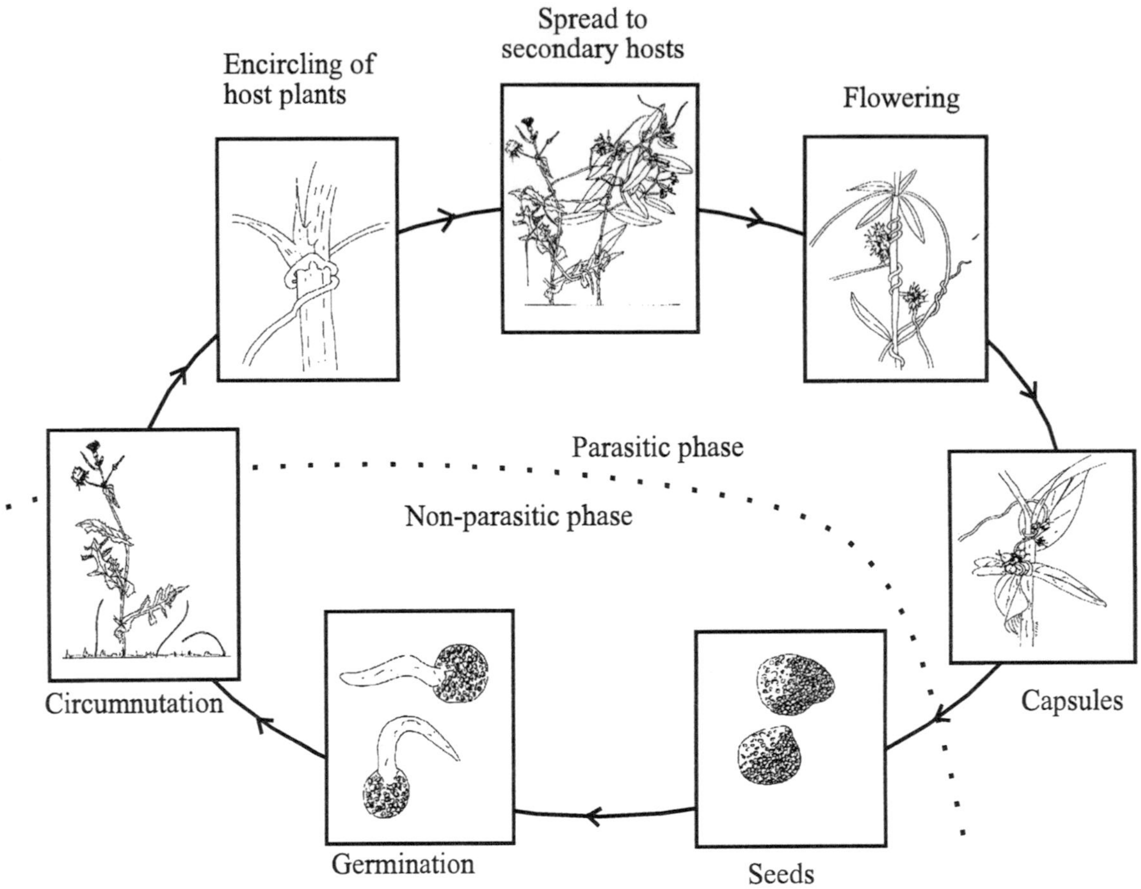

The life cycle of *Cuscuta* spp.

11.6 List of trap crops for *Orobanche* spp. and *Striga* spp.

Family	Scientific name	Common name
Orobanche aegyptiaca, Orobanche ramosa		
Allianceae	*Allium sativum* L.	Garlic
Apiaceae	*Carum ajowan* Benth. et Hock	Caraway
Euphorbiaceae	*Ricinus communis* L.	Castor bean
Fabaceae	*Glycine max* (L.) Merr.	Soybean
	Lablab purpureus (L.) Sweet	Hyacinth bean
	Lathyrus ochrus (L.) DC.	Ochrus pea
	Medicago sativa L.	Alfalfa
	Phaseolus vulgaris L.	French bean
	Pisum sativum L.	Pea
	Vigna radiata (L.) Wilczek	Green gram
	V. unguiculata (L.) Walp. spp. *unguiculata*	Cowpea
Linaceae	*Linum usitatissimum* L.	Linseed
Pedaliaceae	*Sesamum indicum* L.	Sesame
Orobanche cernua, Orobanche cumana		
Asteraceae	*Bidens pilosa* L.	Spanisch needle
	Cichorium intybus L.	Chicory
	Guizotia abyssinica (L.f.) Cass.	Noog
	Tridax procumbens L.	Coat buttons
Euphorbiaceae	*Ricinus communis* L.	Castor bean
Fabacae	*Macrotyloma uniflorum* (Lam.) Verdc.	Horsegram
	Medicaga sativa L.	Alfalfa
	Vigna aconitifolia (Jacq.) Marechàl	Moth bean
	V. mungo (L.) Hepper	Black gram
	V. radiata (L.) Wilczek	Green gram
Linaceae	*Linum usitatissimum* L.	Linseed
Malvaceae	*Gossypium* spp.	Cotton
Solanaceae	*Capsicum annuum* L.	Sweet chillies
Orobanche crenata		
Asteraceae	*Helianthus annuus* L.	Sunflower
Fabaceae	*Astragalus boeticus* L.	Vetch
	Glycine max (L.) Merr.	Soybean
	Hedysarum coronarium L.	Sulla
	Lablab purpureus (L.) DC.	Hyacinth bean
	Phaseolus vulgaris L.	Common bean
	Vigna radiata (L.) Wilczek	Mung bean
	V. unguiculata (L.) Walp. spp. *unguiculata*	Cowpea
Linaceae	*Linum usitatissimum* L.	Linseed

Striga asiatica		
Asteraceae	*Helianthus annuus* L.	Sunflower
Euphorbiaceae	*Ricinus communis* L.	Castor bean
Fabaceae	*Arachis hypogaea* L.	Ground nut
	Crotalaria juncea L.	Sunhemp
	Dolichos lablab L.	Hyacinth bean
	Glycine max (L.) Merr.	Soybean
	Pisum sativum L.	Pea
	Vigna unguiculata (L.) Walp. spp. *unguiculata*	Cowpea
Linaceae	*Linum usitatissimum* L.	Linseed
Malvaceae	*Gossypium* spp.	Cotton
Striga hermonthica		
Asteraceae	*Helianthus annuus* L.	Sunflower
Fabaceae	*Cajanus cajan* (L.) Huth	Pigeonpea
	Cicer arietinum L.	Chickpea
	Crotalaria juncea L.	Sunhemp
	Dolichos lablab L.	Hyacinth bean
	Glycine max (L.) DC.	Soybean
	Mucuna pruriens (L.) DC.	Velvet bean
	Pisum sativum L.	Pea
	Vigna subterranea (L.) Verdc.	Bambara groundnut
	V. unguiculata (L.) Walp. spp. *unguiculata*	Cowpea
Malvaceae	*Gossypium* spp.	Cotton

11.7 Form for a site description

(see chapter 3.1.1)

1	Date						
2	Field number						
3	Field size						
4	Altitude						
5	Relief						
6	Location at the slope						
7	Exposition						
8	Soil moisture						
9	Soil structure						
10	Friability						
11	Soil texture						
12	Soil reaction						
13	Organic matter						
14	Stone- and gravel content						
15	Crop						
16	Growth stage						
17	Planting density						
18	Parasite						
19	Growth stage						
20	Weeds						
21	Remarks						

Explanations to annex 11.7:

No.

1. Date
2. Field number
3. Field size estimated size in m^2
4. Altitude
   (m above see level)
5. Relief

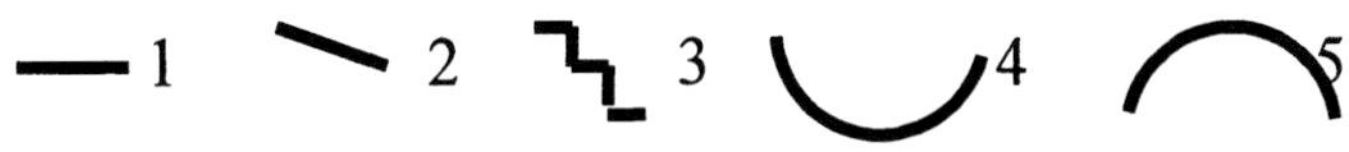

6. Location at the slope
   ridge (1), middle (2), foot (3),
7. Exposition
   sky direction (NW, SE, NE, SW,...)
8. Soil moisture
   arid (1), dry (2), fresh (3), moist (4), wet (5), flooded (6)
9. Soil structure
   very compact (1), compact (2), moderate compact (3), loose (4)
10. Friability
    good (1) , medium (2) , bad (3)
11. Soil texture
    manual field tests
12. Soil reaction
    ph-colorimetric
13. Organic matter
    color of soil when moist: gray (1), dark gray (2), black gray (3), black (4)
14. Stone- and gravel content
    without (1), low (2), middle (3), high (4)
15. Crop
    sorghum (1), millet (2), maize (3), cowpea (4), mixed (5)
16. Growth stage
    shoot (1), flowering (2), green (3), semi matured (4), matured (5)
17. Planting density
    plants / m^2, estimated from 2m x 5m
18. Parasite
    Striga hermonthica (1), *S. aspera* (2), *S. asiatica* (3), *S. gesnerioides* (4)
19. Growth stage
    green shoot (1), single flowers (2), multiple flowers (3), matured (4), dry (5),
20. Other weeds
    estimated density
21. Remarks
    mixed croppings, pests, ...

11.8 Data sheet for the quantification of *Striga* infestation and yield loss assessments

(see→ chapter 3.1)

Date:_______

Field No.______

District / village:__________/____________

Planting hole No.		**1**	**2**	**3**	**4**	**5**	**6**	**7**	**8**	**9**	**10**
Crop[1]											
Growth stage[2]											
No. of plants											
Plant height	1 / 1 2 / 2 3 / 3 4 / 4										
Cob or panicle weight	1 / 1 2 / 2 3 / 3 4 / 4										
Seed weight	1 / 1 2 / 2 3 / 3 4 / 4										
Striga species[3]											
No. of plants											
Growth stage[4]											
Max. height (cm)											

[1]maize (1), sorghum (2), pear millet (3)

[2]Vegetative phase (1), inflorescence emergence (2), anthesis (flowering) (3), milk development (4), dough development (5), ripening (6)

[3]*S. hermonthica* (1), *S. asiatica* (2), *S. aspera* (3), *S. forbesii* (4)

[4]Vegetative phase (1), beginning of flowering (2), flowering complete (3), ripening (4)

11.9 Example of a questionnaire for the analysis of farming systems

(→ chapter 3.2)

Questionnaire on *Striga asiatica* in South Africa

Name: ______________________

Date: ______________________ *Time:* ______________________

Province: ______________________ *District:* ______________________

Village: ______________________

Which is your biggest problem at the moment? (in farming)

____

A: Plant Production

1. What farm size do you have? __________

2. How many hectare do you cultivate this season? __________

3. What crops do you cultivate?

	maize	1	sorghum	2
	sugar cane	3	cowpeas	4
	groundnuts	5	bambaras	6
	pumpkins	7	cotton	8
others	__________	9	__________	10
	__________	11	__________	12
	__________	13	__________	14

4. Number of family members in this resident: men ______
women ______
children ______ entire: __________

5. What level of education do you/your wife have? (years of school)
farmer: __________________
wife: __________________

6. Do you have transport facilities? No 1 Yes 2

⇒ if yes:

6.1. bicycle 1 barrow 2 cart with draught animal 3
bakkie 4 tractor 5

7. What are the main crops on your fields at the moment? (write down the name of the variety if known, the field size and sign if intercropped)

	7.1. name of variety	7.2. hectare		7.3. intercropped
maize	__________	______	1	
sugar cane	__________	______	2	
groundnuts	__________	______	3	
cotton	__________	______	4	
__________	__________	______	5	
__________	__________	______	6	

8. Which are your usual crop mixtures?

____

____

9. Did you buy / get any new seed varieties for each season? No 1 Yes 2

⇒ if yes: name of variety

9.1. maize	____________	1
sugar cane	____________	2
groundnuts	____________	3
cotton	____________	4
____________	____________	5
____________	____________	6
____________	____________	7

10. Do you practice rotation? No 1 Yes 2

____

____

11. Do you have fallow periods? No 1 Yes 2

12. What do you do with the crop residues after harvest?

remaining of the field	1	ploughing in before sowing	4
burning after harvest	2	grazing	5
burning before soil preperation	3	fodder use	6

13. Do you have live stock? No 1 Yes 2

cattle	1	________
sheep	2	________
goat	3	________
poultry	4	________

others: ____________

14. Do you apply organic manure? No 1 Yes 2

if not go to No 15!

⇒ if yes, on which crops?

14.1. crop: quantity:

crop	quantity
____________	____________
____________	____________
____________	____________
____________	____________

14.2. From where do you get the organic manure?

own livestock / fields	1
other farmers	2

value: ____________

14.3. How do you apply it?

rotation of kraal	1	applying crop residues	2
collecting and carrying	3		

15. Do you apply mineral fertilizer? No 1 Yes 2

⇒ if yes: full name of fertilizer bags / hectare

15.1. maize ____________ 1 16.2. ________ 1
sugar cane ____________ 2 ________ 2
groundnuts ____________ 3 ________ 3
cotton ____________ 4 ________ 4
____________ ____________ 5 ________ 5
____________ ____________ 6 ________ 6
____________ ____________ 7 ________ 7

B. Plant Protection

1. What are your main plant protection problems (weeds, diseases, pests, birds, rats) last season in:

maize ____________ 1
sugar cane ____________ 2
groundnuts ____________ 3
cotton ____________ 4
____________ ____________ 5
____________ ____________ 6
____________ ____________ 7
____________ ____________ 8

2. Do you use any pesticides? No 1 Yes 2
⇒ if yes: full name quantity used last year
2.1. maize ____________ 1 2.2. ________ 1
sugar cane ____________ 2 ________ 2
groundnuts ____________ 3 ________ 3
cotton ____________ 4 ________ 4
____________ ____________ 5 ________ 5
____________ ____________ 6 ________ 6
____________ ____________ 7 ________ 7
____________ ____________ 8 ________ 8

3. How often and when do you weed?
3.1. How often 3.2. when (weeks after sowing)
maize ____________ 1 ____________ 1
sugar cane ____________ 2 ____________ 2
groundnuts ____________ 3 ____________ 3
cotton ____________ 4 ____________ 4
____________ ____________ 5 ____________ 5
____________ ____________ 6 ____________ 6
____________ ____________ 7 ____________ 7

4. Who is doing the weeding?
number number
of the family: men ________ 1 from outside: hired labourer ________ 4
women ________ 2 neighbour ________ 5
children ________ 3
4.1. if hired labourer / neighbour:
wage ________
value in kind ________

5. Do you weed all your fields? No 1 Yes 2
if not: Reasons:
____
____

6. Which crops have priority in weeding?

____
____

C. *Striga*

1. Do you know the weed *Striga* (Roiblom)? (photo) No 1 Yes 2
1.1. Do you know the weed Tsono (Alectra)? No 1 Yes 2

MHILIWANE = *Striga* as well as *Alectra*

2. Have you ever found it on your fields? No 1 Yes 2
Roiblom 1
Tsono 2

3. Do you know if *Striga* (Mhiliwane) exists on fields of other farmers in this area?
No 1 Yes 2

4. How does it damage the crop?

____

5. How does it multiply itself?

____

6. Which crops are affected by *Striga*?

____
____
____
____

7. If you compare the *Striga* problem with other problems in your fields, do you think it is:
very serious 1
serious 2
moderate 3
slight 4

8. Can you estimate for how many years *Striga* exists on your fields? _______ years

9. Do you find the *Striga* infestation has increased? No 1 Yes 2
(compared with the seasons before)

10. Do you have a lower harvest because of *Striga*? No 1 Yes 2

11. Can you see an other damage caused by *Striga* (beside the lower harvest) ?
No 1 Yes 2
⇒ if yes:
11.1. Damage: ____

12. Which fields are most attacked by *Striga* ?
(give the alternatives to the farmer and write down the more attacked alternative)

			no difference:
monocropped / mixed croped	__________________	1	
sandy / loamy soils	__________________	2	
dry / humid soils	__________________	3	
(flat / ridges fields)	__________________	4	
no weeding / weeding	__________________	5	
no manure / manure	__________________	6	
no fertilizer / fertilizer	__________________	7	
early / late sowing	__________________	8	

13. Do you weed because of Mhlilwane? No 1 Yes 2
⇒ if yes:

13.1. Is there a special late weeding of Mhlilwane? No 1 Yes 2
(after harvest)

14. What do you do with the weeded Mhlilwane plants?

remaining on the field	1	ploughing in	2
collecting besides the field	3	fodder use	4
burning	5		

15. Do you have any measures to combat it besides weeding? No 1 Yes 2

⇒ if yes:

15.1. fallow periods	1	intercropping	2	others	7
rotation	3	late sowing	4	______________	
herbicides	5	new varieties	6		

⇒ if rotation ask No. 16, if not go to No.17!

16. Did you change crop rotation due to the parasite? No 1 Yes 2
⇒ if yes:

16.1. Changed crop sequence ______________________________

17. Do you know plants, which seem to lower the *Striga* infestation? No 1 Yes 2
⇒ if yes

17.1. cotton	1	sunflower	2
groundnuts	3	root crops	4
pulses	5	others ____________	6

18. Do you know tolerant varieties of maize / sorghum / sugar cane? No 1 Yes 2
⇒ if yes: name of variety

____
____
____

19. Does Mhlilwane affect amount of hired labourer? No 1 Yes 2

20. Did you get recommendations from other farmers concerning Mhlilwane? No 1 Yes 2
⇒ if yes:

20.1. What did they recommend?

____

21. Did you give any advice to your neighbours depending *Striga*? No 1 Yes 2
⇒ if yes:

21.1. Which advice? ______________________________
____

22. Do you receive information from

extension officer	1	magazine	2
radio	3	others ____________	4

⇒ if yes:

22.1. What information do you receive?

____
____

D. Control Measures:

1.I will propose you some possibilities for *Striga* control, please comment on them:

regular hand pulling	______________________	1
weeding after harvest	______________________	2
burning of residues	______________________	3
crop rotation	______________________	4
3 t of organic manure / ha	______________________	5
6 bags LAN (>80 kg /ha)	______________________	6
herbicides	______________________	7
transplanting	______________________	8

Do you have any wishes / cares in future concerning *Striga*?

____

____

What are in future the important desires and worries for your family and your farm?

____

____

Time at the end of the interview: ____________

Length of interview: ____________

Remarks:____

____

11.10 Programme of a three days training workshop

(→chapter 7.1.2)

Training workshop on the
"Biology and Control of *Striga* and *Alectra* in Malawi"

organised by the Department of Agricultural Research and technical Services for extension personnel and trainers/supervisors of extension staff in co-operation with the Bunda College of Agiculture, the Malawi-German Biocontrol and Post-Harvest Project, the supra-regional GTZ-UH project "Ecology and Management of Parasitic Weeds".

Goals of the workshop

* Participants should know and be able to explain the biology and ecology of striga and alectra (to farmers).
* Participants should know different control methods and be able to select and combine adapted control methods.
* Participants should know about appropriate extension materials and their use.
* Participants should discuss on-farm trials with farmers and researchers in the field.

Date:	March 1997
Moderator:	Dr. S. Kachelriess
Resource persons:	Dr. C. Nyirenda (Bunda College)

Programme of the workshop

__1st day__

7.30 a.m.	- 8.30 a.m.	*Breakfast*
9.00 a.m.	- 10.30 a.m.	Welcome Opening address Introduction Expectations Questionnaire for the participants
10.30 a.m.	- 11.00 a.m.	*Tea break*
11.00 a.m.	- 1.00 p.m.	Host plants of striga Striga and alectra species Distribution of striga and alectra Differences to normal weeds Life cycle Seed dispersal
1.00 p.m.	- 2.00 p.m.	*Lunch break*
2.00 p.m.	- 3.30 p.m.	Damage caused by striga and alectra Working groups on striga and alectra biology (Which aspects are most important for extension?)
3.30 p.m.	- 4.00 p.m.	*Tea break*
4.00 p.m.	- 5.30 p.m.	Group presentation

6.00 p.m.		*Supper*

2nd day

7.30 a.m.	- 8.30 a.m.	*Breakfast*
9.00 a.m.	- 10.30 a.m.	Why are parasitic weeds a problem in agriculture? - Changes in farming systems in Malawi Preventive methods
10.30 a.m.	- 11.00 a.m.	*Tea break*
11.00 a.m.	- 1.00 p.m.	Control methods and their combinations Crop rotation, trap crops, catch crops Green manure, organic manure, improved fallow, fertiliser Cover crops Weeding practices "Serwala" method
1.00 p.m.	- 2.00 p.m.	*Lunch break*
2.00 p.m.	- 2.30 p.m.	Gender related issues relevant to striga control
2.30 p.m.	- 3.30 p.m.	Group work: Integrated control
3.30 p.m.	- 4.00 p.m.	*Tea break*
4.00 p.m.	- 5.30 p.m.	Presentation of group work results
6.00 p.m.		*Supper*

3rd day

7.30 a.m.	- 8.30 a.m.	*Breakfast*
8.30 a.m.	- 10.30 a.m.	Field visit
10.30 a.m.	- 11.00 a.m.	*Refreshments (in the field)*
11.00 a.m.	- 1.00 p.m.	Discussion with farmers and evaluation of experiences with integrated striga control
1.00 p.m.	- 2.00 p.m.	*Lunch break*
2.00 p.m.	- 3.30 p.m.	Extension and communication aids Exam for the participants
3.30 p.m.	- 4.00 p.m.	*Tea break*
4.00 p.m.	- 5.30 p.m.	Evaluation of the workshop Distribution of certificates Cost reimbursement
6.00 p.m.		*Supper*

11.11 Example for a benchmark test

(→chapter 7.1.2)

Name: ____________	Score:
Date: ________________	

Training Course

Biology and Control of *Striga* and *Alectra* in Malawi, March 1997

Benchmark test

Please mark ☒only the correct statement (s). There can be more than one correct answer.

1. Which of the following weeds are parasitic weeds?

a) *Cynodon dactylon*.......................................x
b) *Striga asiatica* x
c) *Commelina benghalensis*.......................................
d) *Cyperus rotundus* ....
e) *Alectra vogelii* ....
f) *Striga forbesii*.......................................

2. Which are parasites of cereals (sorghum, millet, maize)?

a) *Striga asiatica* ....
b) *Striga gesnerioides*.......................................
c) *Striga forbesii*.......................................
d) *Alectra vogelii* ....

3. Which of the following crops are host plants of ***Striga asiatica*** ↓ ***Alectra vogelii*** **?** ↓

	Striga asiatica	*Alectra vogelii*
a) cotton	/	/
b) soybean	/	/
c) maize	/	/
d) groundnut	/	/
e) sorghum	/	/
f) millet	/	/
g) cowpea	/	/
h) bambara groundnut	/	/
i) yam	/	/
j) rice	/	/

4. Do you know the life cycle of *Striga*? Please put the following terms of the *Striga* life cycle in the right order (from 1 - 8). Start with seeds in the soil (1).

( ) emergence of *Striga*
( ) attachment
( ) germination
( ) dissemination of seeds
( ) capsule
( ) flowering
(**1**) seeds in the soil

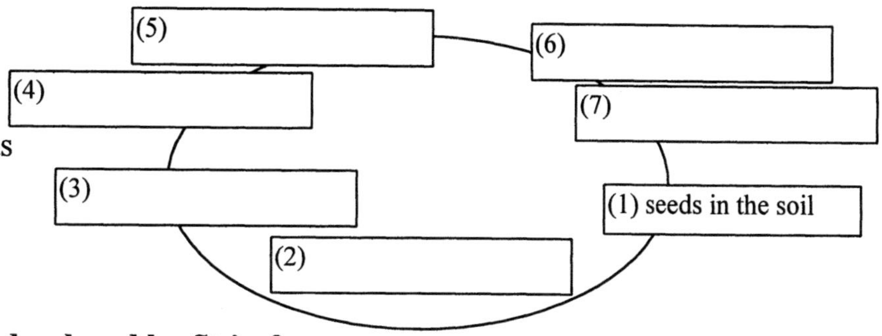

5. How is the root system developed by *Striga*?

a) few roots .... /
b) a dense root system .... /
c) a dense and deep growing root system .... /

6. How does *Striga* reduce the yield in cereals?

a) shading of the host plant .... /
b) competition for space and light .... /
c) uptake of water and minerals from the soil .... /
d) reduction of size and weight of the host .... /
e) uptake of water and minerals direct from the host .... /

7. How is the reproduction of *Striga* and *Alectra*?

a) by roots .... /
b) by seeds .... /
c) by tubers .... /
d) by rhizomes .... /

8. How is the distribution of from field to field?

	Striga ↓	*Alectra* ↓
a) by animals	..../	..../
b) by wind	..../	..../
c) by water	..../	..../
d) by men	..../	..../
e) by crop seeds	..../	..../
f) by tools	..../	..../

9. Can you imagine the size of a *Striga* seed?

a) 1 cm .... /
b) 0.5 cm .... /
c) 0.2 mm .... /

10. How many seeds can be generated by one single *Striga* plant?

a) 50 .... /
b) 500 .... /
c) 5,000 .... /
d) 50,000 .... /

11. How many years can *Striga* seeds survive in the soil?

a) 1 year ..../
b) 5 years ..../
c) 10 years ..../
d) 50 years ..../

12. On which fields is *Striga* a serious problem?

a) fields with high soil fertility ..../
b) monocropping of cereals for many years ..../
c) good crop rotation between sole cereals and sole legumes ..../
d) after a long fallow period ..../
e) fields with low soil fertility ..../

13. Why is weeding an important control method for *Striga*?

a) to prevent the yield reduction ..../
b) to prevent the reproduction of *Striga* during the vegetation period of cereals ..../
c) to prevent the reproduction of *Striga* after harvest of cereals ..../

14. When would you recommend to weed *Striga*?

a) immediately after *Striga* emerges ..../
b) at the beginning of the flowering stage of *Striga* ..../
c) after crop harvest ..../

15. What should be done with weeded *Striga* plants?

a) leave the *Striga* plants on the fields as green manure ..../
b) feeding animals ..../
c) burying/burning at the field border ..../

16. Which crops in the rotation reduce *Striga* infestation?

a) maize ..../
b) groundnut ..../
c) yam ..../
d) sorghum ..../
c) cotton ..../
f) soybean ..../

What is a trap-crop?

a) A plant which stimulates the parasite's seeds to germinate but does not allow their development and reproduction. ..../
b) A plant which stimulates the parasite's seeds to germinate and ensures its development ..../
c) A plant which does not produce a stimulant for the parasite to germinate ..../

17. Please, give an example for a good crop rotation to control *Striga*.

11.12 Example for a questionnaire used for the evaluation of a training course

(→ chapter 7.1.2)

Striga Workshop Questionnaire

Dear Madam, dear Sir,

in order to evaluate the workshop the organisers would like to ask you to answer the following questions.

Please complete the questionnaire and return it to one of the moderators during the last day of the workshop.

Where are you employed and what is your position?

....

....

Have you ever done extension work concerning *Striga* before?

Yes:..... No:......

If yes, when, how, and what target group was involved?

....

....

....

Ranking

Please give a score from -2 (very bad) to +2 (very good) for the following aspects of the workshop (organisation, programme, content, moderation)

Organisation (venue, food, accommodation...)

Score	**-2** (very bad)	**-1** (bad)	**0** (moderate)	**1** (good)	**2** (very good)

Programme (structure and schedule of the workshop, working groups, lectures...) not the content.

Score	**-2** (very bad)	**-1** (bad)	**0** (moderate)	**1** (good)	**2** (very good)

Content (quality of information...)

Score	**-2** (very bad)	**-1** (bad)	**0** (moderate)	**1** (good)	**2** (very good)

Moderation (stimulating discussions, exchange of ideas...)

Score	-2 (very bad)	-1 (bad)	0 (moderate)	1 (good)	2 (very good)

What did you like at the workshop?

....

....

....

....

What did you not like at the workshop?

....

....

....

....

Which suggestions for improving workshops in future do you have?

....

....

....

....

What aspects should have been explained more in detail or what could be left out or shortened?

....

....

....

....

Where and how do you think you can use the contents of the workshop?

....

....

11.13 *Striga*, an enemy to all farmers – a radio play

(→ chapter 7.2)

Key Issues to be delivered:

⇒ Clear distinction between bochaa and wublum; clarify that the broadcast deals with the pink flowering wublum only.

⇒ *Striga*'s reproduction through seeds.

⇒ *Striga* sits on the roots of the host crop (just like a tick), sucking on it and taking up nutrition and water. The reason for the severe effect for *Striga* infestation.

⇒ Ways of seed dispersal and distribution, particularly by crop seeds.

⇒ *Striga*'s relation to soil fertility issues - *Striga* being an indicator for poor soils.

⇒ Different types of soil fertility management that are enhancing soil fertility and thus combating *Striga*. In particular manuring and rotation with leguminous crops.

⇒ Clear distinction between host crops of *Striga* (maize, guinea corn, millet and rice) and trap crops (cotton, soybean, groundnuts, bambara bean, pigeon pea and cowpea).

⇒ Neutral crops in regard to *Striga*, i. e. yam and cassava.

⇒ Weeding practice to be carried out at the beginning of flowering stage and the need to bury the uprooted *Striga*.

⇒ The long term approach needed for the control of *Striga* and soil fertility management, respectively.

The setting: Discussion between Front Line Staff (FLS) and Farmer. After the usual greetings.

Farmer: These days I get very sad when I go to my farm. The fields are full of *bochaa* (indefinite expression for *Striga*). It is destroying all my crops. How can I fight it?.

FLS: What in particular are you talking about?

Farmer: I am talking about that vicious weed that is flowering pink. It is also known by the name *wublum* (definite expression for *Striga*).

FLS: Okay that is making it more clear. We are going to talk about *wublum* and not any other sort of soil born plant disease.

First of all you need to get to know your enemy well as to learn about its weak points. Let us start with some essential information on *Striga*. How does the weed multiply / reproduce?.

Farmer: I see it flowering but I have never seen seeds.

FLS: In fact *Striga* disperses very tiny seeds, one plant can produce several thousands of dust like seeds. Therefore one of the most important measure is stop *Striga* from dispersing its seeds. How can you achieve that?

Farmer: I have to weed or hand-pull it before it disperses seeds.

FLS: Have you been doing it and did you observe any improvement?

Farmer: I have always been weeding it but I haven't seen any great impact. Besides, it is impossible for me to weed all my fields at the time of *Striga* occurrence. Moreover, the crop is already very much affected and it would not give me much of a better yield.

FLS: That is true, but in order to achieve better yields in future we have to reduce the seeds in the soil.

Farmer: I understand that, but how comes that the crop is so seriously affected after it has germinated nicely?

FLS: *Striga* is different to other weeds, it does not grow towards the light but towards the roots of particular plants, i.e. maize, guinea corn, millet and rice. As the *Striga* seedling reaches the crop root it will attach itself to it and sit on it, sucking all what it needs directly from the crop root. Then it takes a number of weeks before *Striga* will appear above ground in the field. Before that time it has already done the damage to the food crop.

Farmer: Now I understand why *Striga* is such a severe weed as it steals all what it needs from the crop. But you haven't mentioned groundnuts and other crops that I have seen being attacked by *Striga*.

FLS: Believe me it is only maize, guinea corn, millet and rice that *Striga* can strive on. Have you ever seen *Striga* in a sole stand of groundnuts, cowpea, soybean or other crops?.

Farmer: No, only when I intercrop it with staple crops.

FLS: You see, sometimes *Striga* is a few feet away from its food provider, the food crop, yet it is sitting and sucking on its root. However, sometimes you can also see it growing on wild grasses.

Farmer: Okay, but the groundnuts and other crops don't perform well in *Striga* infested fields.

FLS: Yes that may be true, because *Striga* is generally an indicator for poor soil conditions Therefore other crops may not perform well in such fields.
Is there any *Striga* problem in your compound fields?.

Farmer: No, I don't see much *Striga* in my compound fields.

FLS: Why is it that you don't see it much there?

Farmer: Maybe because the compound fields are more fertile.

FLS: That is true. In general soil fertility and any measure contributing to it, is a major help in combating *Striga*. If you manage to make your bush fields more fertile it will also fight *Striga*.

Farmer: I don't have the money to buy enough fertiliser!

FLS: I understand, but there are other ways of enriching the soil and combating *Striga*, respectively.
One very effective way is the application of cow dung.

Farmer: My cattle is herded by Fulanis so it is difficult for me to get cow droppings in a large quantity.

FLS: So you should come to an agreement with the herdsmen to move the cattle around your *Striga* infested fields in the dry season, so that they can rest there during the night, disposing their dropping to the soil.

Farmer: I am not sure whether I can arrange that.

FLS: Don't you have other animals?

Farmer: Yes, some goats, sheep and fowls.

FLS: You can also use their droppings.

Farmer: I don't think that the quantity is going to be sufficient.

FLS: Why don't you concentrate on one field in this season and you will go on with other fields in the next season.

Farmer: That sounds reasonable but how should I convey the manure to the far away field?

FLS: That is indeed a problem, that's why I proposed dynamic kraaling to you. However, there is a very effective method of applying manure, i. e. to apply it directly to the planting hole. You have to make sure that the manure is well decomposed so that it will not harm the crop seedling.

Applying the manure directly to the planting hole will allow you to make best use out of a small quantity. Both in its effectiveness to fight *Striga* and the transport burden.
Due to the manure the crop seedling will be developing quickly and escape *Striga* or will at least avoid the most severe effects of it.

Farmer: I will try the direct application of manure to the planting hole in the next season.

FLS: That is good, but there is more to be done. Have you observed any crops that are some how reducing *Striga* in the following season?

Farmer: Yes, cotton.

FLS: That is true, cotton will stimulate the *Striga* in the soil to germinate but before it can attach itself to cotton it will die off. Therefore the seeds in the soil are reduced. Besides cotton, there are groundnuts, cowpea, pigeon pea, and bambara bean having the same effect on *Striga*, that is reducing the seeds in the ground due to ineffective germination of *Striga*.

Farmer: What about yam and cassava?

FLS: They don't have any effect on *Striga*. Yet, they are not affected, but will also not lead to a big reduction of the seeds in the soil. Once you crop them, you only give *Striga* seeds the time to die off naturally. But this takes many years. *Striga* can lie in the soil for years until it smells a root to grow towards it.

Farmer: Now I see the how harmful the weed is. It grows best on soils that are already less fertile and then it comes to attack the food crop. It will sit on the root and steal water and nutrition from the crop. I really need to fight that *Striga*.

FLS: That is good, but keep in mind that it takes a while. You can not get rid of *Striga* in only one or two years. However, you can gradually improve the soil fertility with manure and by planting crops that will reduce the *Striga* seeds in the soil.
Can you recall these beneficial crops for me?

Farmer: Yes, they are cotton, groundnut, cowpea, pigeon pea and bambara bean.
But all this can only be applied in the next season. What can I do now?.

FLS: Talk to your friends and inform them about *Striga*. Get together and uproot and bury *Striga* today for a better tomorrow.

Farmer: Did you say bury?

FLS: Oh yes, I forgot that earlier on, it is necessary to bury *Striga* since it can still produce seeds even when uprooted and left in the field. Therefore it is advisable to dig a hole (about 2 feet deep) and dump all uprooted *Striga* in there. So that it can decay without producing any seeds.

You can only succeed in controlling *Striga* effectively if you go hand in hand with your community. As I mentioned *Striga* reproduces by seeds and they are easily carried from one to another field by wind, water, tools, animals and most important by crop seeds.

Farmer: Why by crop seeds?

FLS: When the crop gets in contact with contaminated soil during harvest, the tiny seeds will stick to the crop. Thus you have to prevent the crop from getting in touch with *Striga* or *Striga* infested soil during harvest.

Farmer: How can I prevent that? And what do I do when I buy some seeds from the market?

FLS: At harvest, don't allow the crop to come in contact with *Striga* infested soil. It is best to uproot and burn all *Striga* before harvesting. When you are getting your seeds from the market, it is a good idea to clean the crop seeds by putting them in a pot and pouring water on them. Stirring and putting a bit of soap in the water enhances the cleaning process. The *Striga* seeds and other weed seeds as well as bad crop seeds and dirt will float on top and will be poured away while the good seeds remain on the bottom.

Farmer: Thank you very much for all this information. I will inform my fellow farmers about it.

FLS: Don't mention, let us recall the most important aspects. *Striga* attaches itself only to which plants?.

Farmer: Food crops, i.e. maize, guinea corn, millet and rice.

FLS: Who are our allies in combating *Striga*?

Farmer: Those crops that will seduce *Striga* to germinate but will not allow it to attach and to reproduce. Our allies in the battle against *Striga* are: Cotton, groundnuts, cowpea, pigeon pea and bambara beans.

FLS: Very good and when is the right time to for hand-pulling *Striga*?

Farmer: Before *Striga* disperses its seeds.

FLS: Yes, from the start of flowering there are about two weeks until it disperses seeds. It is most efficient to hand-pull them at this point in time. In other words whenever you see *Striga* flowering you should be prepared to hand-pull and bury.
Why did you agree to hand-pull *Striga* now, even though not much positive impact can be expected for the current season?

Farmer: Because I want to stop the built up of *Striga* seeds in the soil. Therefore I have to uproot and bury *Striga* today for a better tomorrow.

FLS: Very much so. Can you recall how *Striga* seeds get from one place to another?

Farmer: Yes, by crop seeds, wind, water, animals, and tools.

FLS: Yes that's right. Now, here is the last question for today: What other ways of making your field more fertile and combating *Striga* do you know now?

Farmer: I can practice dynamic kraaling, the application of cowdung, and other manure as well as direct planting hole application of manure.

FLS: Very good, you're now well prepared to fight *Striga* effectively and raise soil fertility, too.

(From Fischer 1999)

12 Color Plates

1a

1b

2a

2b

Economic important parasitic weed species of the family Scrophulariaceae:

1a, b *Striga hermonthic*

2a, b *Striga asiatica*

Plate 2

1a

1b

2a

2b

Economic important parasitic weed species of the family Scrophulariaceae:

1a, b *Striga gesnerioides*

2a, b *Alectra vogelii*

1a

1b

2a

2b

Economic important parasitic weed species of the family Orobanchaceae:

1a, b *Orobanche crenata*

2a, b *Orobanche aegyptiaca*

Plate 4

1a

1b

2a

2b

Economic important parasitic weed species of the family Orobanchaceae:

1a, b *Orobanche cernua*

2a, b *Orobanche foetida*

1

2

3

4

Economic important parasitic weed species of the family Cuscutaceae:

1 *Cuscuta campestris*, encircling a young weed
2 *C. campestris* on faba bean
3 *Cuscuta monogyna* on citrus
4 *Cuscuta* spp. in the natural vegetation

1

2

3a

3b

3c

Economic damage by parasitic weeds:

1 *Striga hermonthica* infestation in sorghum, Sudan

2 Stunted growth of maize as a cause of *Striga* infection

3a-c *Orobanche crenata* infestation in Morocco and wilting symptoms of faba bean

1

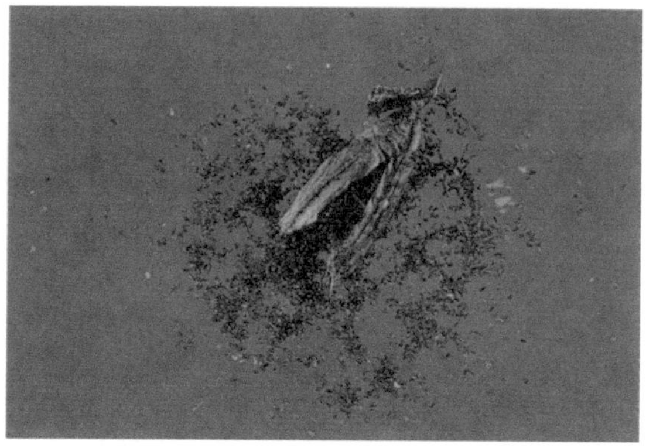

2

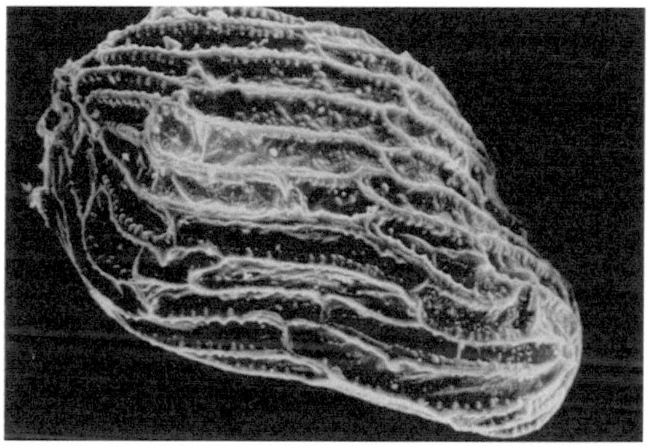

3

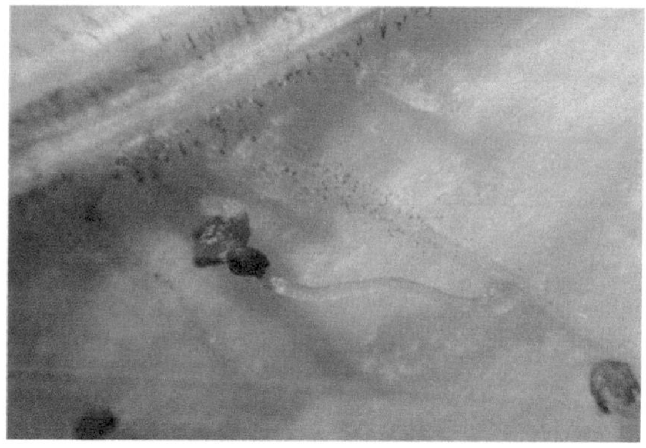

4

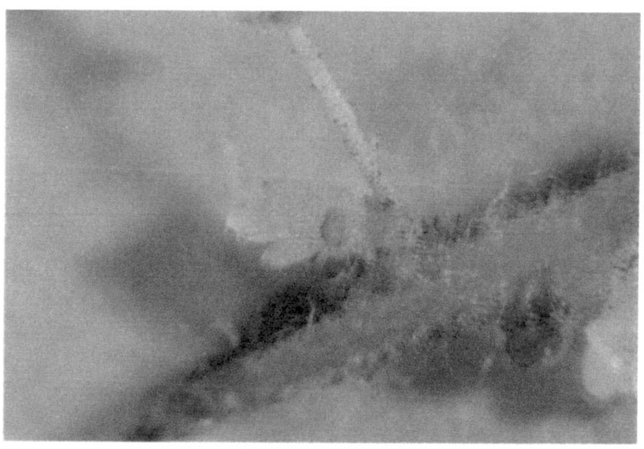

5

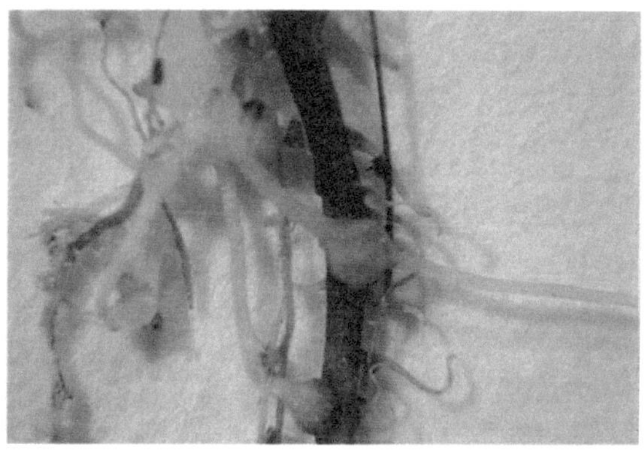

6

Developmental stages of *Striga* spp.:

1 The size of *Striga* seeds (left) compared to crop seeds
2 Seed capsule of *S. hermonthica*
3 Electron microscope picture of a *S. asiatica* seed
4 Germ tube and attachment of *S. hermonthica*
5 Underground shoot of *S. hermonthica*
6 Secondary haustoria of *S.hermonthica*

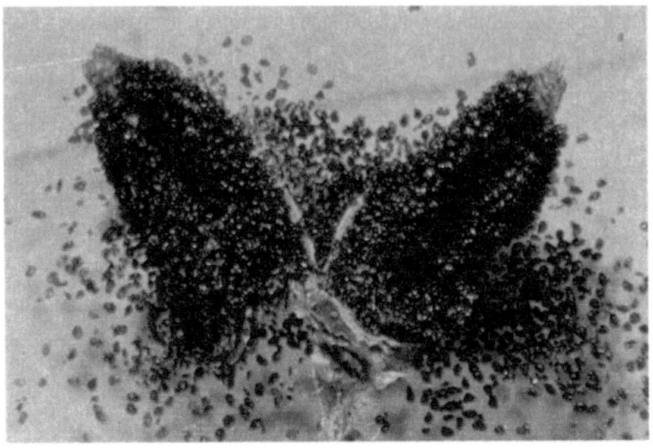

1

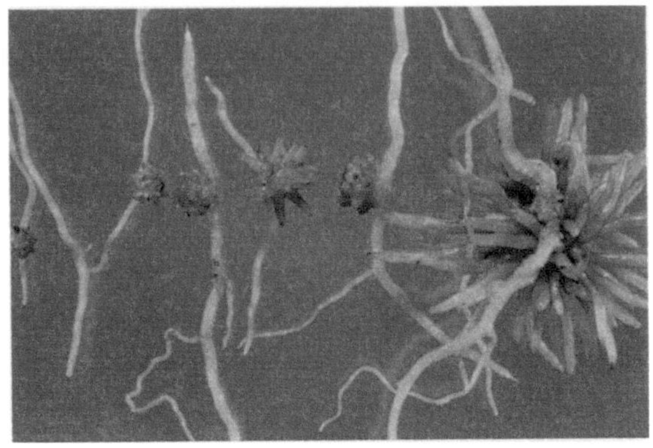

2

3

4

5

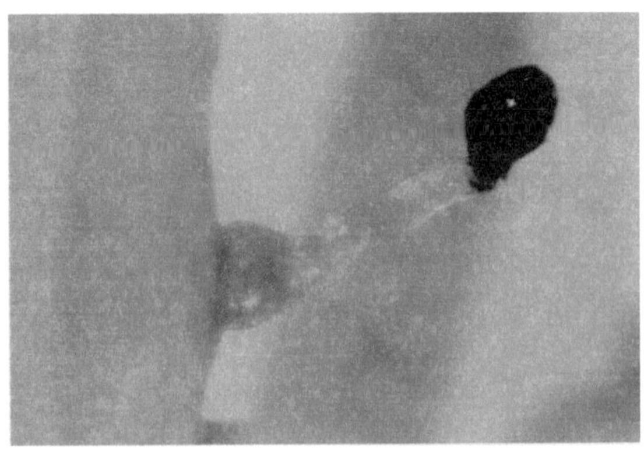

6

Developmental stages of *Orobanche* spp.:

1 Attachment of *Orobanche crenata*
2 Tubercles of *O. crenata* of different age
3 *O. crenata* shoots of different age on faba bean roots
4 Seed capsules of *O. crenata* (petals has been removed)
5 Unripe seed capsule of *O. crenata*
6 Opened seed capsule of *O. crenata* with ripe seeds

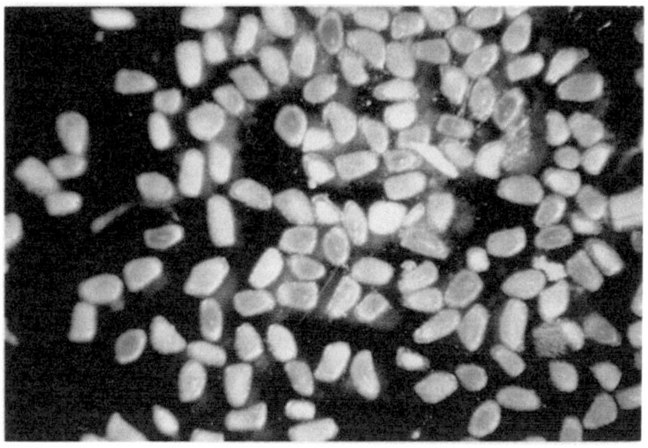

1

2

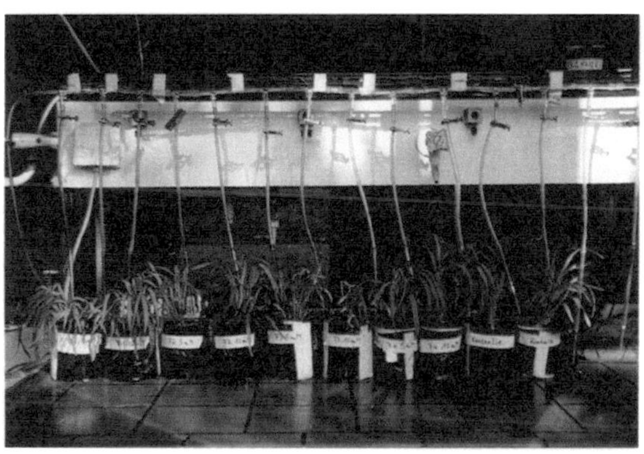

3

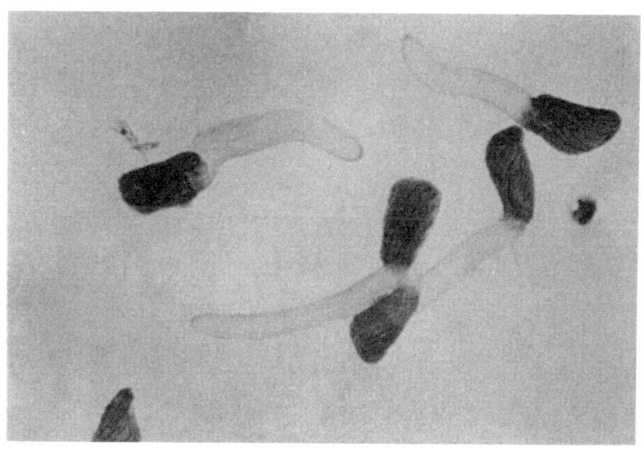

4

Research techniques: Seed viability, germination and underground development

1 Viability test: Viable seeds are red, reddish or yellow, non-viable seeds are white
2 Sorghum seedlings growing in sand to be used for the production of root exudate
3 Sorgum seedlings immersed in deionized water. The solution in the vessel is used as germination stimulant
4 Germinated *Striga* seeds as observed in a germination test

Plate 10

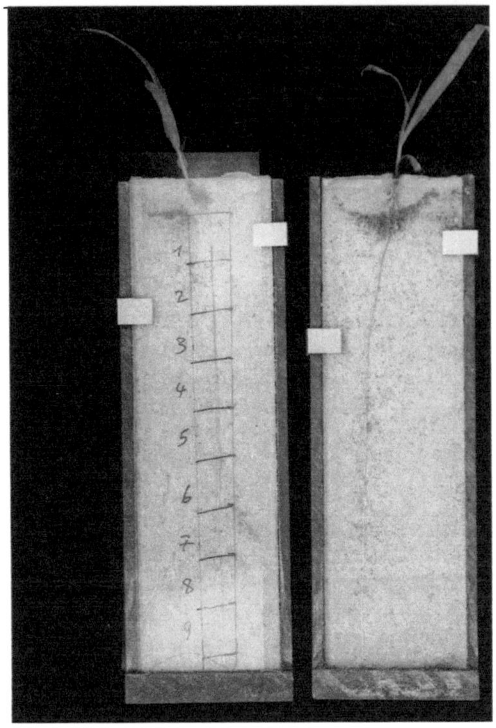

1a

2a

2b

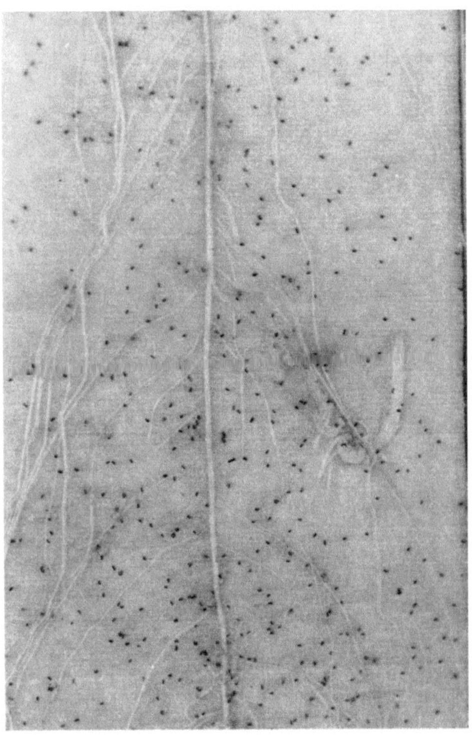

1b

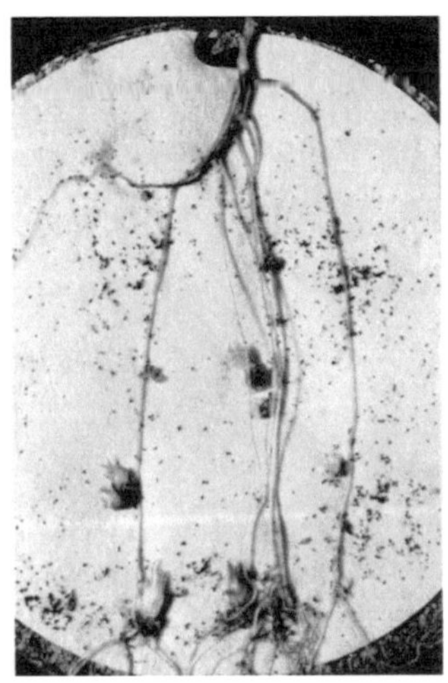

2c

Research techniques: Observation of underground development

1a, b Root chamber technique: The root area is marked where observations will be made
2a-c Petri dish method: Screening of lentils for resistance to *Orobanche crenata*

1a

1b

2a

2b

3

4

Application of control methods:

1a,b Field testing of trap and catch crops in micro plots for the control of *Orobanche* spp.

2a,b Testing of the effectivity of soil solarisation: Besides of *Orobanche* spp. also other weeds are controlled.

3 Screening for *S. hermonthica* resistance in sorghum

4 Field testing of different methods to control *O. crenata* in faba bean. The effect of the use of imazethapyr is clearly visible in the „green" plots

1

2

3

4

5

6

Phytophagous insects of *Orobanche* spp. in Morocco

1 Egg laying female of the monophagous fly *Phytomyza orobanchia* (Diptera, Agromyzidae)
2 *Smycronyx cyaneus* (Coleoptera, Curculionidae) on a flower of *O. crenata*
3 *Meloe majalis* (Coleoptera; Meloidae) feeding on a shoot of *O. crenata*
4 *Mylabris olea* (Coleoptera; Meloidae) feeding on flowers of *O. crenata*
5 *Tropinota squalida* (Coleoptera, Scarabaeidae) occurs on flowers of *O. crenata*
6 *Pachychila salzmanni* (Coleoptera, Scarabaeidae) feeds on young *Orobanche* shoots

Plate 13

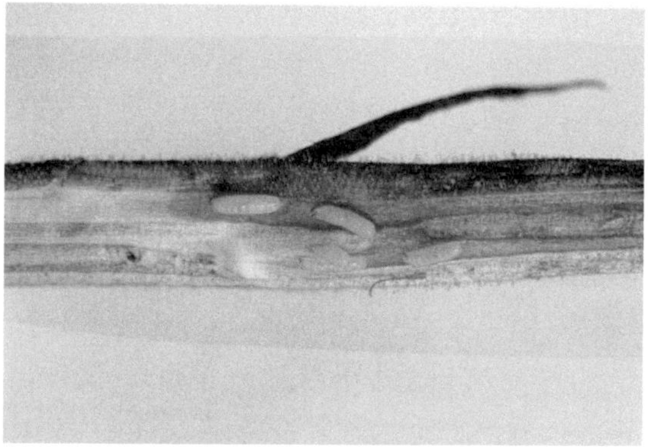

1

2

3

4

5

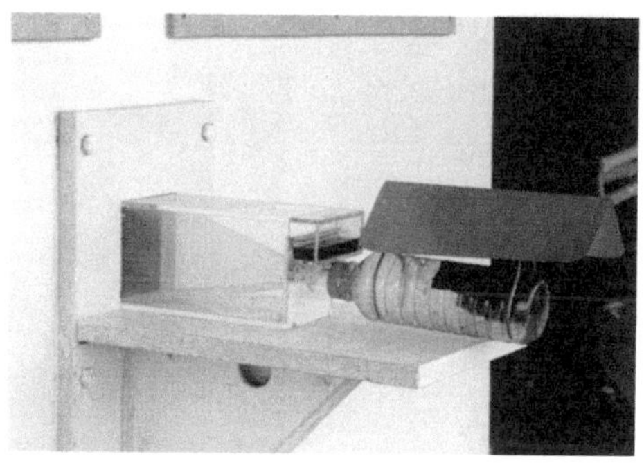

6

Biology and rearing of *Phytomyza orobanchia*:

1 Before flowering eggs are deposited on shoots and larvae mine under the epidermis
2 As a cause of the feeding damage (in combination with larvae of *S. cyaneus*) and of the secondary infestation with micro-organisms the shoot basis of *Orobanche* starts rotting
3 Shoot of *O. crenata* with infested seed capsules
4 Infested seed capsule. Infestation leads to a seed reduction of more than 96%
5 Phytomyzarium for storing of *Orobanche* shoots and for the rearing of *P. orobanchia*
6 The separator, which is fixed at the Phytomyzarium during the time of hatching ensures that only adults of *P. orobanchia* enter the collecting box

1

2

3

4

5

Monitoring the population dynamics of *P. orobanchia* and the efficacy of releases:

1 Malaise trap
2 Window trap
3 Emergence cages
4 Testing the efficacy of releases in field cages
5 Phytomyzarium established in farmers fields' for testing of field releases

Plate 15

1a

2a

1b

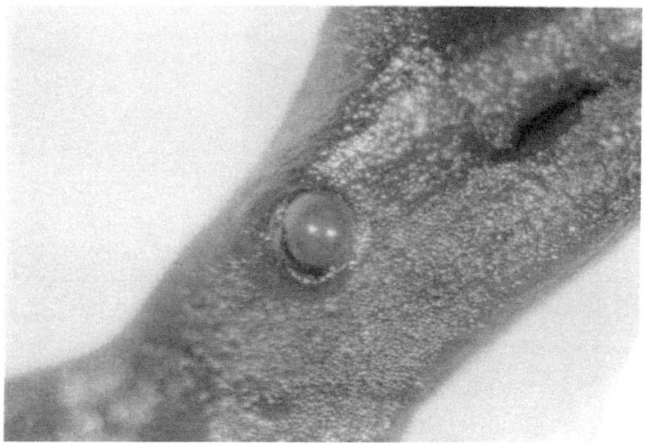

2b

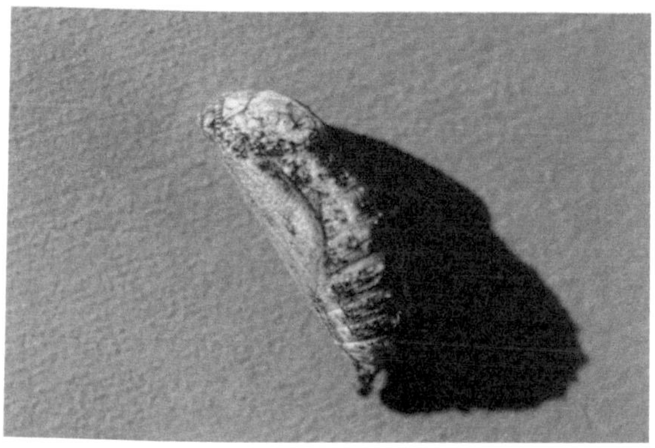

1c

2c

Phytophagous insects of *Striga hermonthica* occurring in West Africa:

1a-c *Junonia orithya* (Lepidoptera, Nymphalidae): a) adult, b) larva, which defoliates *Striga* and c) the pupa

2a-c *Smicronyx umbrinus* (Coleoptera, Curculionidae): a) copulating adults, b) usually one egg is deposited in a single ovary, c) seed capsules transformed into galls

Plate 16

1

2

3

4

5

6

Biological control of parasitic weeds with fungal pathogens and phytotoxins:

1 *Fusarium oxysporum* f.sp. *orthoceras* applied in pot trials to control *O. cumana*
2 *F. oxysporum* f.sp. *orthoceras* infected *O. cumana*
3 *Fusarium oxysporum* infected *S. hermonthica* in Northern Ghana
4 *F. oxysporum* infected *S. hermonthica* in pathogenicity tests
5 Host range screening for *F. oxysporum* in pot trials
6 Post emergence application of fumonisin B_1 on *S. hermonthica*

1

2

3

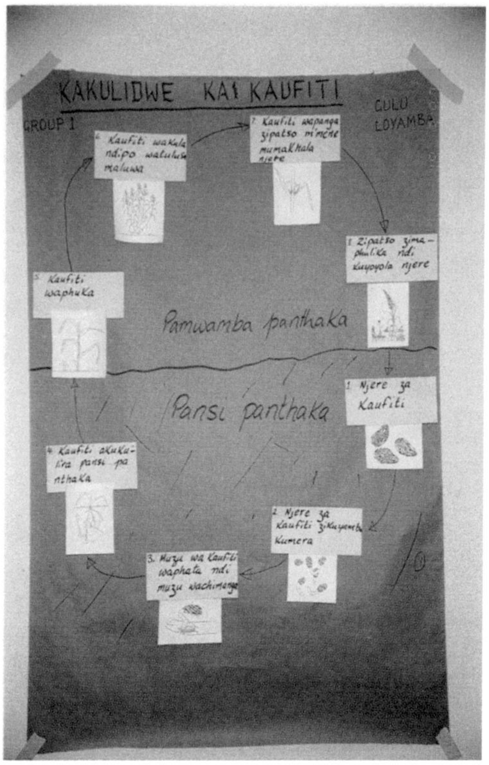

4

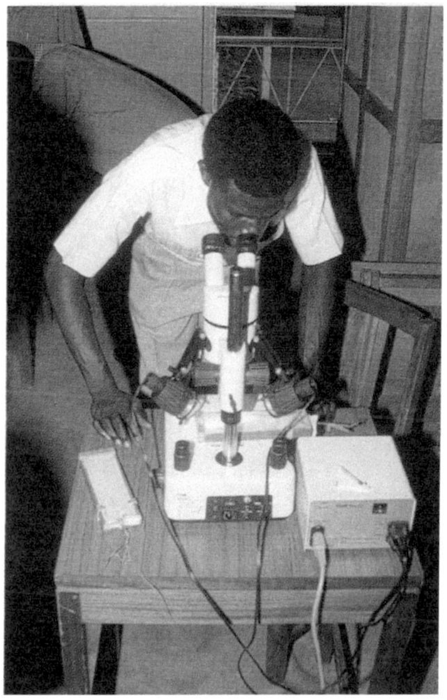

5

Training of researchers, extension staff and farmers:

1 Lectures as part of national training courses for extension staff
2 Presentations by local trainers and participants
3 Visits of demonstration plots
4 Translation of scientific names into local terms
5 Demonstration of the underground development of *Striga*

1a

1b

2a

2b

3a

3b

Development of visualized extension aids. Pictures of „The *Striga* Control Programme“ as developed for Northern Ghana. Examples of a more detailed (left) and simplified version (right):

1a,b Demonstration of the use of trap crops like cotton

2a,b Demonstration of the collection and use of organic manure

3a,b Hand-weeding of *Striga*

MIX
Papier aus verantwortungsvollen Quellen
Paper from responsible sources
FSC® C105338

If you have any concerns about our products,
you can contact us on
ProductSafety@springernature.com

In case Publisher is established outside the EU,
the EU authorized representative is:
Springer Nature Customer Service Center GmbH
Europaplatz 3, 69115 Heidelberg, Germany

Printed by Libri Plureos GmbH
in Hamburg, Germany